TH 4818 .A3 E93 2002

Evans, Ianto.

The hand-sculpted house

THE HAND-SCULPTED HOUSE

NEW ENGLAND INSTITUTE OF TECHNOLOGY
LIBRARY

Praise for *The Hand-Sculpted House*

A lovely book that will inspire builders and others to open their hearts and minds to the possibilities which exist outside of "normal" building and living standards. —*Ecology Action Newsletter*

This is truly an inspirational book, loaded with beautiful ideas and packed with practical knowledge and skill. —*Earth Garden*

The Hand Sculpted House combines technical detail with the greater philosophical and psychological significance of building directly with Earth's abundant and elemental ingredients. From the joy of mixing by hand (or by foot as the case may be in "mud dancing") to literally sculpting a home that fulfills our great need for meaning within our daily lives, the authors expand "notions of what is possible aesthetically, ecologically, and in terms of the spirit of the building" with "the least technical, safest, most forgiving of natural building materials." —*HopeDance*

The Hand Sculpted House goes beyond the practical. It covers the things you don't always see in building texts: an encouragement to the understanding of the site, comprehension of the value and inclusion of the natural, a whole chapter on space and intuitive design, and thoughts on warmth, comfort, and natural lighting. —*BackHome Magazine*

Written with a personal and honest pen, *The Hand Sculpted House* is approachable and fun to read—even for those with no intention of building a cob cottage. —*Environmental Building News*

THE HAND-SCULPTED HOUSE

A Philosophical and Practical Guide to Building a Cob Cottage

IANTO EVANS, LINDA SMILEY, AND MICHAEL G. SMITH
ILLUSTRATED BY DEANNE BEDNAR

Chelsea Green Publishing Company
White River Junction, Vermont

NEW ENGLAND INSTITUTE OF TECHNOLOGY
LIBRARY
#49493511
3/04

Copyright 2002 by Ianto Evans, Linda Smiley, and Michael G. Smith.
Illustrations copyright 2002 by Deanne Bednar.
Unless otherwise noted, photographs are copyright 2002 by Ianto Evans.

All rights reserved. No part of this book may be transmitted in any form by any means without permission in writing from the publisher.

Designed by Jill Shaffer.
Edited by Gayla Groom.

Printed in the United States.
First printing, June 2002.

07 06 05 04   3 4 5 6 7

Printed on acid-free, recycled paper.

Due to the variability of local conditions, materials, skills, site, and so forth, Chelsea Green Publishing Company and the authors assume no liability for personal injury, property damage, or loss from actions inspired by this book. Remember that any construction process can be dangerous, and approach the work of building with due caution, care, and a sense of responsibility.

Library of Congress Cataloging-in-Publication Data

Evans, Ianto.
The hand-sculpted house: a philosophical and practical guide to building a cob cottage / Ianto Evans, Linda Smiley, and Michael Smith.
p.   cm. (A Real Goods solar living book)
Includes bibliographical references and index.
ISBN 1-890132-34-9 (alk. paper)
1. Earth houses—Design and construction—Amateurs' manuals. 2. Cob (Building material) I. Smiley, Linda, 1952- II. Smith, Michael, (Michael G.), 1968- III. IV. Series.

TH4818.A3 E93 2002
693'.22—dc21   2002025909

Chelsea Green Publishing Company
Post Office Box 428
White River Junction, VT 05001
(800) 639-4099
www.chelseagreen.com

We dedicate this book
with our deepest love and gratitude
to our parents:

BECKY AND BILL SMILEY

JANICE AND MAURICE SMITH

BERT AND GENE BEDNAR

SHEILA EVANS

and in loving memory of

HARRY EVANS

without whom you would not be
reading this book.

Table of Contents

Acknowledgments

THIS BOOK IS A FRUIT OF THE COLLABORATION of literally hundreds of people, working together in various ways for nearly a decade. It represents the work of the Cob Cottage Company, which was founded by Ianto, Linda, and Michael in 1993 but could not have achieved anything like the scope of what is described herein without the support and active assistance of a host of apprentices, volunteers, colleagues, friends, family members, workshop sponsors, and students. Dozens of people contributed directly to the writing of this book—either by teaching us techniques or information which found their way into the manuscript; by sending stories or photos for our use; or by giving us advice on the business of writing and publishing. We'd also like to acknowledge the cob builders of Devon, England, and Wales—both modern-day revivalists and their distant ancestors—for providing inspirational examples of cob's potential. A complete list of the people to whom we are indebted would require a small book of its own; we hope most of you will accept our sincerest gratitude anonymously. However, there are a few people who have given so much to this project that we could not in good conscience fail to mention their names.

The patron saint of this book was surely Gayla Groom, who volunteered countless hours of priceless editorial assistance just when the project had reached its darkest hour. Without her help, we might never have pulled the manuscript into publishable condition. Deanne Bednar has been an invaluable accomplice in nearly every step of the process, not only creating the beautiful illustrations but offering editorial advice and helping to keep communications clear among the many parties involved. Likewise, the staff, apprentices and volunteers at Cob Cottage Company have gone far beyond the call of duty in bringing this book together. In this regard we'd especially like to thank Susan and Hop Kleihauer and Janine Björnson. Our gratitude goes also to the good people at Chelsea Green for their experienced guidance, especially to Jim Schley, Hannah Silverstein, and Rachael Cohen. And finally, to our dearest friends, community members, and families, thanks so much for helping to keep us balanced and healthy throughout a long, demanding and sometimes frustrating process. As the saying goes, here's mud in your eye!

Introduction

BY IANTO EVANS

What is cob? Cob is a structural composite of earth, water, straw, clay, and sand, hand-sculpted into buildings while still pliable. There are no forms as in rammed earth, no bricks as in adobe, no additives or chemicals, and no need for machinery. Cob is not new and not untested. Its viability has been thoroughly proved, all over the world, for centuries and probably millennia. But despite great public enthusiasm for natural building, most how-to building books and architectural histories contain hardly a mention of this common, almost universal, building technique. This book was written to fill that gap.

What is Earth's most common building material? Why, earth itself of course! Even today between a third and a half of us humans live in houses of unbaked earth. Must all earth buildings be mud huts in Africa? Well, no, they're also lavish adobe haciendas in Latin America, rammed earth mansions in France, earth-brick palaces in China. They're ten-story apartments in Yemen, old fortified monasteries in the Middle East, puddled and hand-shaped and pressbricked and foot-stomped earthen buildings from near the Arctic Circle in Norway to the tip of Chile and the polar end of New Zealand. Millions of them.

Adobe is well known to most of us. Rammed earth is increasingly in the news again. Even *pressbrick, poured adobe,* and *wattle and daub* are terms we have heard over the years. Yet the simplest, most accessible, and most democratic earth-building technique was until very recently almost unknown.

Ten years ago there were no cob builders in the United States. Nobody had built cob in North America for 150 years, or in the British Isles since the 1920s. There was no guidebook to cob construction, almost no descriptive writing, and no public awareness of the possibility. The continuity of master builders had, like a dead language, been lost. The revival of cob in the late 1980s depended on assumptions, deductions, and flimsy scraps of outdated information. In a single decade this has changed. More than a thousand students have passed through The Cob Cottage Company's trainings; many are themselves already teaching. Michael Smith's book, *The Cobber's Companion,* the first work exclusively devoted to cob building, has been a remarkable success. Now the mainstream media are involved, and cob building looks as if it may be here to stay.

I've been an architect involved in building since the early 1960s. In these thirty-five years I've seen dozens of "alternative" building fads flit by, many of them touted as more ecological, less energy-intensive, suitable for owner-builders, more occupant-friendly, and so on. Many were utterly frivolous attempts at publicity and profit, or obvious dead-ends. Some were structural liabilities. Others were basically good ideas that got out of hand. Most of them I've tried, one way or another; few left any permanent inspiration. I'm cynical now about any new, heavily publicized "ecological" building system. Why would cob be any different? I have regularly had to ask myself whether at sixty-one I'm squandering what's left of my productive life on yet another irrelevant whim.

The evidence suggests that with cob building we may be tweaking the tip of a big and very solid iceberg, the whole impact of which has yet to be seen. Cob construction seems to satisfy its builders in very profound ways. Our files are stuffed with letters of encomium, extravagant appreciation of how good it feels to build a house of mud pies, to involve yourself in building in such a primal way. You don't get ecstatic about building with concrete blocks or drywall, but with cob there seems to be universal enthusiasm. As a specific remedy for what ails our buildings, cob is unlikely to cure the epidemic, but it seems to be having a catalyzing effect as an inspiration and a tool for considering all the crucial issues.

WHAT THIS BOOK OFFERS YOU

This book is a groping toward an Ecological Architecture. It's not about a return to the past; it's an exploration of where we can go in the 21st century, creating buildings we will still love in the 31st. We spend most of our time in buildings; any way we can enjoy them more must surely be worth our effort. We have written this book to expand your notions of what is possible aesthetically, ecologically, and in terms of the spirit of the building—how it makes you feel. You still may never build, or you may choose to build with materials or methods other than cob, but this book can help you reflect on what you really need from the house where you will spend so much time. It should help you establish quality standards for where you live.

We are surrounded everywhere by free building materials. This book can help you find them, adapt them to what you need, and build with them. A new world of possibilities opens up. Cob is the least technical, safest, most forgiving of natural building methods. If you don't feel competent to begin building with logs, rammed earth, straw bales, or other more intimidating materials, start with cob. You will gain self-confidence in a completely safe environment.

Part 1 explains what you should consider before ever picking up a tool—*why* and *where* to build, with two chapters presenting for the first time a comprehensive and unique design process specially prepared for natural builders. For simplicity, we have addressed ourselves mainly to persons considering construction of what we would call a cottage, a humble little house in which a small family could live joyfully. We have assumed an intelligent reader who may never have built anything before but will no longer settle for less than the best.

Part 2 is the how-to-do-it section, for people who want or need to be able to build their own house, who won't support lending institutions, who are determined to tread more lightly on Earth, and who want a house perfectly suited to their own unique needs. It will also help encourage those whose creative talents have long been repressed to go out and *build*! The technique sections will lead you through this very peaceful process.

The final "Onword" talks about the sheer joy of building, and provides examples of how cob can help heal your body and spirit, by creating a sacred space, which is the quality of shelter that so often results from building with this medium. The color section is included for inspiration.

The Hand-Sculpted House is a practical guide based on the experience of The Cob Cottage Company and our associates. The ideas and techniques in this book have been developed and tested during the construction of more than a hundred demonstration cob buildings. In less than a decade a small group of enthusiasts has resurrected a noble building tradition and adjusted it to our own era, inventing tools and techniques, improving the process, and questioning our every move. This process continues as we write, so your book is already out of date; our opinions on many minor technical issues will probably have changed even before you read this.

The Hand-Sculpted House is not a recipe book. Because earth is such a variable material, any attempt to provide standard recipes and proportions for mixing cob and mud plasters often leads to frustration. Rather than codifying the process, we prefer to give you a thorough understanding of what makes cob buildings work, so you can make your own decisions based on our practical experience and your own sound judgment. The best experimental builder is a skeptic. Consider this book a widely experienced but potentially fallible advisor. Always do your own tests, and trust your own experience and intuition over the advice of the "experts," including us. Build something, experiment, ask questions, then push the known limits of the materials, systems, and techniques described herein. This book is intended as a companion to, rather than a substitute for, workshops and other hands-on learning. At times we've struggled to explain in words what can be much more easily demonstrated in practice.

We hope this book will be a bridge to your greater involvement in the natural building movement. Societal change can only be brought about by many people acting together, so don't keep your ideas, enthusiasm, or discoveries to yourself. Take a workshop, write an article, invite friends to help you build, teach your neighbors, get on the radio, attend a Natural Building Colloquium. Together we can make our voices heard. As the old English proverb says, "If you throw enough mud, some of it will stick."

Congruent with developing an essentially new construction system and a gentler approach to creating buildings, we have attempted to manage our individual and collective lives in a manner consistent with our values. All of this needed to be reflected in this book, otherwise none of us would feel honorable. In committing to publication we must take responsibility for every tree that was cut to make this paper, the fuel for transport, chemicals used in printing, and the seemingly inevitable toll of tiny lives that are extinguished by commerce and industry. We hope this book is valuable enough to more than offset these costs. The dividends in changes of attitude, creation of more opportunities for ecological buildings, and diminution of environmental damage all need to be great, or we have merely contributed to the problem. So, borrow a copy if you can (ask your local public library to order it) or if you can't, buy a copy *for* your library to lend out, knowing that this way you can read the book whenever you want.

Most importantly, building is not something you should try to do quickly to get a finished structure. Both building and living in your house are spiritual processes of daily joy, reflection, and connection with Nature. Taken slowly, the experience of building for yourself will be a high point of your life, immersed in meaning and saturated with joy. Building your own natural house may be the threshold to reinhabiting your own part of this small planet in a way that gives you great satisfaction and rests comfortably with your conscience. We hope this book will help you summon the courage to take charge of your own life, to build your harmless house and live in it in harmony with the cosmos.

ABOUT THE AUTHORS

The Hand-Sculpted House is a result of eight years of close partnership in The Cob Cottage Company and the accumulated experiences of three lifetimes.

Ianto Evans is an applied ecologist, landscape architect, inventor, writer, and teacher with building experience on six continents.

COURTNEY ROGMANS

LEFT TO RIGHT: *Michael, Linda, and Ianto.*

Cob is traditional in Wales, his homeland. He teaches ecological building and has consulted to indigenous nations, USAID, the World Bank, the U.S. Peace Corps, and several national governments.

Michael G. Smith has formal training in environmental engineering and ecology. He is the author of *The Cobber's Companion: How to Build Your Own Earthen Home* and co-editor of *The Art of Natural Building: Design, Construction, Resources.* He teaches, writes, and consults on cob, natural building, and Permaculture.

Linda Smiley is a director of The Cob Cottage Company, as well as a master cobber and recreational therapist. She also teaches workshops in cob, sculpting sacred spaces, intuitive design, and natural plasters and finishes.

Part 1 was mostly written by Ianto, part 2 mostly by Michael, and the Onword mostly by Linda. You will hear each of our voices in these three sections, but we all had a hand in all of it: drafting, editing, circulating, rewriting. In cases where a section or chapter is very much in one author's voice, offering that person's particular viewpoints on an issue, we have used a by-line to indicate which of us is addressing you.

QUESTIONS & ANSWERS ABOUT COB

At The Cob Cottage Company, we have found that many visitors and workshop participants have similar questions about cob, as no doubt will many readers. Here are short answers to some of the most common questions.

What Is Cob? Is It Adobe?

Cob is a composite building material, a mixture of earth, clay, sand, straw, and water, hand-worked into monolithic earthen walls. No forms, no cement, no ramming, no machinery. Adobe means sundried earthen bricks, the material they are made of, and the resultant earthen buildings. The word *cob* comes from an Old English root, meaning "a lump or rounded mass." This term refers to the material itself, to buildings made of it, and to a traditional building technique used for centuries in Europe, in rainy, cold, windy climates as far north as the latitude of Alaska.

Cob is one of many methods for building with raw earth, the world's most common construction material. In simplicity of construction and freedom of design, cob surpasses adobe and related techniques such as rammed earth and compressed earth bricks. Since you don't need straight forms or rectilinear molds, cob lends itself to organic shapes, curved walls, arches, and vaults. Building with cob is a sensory and aesthetic experience much like sculpting with clay. Unlike adobe, cob can easily be built in cool, wet climates such as the Pacific Northwest; its resistance to rain and cold makes cob well suited to all but the coldest parts of America, where it may need additional insulation.

Why Hadn't I Heard of Cob until Recently?

Cob was little known in North America until The Cob Cottage Company reintroduced it in the 1990s. However, the technique was once common in the Southwest, where it was called *puddled, coursed,* or *monolithic adobe.* Some early settlers in the Northeast also built with cob. More recently, government regulations have supported the lumber industry's sales needs by not recognizing earth as a suitable building material.

Is Cob Durable? What about in Rainy Places?

In England there are tens of thousands of comfortable cob homes, many of which have been used for more than five centuries. Yemen's medieval 10-story skyscrapers are part cob, as is Taos Pueblo, continuously inhabited for 900 years. A large part of both the Great Pyramid and the Great Wall of China are earth, and the oldest known human dwellings (earthen, of course) in Jericho have survived 9,000 years. So your cob house should easily outlast neighboring studframes, designed for only 50 years' service.

Like any other kind of structure, cob buildings need good roofs and adequate foundations to protect them from water damage. Traditionally, cob walls are protected from driven rain with a lime plaster or stucco, though in protected places in England, unplastered walls have stood for centuries.

Aren't Cob Houses Cold and Damp Inside?

Winter visitors to our cob buildings in the Oregon rain forest often comment on how warm and dry they feel. Cob walls 1 to 2 feet thick provide immense thermal mass and adequate insulation, ideal for passive solar construction. Cob structures require little additional heating in winter and remain cool and comfortable on hot summer days. As it is fireproof, cob can be used for building ovens, stoves, and chimneys, and is ideal for completely unburnable houses in forest fire areas. One of our favorite designs is a cob bench or bed heated by the flue of a wood-burning stove.

What about Earthquakes?

No building system is guaranteed against earthquakes, but cob buildings have a good survival record in seismic zones. Unlike adobe, held together mostly by gravity, cob buildings are locked together by an invisible three-dimensional textile of interwoven straw, with thousands of individual stems giving great overall strength. A cob mansion in Nelson, New Zealand, has survived 150 years in one of the world's most active seismic zones, enduring two major earthquakes without a crack, while the town around it collapsed. The curve and taper characteristic of Oregon cob walls make them even stronger.

Will Cob Make a Normal, Big, Square House?

Absolutely. It can be as big as you need it. But don't sacrifice quality for size. Do you want a work of art or just square footage? Yes, you can make your walls square, but it will cost extra in time, effort, or money. Nature doesn't provide Square; we have to carefully true it up, whether we mill the round tree into square boards or cut round stones square for a rock wall. Earth's yearning to be curved and sculptural gives cob walls an aesthetic quality we can't otherwise easily achieve.

What Does a Cob Cottage Cost?

Cob is one of the least expensive building materials imaginable. Often the soil removed during sitework is enough to build the walls. With inventiveness and forethought, the costs of other components (doors, windows, floors, and so forth) can be extensively reduced. Total building expenses will depend upon size, design, your creativity and organization, and how much you want to be involved. Most of us go to work to pay someone else to build our houses. We also pay for their profit, mistakes, transport, overheads, and bad luck. Above all, we pay loan institutions to borrow their money, then build under their constraints. By being your own builder, you can manufacture your own materials, and by going slowly and carefully, your house could cost one-tenth of what you'd buy one for.

How Fast Can I Build a Cob House?

It takes a long time to make a good house whatever the materials, but in dry weather you could build a two-story wall in a month. A determined owner-builder can move into a modest cob house in less than a year. In wood construction, the frame is a tiny part of the work, but a cob wall once built is finished, apart from the plaster. Pipes and wires are laid directly in place, and there's no need for drywall, tape, spackling, sanding, painting, sheathing, or vapor barriers. Yet racing to build fast is missing the whole point and half the fun. Unlike conventional modern building with its frenetic pace, power tools, and scope for errors and accidents, cob-making is a peaceful, meditative, and rhythmic exercise. Building cob is faster, easier, and more enjoyable with a team, so it lends itself to community projects, building parties, and workshops.

What about Building Codes?

Codes today protect the corporate manufacturers of building components better than homeowners. Not surprisingly, there is no code for cob, though nowhere is earthen building prohibited. Many cob builders choose not to involve building officials and have had few problems. Legally permitted cob buildings are beginning to appear, but there is considerable expense and paperwork involved, as with any permit. You have several alternatives to these inconveniences; see appendix 2.

Can I Find a Competent Builder?

With cob construction now a decade old in North America, professional contractors are beginning to appear. We keep a roster at the Cob Cottage Company; contact us. But rather than hiring a builder, consider building for yourself. If you are personally involved, you'll likely get a better job. Many cob builders we know are middle-aged or elderly, or are women. You don't need expensive equipment or lengthy training to learn the basics of this friendly, safe, and forgiving medium.

How Can I Learn More about Cob?

- The only proven way to learn about cob building is to try it!
- The Cob Cottage Company offers hands-on construction workshops in most parts of North America. Cob building is amazingly simple. In a week-long workshop, you can learn how to select materials, prepare a mix, and build a wall. We have taught men, women, and children of all ages and abilities everything needed to build cob parts of a small cottage, including site selection, foundations, windows and doors, floors, attachment to wood and other materials, detail work, and finishing. Course graduates with no previous building experience leave feeling confident and enthusiastic about building their own cob cottage.
- At the new North American School of Natural Building in the rain forest of southwest Oregon, we sponsor training in cob and other aspects of natural building, including roundwood carpentry, natural plasters, earthen flooring, passive solar design and siting, earth stoves and built-in furniture, living roofs, etc.
- For those serious about a livelihood in natural building, apprenticeships are available with a year-long certificate program. Contact us for our current schedule, or check our visually inspiring Web site at www.deatech.com/cobcottage.
- We supply videos and slides of finished cob buildings, *The Cobber's Companion* (a step-by-step, how-to-do-it manual), *Earth Building and the Cob Revival* (an informative illustrated reader), and other literature.
- *The CobWeb,* the only newsletter mostly devoted to cob, is available by subscription.
- We are available for public slideshows and demonstrations.

PHILOSOPHY, BACKGROUND, AND DESIGN

I

(What You Need to Consider Before You Start to Build)

BY IANTO EVANS

Ianto's Story

Although I never met him personally, Adolf once did me a tremendous favor. Even without knowing I was there, he sent his planes over Liverpool, where that night in August 1940 I was born to the height of the Perseid meteor shower and a hail of German bombs. My mother, though not normally at all timid, had the good sense to flee as soon as she could, leaving our ruined street, and she took us to live in the countryside of Wales, the only home I would know until my twenties. We survived, and throve; poor Liverpool hardly did.

COURTNEY ROGMANS

Wales is a cold, windy country full of mountains, sheep, and beer, and we lived in a series of tiny, ancient stone houses. You will not be surprised that the first phrase I can remember hearing was CAE'R DRWS, MWDDRWG! meaning "Shut the door, you scamp!" I lived my solitary childhood happily marauding the empty landscape, digging muddy holes, and collecting wildflowers. School was a constant background irritation to a real life of holidays and weekends. Daily I made the six-mile round-trip trudge to my prison, down narrow country lanes, squeezing through prickly gaps in hedgerows, birds-nesting, building forts.

Somehow, despite the chastisements of enforced education, I came to be a government landscape architect and graduated in the late 1960s to a desk with four telephones. Mercifully, Harold Wilson was unreasonably obdurate about keeping Britain out of Vietnam, thereby keeping me from the horrors of conscription. To maintain professional mystique, my licensing board made it clear that no real architect (and gentleman) would ever lower himself to pick up shovel or hammer, so I became an expert at many aspects of building about which I knew nothing at all. Liberation came only by fleeing to the United States in the 1970s and mailing back my license, only after which did I feel the freedom to actually build anything.

Three years in rural Guatemala refined some skills and taught me how to make sand-clay stoves. I built hundreds, experimentally, and wrote a book on them, *Lorena Stoves,* naively precipitating myself into the crazy world of international consultancy, parts of which I would rather forget. But I learned how to manipulate natural materials. Mostly, I got an abrupt course in how the world really works—the catastrophic impact of consumerism on traditional cultures and on the wild ecologies that support us all.

In 1978, teaching a workshop on mud stoves for development workers, I met Linda Smiley. We trod mud together, a sticky procedure that has glued together our lives ever since. In the 1980s cob buildings demanded our attention during a bicycle tour of Ireland and Wales. In 1989 we built the first cob cottage constructed in the United States in the past 150 years, and in 1993 we formally set up The Cob Cottage Company. This book is one result. Read on.

It has been my privilege to have spent much of the past twenty-five years in villages in conserver cultures, all over the world. I went not as a tourist, to gaze at them, but as an invited specialist, to help communities deal better with the daily strain of fuel shortages, smoky kitchens, and cooking in primitive conditions. I lived with the people. I worked in the inner sanctum of the home. Often people had never before met a foreigner, and in Muslim countries I was sometimes the first man ever to set foot in the kitchen. They taught me far more than I ever taught them, those tireless women with flocks of children. Though they may have lacked running water or a table or a finished floor, they were endlessly cheerful, exceptionally hospitable, always laughing.

My conserver-culture friends showed constant curiosity about where I lived, how many wives I could afford, what gets cooked in my kitchen. In answering their questions I reflected on the contrasts between our lives, and so saw more clearly how their days unfolded. Time is not money in traditional places, so everything is done at a dignified pace, done until it's done. Our own language lacks the finesse to describe their daily rhythms; there are no words in English. In how they cook, in the buildings they make, there is a completeness and immediacy we seldom know. It feels almost religious. Everything is done with care and respect and humor. *In every action is sacred process.*

PEASANTS AND FARMERS

I once heard a Chilean named Ana Stern give a speech on "The Difference Between Peasants and Farmers in Mexico." Peasants, she said, satisfy their own basic needs: they grow their food, build the houses they live in, often make their own clothes. Most peasants collect medicinal herbs, treat medical emergencies, supply their family entertainment. They experience fully what they do every day; they have time—they feel joy. Their culture is integral, it makes sense. Farmers by contrast grow things to sell. With what they earn from their products, they *buy* their groceries, building materials, clothes, entertainment, and medical insurance. They must also buy into a system which demands that they drive to market, pay taxes, perhaps send their kids to agricultural college. Increasingly they must buy machinery, seeds, farm chemicals. Farmers have no time to directly enjoy satisfying their own needs, so they purchase their satisfactions; they buy ready-made clothing and "convenience" foods.

I've thought a lot about Ana's presentation. Her definition shook my worldview. In her terms we are all farmers—there are few peasants in the U.S.A. I'd always felt comfortable in the traditional villages of Africa and Latin America, and now I understood why. The parts of my own life that I truly enjoy are the peasant parts, the parts I don't pay for, the parts that I myself create. A life of working for someone else and paying for basic needs is essentially unsatisfying. Why? Because our links to Nature are severed when we live that way.

Why do we grow garden vegetables? It's not the easiest way to obtain food. The simplest cost-benefit analysis will show that it's hard to make the same money growing lettuces as we do going to the office. Otherwise wouldn't most of us be lettuce farmers?

We grow food (or flowers) for completeness, for the *grounded* understanding that comes from putting seeds in the *ground* and tending them, feeding, watering, picking, eating. To be complete we need to have a constant awareness of our cosmic bearings, where and when we fit into Nature's patterns. If you compost your excrement as the Chinese do, use your own urine for fertilizer, and grow your own vegetable seeds on the plants you raise, the cycle is complete. You have inserted yourself into a completely visible ring of cause and effect. You experience the whole natural process, and the better you observe how that process works, the easier you slide into it.

THE IMPORTANCE OF HOUSING YOURSELF

The peasant/farmer analogy works equally well for houses. For most of history most of us humans have created our own homes. The whole family helped when the work was too heavy or too slow; the entire community assisted when the work required it, as with an Amish barn-raising. Only recently have we traded with those outside of our friends and family in order to have homes. At first we traded only for parts or techniques beyond the reach of the homemade; the village blacksmith made the hinges, and we gave him eggs. Later we paid money to skilled local specialists for more durable, better-made work. Then, not long ago, we started to pay complete strangers and distant corporations our hard-earned cash to supply us with skilled trades and pre-assembled components. To earn that money, we had to grow a surplus. The self-sufficient plot was no longer big enough.

Peasants became farmers. Yet small landholders often can't survive in a cash economy: their land is sold to a bigger operator. Not having land, they don't have access to the earth, rock, trees, or straw that were previously at hand for building materials. In order to pay for housing, perhaps they turn to producing artifacts or services to sell.

That's the stage set. We go to jobs doing possibly meaningless work for thirty, forty, or fifty years to pay for a house with which we no longer have any direct connection. How many of us have been in a steel mill or a drywall factory? If we have, do we enjoy what we smell and hear and feel there? When schoolchildren take a field trip to the slaughterhouse, they stop eating meat. When we see how building components are made, we seek better ways to house ourselves.

The natural building movement has helped humans reconnect with our tradition of self-reliant shelter, surely one of our natural rights. With cob, we aren't banging together ready-made components bought at a chain store. With cob, we take the free building materials from the ground beneath our feet—good clean dirt—and shape it to make floors, walls, plasters—homes. Homes so beautiful they make grown men cry.

Most of the buildings most of us live and work in are soulless, anti-ecological, and ugly. We close our senses when we are in them. But there's another kind of architecture, one that feeds the soul and spirit, that helps us feel good, that elevates our daily lives. The old Dominion-Over-Nature days are past; we need an "ecological architecture" that reestablishes us in our place in Nature, where we are constantly reminded of the glory of the world around us. In creating natural buildings, you will create ecstatic spaces, a place where your spirit can soar.

A shift in attitude comes of making what you need for yourself. You change your outlook from "I want, so I have to buy . . ." to "What's here? What can I best do with it?" The first attitude is how a consumer society approaches life. The second is how people in traditional societies have always looked at their world. It's called "creativity," and it's enormously satisfying.

Once you learn to create your basic building materials from the ground beneath your feet, your vision opens up. Now you see the role of roundwood thinnings in framing a roof and realize how easy it is to build door frames from poles, to shovel sod onto your roof, to set frameless glass shards for windows into a cob wall. A world of possibilities opens.

"But I don't have time for all that. When you have a REAL job and work for a living and raise three kids there just isn't time for such a laid-back attitude. Someone's got to pay the bills. Anyway, we don't own a building site." In this book we'll address all these concerns and tell the stories of people who have overcome them. We'll show that it's not the housebuilding that many of us can't afford, but rather the drive to work to pay for the gas to get to work to pay income tax so we can pay the remainder to the bank to borrow *their* house to live in.

Central to building your own natural house is the lifestyle change that frees you from tedium and feeling trapped by debt. If you follow the thought processes and building principles explored in this book, your housing costs may almost disappear, creating an opportunity for you to take a sabbatical and create a house that really inspires you.

Most importantly, remember that building is not something you do in a rush to get a finished structure. Both building and living in your house can be spiritual processes of daily joy, reflection, and connection with Nature. You are not producing a product—take the time to enjoy all of the process.

Natural Building 1

IN A SOCIETY WHERE MOST OF US HAVE never been in a natural building, people are hungry for any refuge from the right angle. Nature's geometries, though frequently incomprehensible to Industrial Man, were well known to our ancestors, who understood that Nature is tolerant of our aberrations only briefly, and then only if we continue to expend energy in maintenance of our own unnatural versions.

The rigidly rectilinear building components of our industrial system stay that way for a very brief time. The moment they leave the mill or the mold, Nature's forces begin working to reassimilate them into the cycles of life and death they were shanghaied out of. Square corners rapidly become rounded. Flat boards warp, glass shatters, plastics degrade in ordinary sunlight. Entropy works fastest on the most unnatural shapes. Each of our carefully standardized identical productions deteriorates differentially, gradually reestablishing the diversity that is universal.

Natural materials occur in the state nature provides, irregularly curved and knotty, or plastic, heterogeneous, and without Cartesian geometries. Using natural materials, it takes less effort to make buildings that fit the spatial needs of the occupant, where each space is distinctively different, free of the deadness and sameness imposed by manufacture and codified standards.

> ❧ TRADITIONAL PEOPLE TEND TO DO EVERYTHING MINDFULLY AND CAREFULLY . . . WITH NO DISTINCTION BETWEEN ART AND WORK AND MEDITATION AND EDUCATION. IN EVERY ACTION IS SACRED PROCESS. ❧

But natural building is more than just materials. Stuffing the gaps in a 2 × 6 frame with straw bales rather than fiberglass addresses only a tiny part of the challenge. Natural building implies profoundly different attitudes to places, building sites, ecology, work, and how we live in buildings. Natural building means paying more attention to all the details of how the world really works.

Natural materials fall neatly into two groups: geological and biological, connected by water.

LEARNING FROM NATURE

We need to do more than avoid toxic materials. When we take on construction in a natural way, we will assume a whole range of new activities and ways of interacting with a building. Above all is an overarching respect for Nature, respect for the place where your building stands, and respect for distant places where the cumulative effects of each of our activities will be felt.

Involving Nature in building doesn't mean adding more money or work; it's more like relaxing. Following Nature's way means doing less, buying less, building less, but thinking, feeling, and observing more. It's what you *don't* do that makes natural building easy.

Because of the consumer paradigm most of us hold, it may be tempting to assume that ecological building needs to cost more, or that it is nonecological building with green products tacked on. For instance, "solar architecture" can have its black boxes, fans, timers, pipes, and valves, all of them additions to the basic house, which may be oriented completely wrong, be built of inappropriate materials, or fail to let in sunshine. After centuries of industrial thinking, it will take time, effort, and careful observation to learn to build in harmony with Nature, and to make house construction easy.

Your main task is to observe, simplify, and involve yourself.

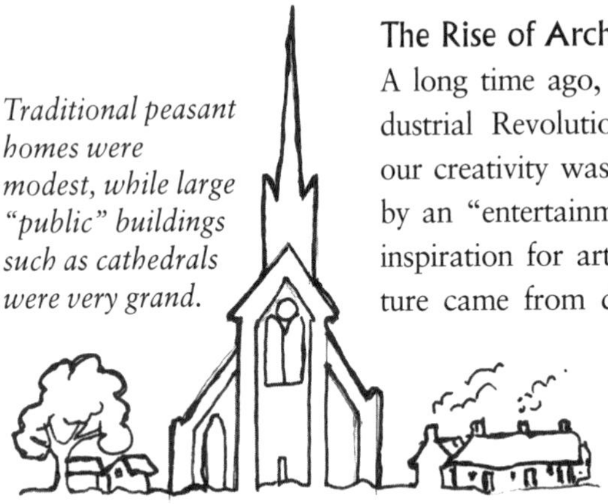

Traditional peasant homes were modest, while large "public" buildings such as cathedrals were very grand.

The Rise of Architects

A long time ago, before the Industrial Revolution and before our creativity was compromised by an "entertainment industry," inspiration for art and architecture came from direct observation of Nature. Nature was the omnipresent matrix within which our little effects were small islands. We didn't need to travel *to* it, for it surrounded us. We ate Nature daily, as most of us provided our own basic foodstuffs. Long days tending gardens, fishing, and collecting wild things were saturated with watching, observing, and perceiving the cause and effect of the natural world. Daily we noticed, commented on, and discussed wildlife, wind, and weather. The fabric of our lives was woven on the warp of natural happenings.

When we built, we built in the image of Nature's constructions: standing poles, rock upon rock, reconstituted earth, matted straw thatch on roofs. Most people knew intuitively and by practice from childhood how to follow Nature's basic rhythms, shapes, processes, and timescale. Our native wisdom came from daily immersion in the functions of the natural world, because we had no other model.

Even today, nonindustrial traditional societies build with pragmatism, as do beavers, bees, or bluebirds. They build only big enough to satisfy their shelter needs, with a crisp efficiency in using materials. Traditional shelters tend to be compact, minimal, and simple, and only occasionally, more extravagant in scale, finish, and handwork. The village of peasant huts sported the Norman cathedral, or a bright and turreted mosque rose above the somber mud houses of the surrounding town.

A new class arose in Europe during the Middle Ages, a class of merchants and professionals, and among these were architects. These men lived in towns and forgot to watch the clouds in the sky. They were busy; they were ambitious; they knew better than Nature how to build. The grandeur of their stone cathedrals and gigantic castles gave them arrogance and a belief that they could dominate Nature's laws, mold natural stuff into the precisely rectilinear shapes of their drafting tools. With the Industrial Revolution, they could forge iron in quantity, then aspire to outdo each other in height, length, and breadth.

Now, for two hundred years or so, we have chiefly competed with and copied one another. Unlike the most influential poets, painters, or composers, who more often have taken inspiration from Nature, building designers have incestuously idolized their own kind. The influences on architecture are almost entirely human, mostly emulations of other architects, the media, and industry. Our buildings have moved ever further from reflecting the wonders of the natural world. Now we need a way to connect our buildings with those principles and wonders again, because after all, the only true constants are in Nature.

How we build at the start of this millennium will affect the chances for survival of our grandchildren. Our recent record at building—architect-impelled and industry-mandated through government decree—has been an unprecedented drain on planetary resources. *In the 1990s, 10 percent of global energy/resource use went into buildings in the United States alone.* Even small reductions in this demand could equal the entire energy budgets of other big countries. Tiny reductions in such waste could leave thousands of square miles of forest standing, reduce global air pollution, free millions of people from stunting physical labor, and reduce the need for nuclear power.

Today's typical buildings generate appalling pollution, deforestation, and habitat destruction during construction, in the manufacture of components, and in decades of daily use. They will be the cause of many species' extinctions, especially where their materials are mined, transported, assembled, or processed. Their purchase and upkeep will impoverish our children and their children. By creating such buildings we are limiting the chances for our own descendants to survive. Many of us now acknowledge this truth, yet, trapped by the momentum of our society's suicidal inertia, we fail to recognize the steps that would restore natural equilibrium. Because architects, engineers, and the building industry are responsible for the constructed world we now inhabit, we need to find other, more reliable inspirations from which to create a new building ethic.

Natural Laws

Many of us are now demanding natural buildings. *Natural* means conforming to the laws of Nature, to principles we can extrapolate from the workings of unaltered ecology. How do we most effectively ask Nature for answers?

Nature's laws are not written like human moral codes that take the form, "Thou shalt . . ." or "Thou shalt not . . . ." Instead, they are written in probabilities, connections, and cause and effect. It is not possible to break these laws because they are only descriptions of what happens. When you drop an apple, it falls, every time.

Certain general truths about ecology give clues as to how natural processes work so efficiently, elegantly, and effortlessly. Here are some examples of Nature's tendencies, from which we can extract principles to help us in building.

❧ ***Nothing is ever created or destroyed; it merely changes form.*** This is called the First Law of Thermodynamics.

❧ ***Everything gradually falls apart*** (this is the Second Law of Thermodynamics). On a local level, living organisms and systems appear to be an exception to this rule. Children grow bigger, stronger, and more organized each year. Older ecosystems grow more complex than their predecessors. But these particular occurrences all happen in a solar system where the sun's atoms are continuously decaying, releasing heat and energy, and increasing entropy. All life feeds on death.

ABOVE: *Each daisy is observably different from all others. Nature never repeats.* BELOW: *Shore life in Oregon. Life occupies every available niche.* BOTTOM: *Xerophytes in the Sonoran desert.*

❧***Everything is unique.*** No event repeats itself identically; no physical entity is the same as any other. Even within Nature's basic templates, every rendition is observably unique and subject to its unique circumstances.

❧***There are no monocultures in Nature.*** Living things exist only in the company of other species. Complex systems are more diverse and therefore more stable and self-regulating.

❧***Nothing ever stops moving.*** Change is the only constant. Sometimes, as with the growth of a bamboo plant, that change happens quickly. Sometimes, as with the movement of continents, change is so slow that we can't even see it.

❧***Nature has a series of fundamental geometries, each for a certain set of phenomena at a particular scale.*** Biological forms tend to be irregularly curved. On the scale from humans to homes, natural objects never come in perfect squares, circles, triangles or cubes. Nature produces few straight lines, right angles, flat surfaces, or unbroken colors, and when they accidentally arise, she rapidly sets about diversifying them.

❧***Life quickly occupies any niche it can exist in.*** It takes energy to keep it out, but once one organism has occupied a niche, that space is less vulnerable to invasion by others.

❧***Nature uses just as many resources as are necessary and no more,*** for example, when growing a structure such as a tree trunk or a skeleton. Animals that create their own shelter shape it to snugly enclose their activities. They don't build surplus space or add unnecessary structure.

LEARNING FROM CONSERVER CULTURES

Most of us who can afford to read these pages have been conditioned to think of the world as divided into groups of nation-states, Us and Them. *Us* we call the Western Developed Nations, the North, or the Industrial Countries. *Them* have been variously referred to as the Developing Nations, the Underdeveloped World, the Lesser Developed Countries, the Third World, the South, and, calling a spade a spade, the Poor Countries. (In Mexico I have heard *Us* described as the Overdeveloped Countries.)

Not only is this labeling insulting and damaging, it is inaccurate and it misses the crucial point. National boundaries don't necessarily separate cultural groups, nor should we lump together whole cultural diversities as "China" or "Brazil." There are pockets of resourceful self-sufficiency in Switzerland; there are wasteful industrial elites in almost every nation, worldwide. *Consumer culture is not bound by geographical, political, or even economic borders, but spreads gradually, like an infection, from the cultures that institutionalize waste to those that practice conservation.*

Most traditional cultures, particularly those based on hunting and gathering, lived in one bioregion for a long time. Usually, through careful use of natural resources and because their numbers were well within the ecological carrying capacity, their long-term effect on the ecosystem was minor, perhaps only slowly producing a gradual series of changes. Typically these traditional cultures were forced to conserve local natural resources, move on, or die out. The people who inhabited the Willamette Valley of Oregon for the several thousand years before the coming of Europeans in the 19th century left only a few arrowheads and stone knives; overall they appear to have lived more or less sustainably. We could call them a *conserver culture*.

In complete contrast, consumer culture institutionalizes and glorifies waste at many levels (though a minor part of ours currently pays lip service to conservation). Its "wealth" is dependent on the ceaseless expansion of consumption. A consumer culture demands a continuous acceleration of "throughput," which is the tranformation of raw materials into refined goods and ultimately into waste and pollution. Consumerism depends on regular clearance of old goods to make way for new. Resources are ever more quickly hurried from source to disposal.

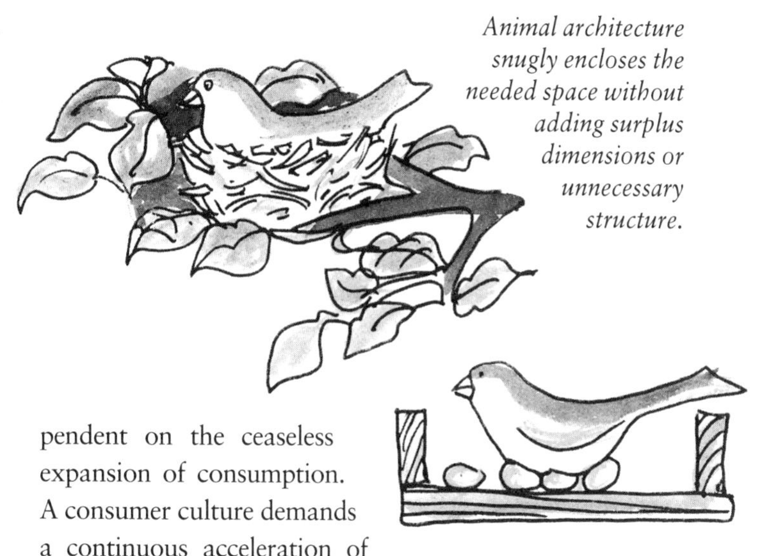

Animal architecture snugly encloses the needed space without adding surplus dimensions or unnecessary structure.

Not long ago, we were all conservers. Just a few generations back most of our ancestors lived in traditional societies where resource use was close to sustainable. Our forebears *didn't* live lives "nasty brutish and short." Their days were filled with mindful application of labor to honored tasks, and their artifacts were fewer, more durable, and cherished.

Today's surviving traditional cultures remain much closer to Nature than is possible in any industrial throwaway society. They still practice the ancestral wisdom that develops in societies where there have not been resource surpluses. They have much to teach us about how to build and how to think about *place*. Here's a sample of some lessons we could learn from them:

❧***Don't build any bigger than you absolutely need.*** Most traditional cultures have built humbly, except for fortifications, ceremonial gathering places, or religious monuments.

ABOVE: *Tiny houses in northeastern Brazil shelter big families.* RIGHT: *Abandoned French cannons built in to protect portals from wagon wheels. St. Louis, Senegal.*

❧***Use whatever materials are close at hand; fashion your architecture around them.*** Rather than transporting heavy materials from far away, traditional cultures usually built stable homes with stone or earth from right under their feet.

❧***Consider the advantages of conjoined and clustered buildings.*** They occupy less space and are easier to heat and cool. They function better, and they *feel* better. They foster cooperation and community; it's hard to be a rugged individualist if your house is nestled in close to family members and neighbors.

❧***Never demolish a serviceable building if you can change it or add to it to accommodate new uses.*** The tangible history of older buildings is essential to your new use; the ghosts are there to make us feel welcome. Let the building carry a cumulative record of its users by the changes they make to it. History can speak from a stone threshold hallowed by centuries of treading feet or in the heights of growing children scored and dated on an ancient doorpost.

❧***Pillage abandoned buildings for components.*** For example, an 18th-century stone house in England contained squared and carved sandstone blocks taken from a ruined church that had been burned by Henry VIII. The church was built in the 13th century with stone quarried by the Romans for a villa which was then abandoned for seven hundred years. Likewise, much of old Cairo was built with stone facing from the pyramids. Good materials are better for having experience.

❧***Accept and encourage the natural decay of natural materials.*** Managed well, decomposition can be very graceful. Ruined castles are "picturesque," but deteriorating mobile homes are squalid.

❧***Understand that straight lines, flat surfaces, and right angles are ecologically expensive.*** So are expanses of only one color or texture, or any mechanized geometries. Nature will steadily sabotage any attempt to introduce monotony and consistency, so if you try to impose them, understand there are construction and maintenance penalties. Refashioning Nature costs energy and usually results in a loss of structural strength and durability, as happens when we mill roundwood into square timbers, losing the strength of its inherent geometry.

RIGHT: *Mahogany dugouts, palm and log houses, each humbly distinct, on Guatemala's Caribbean coast.*

ABOVE: *This large, complex farmhouse in Wales began tiny, a single story (at extreme right), then grew with need over centuries.* BELOW: *Plan of a village in West Africa. Note repeating but constantly changing pattern of long, shallow curves connected by short, steep curves. There are no right angles.* BOTTOM: *House in Guatemala made of motorcycle crates.*

❧***Make warm, cool, or dry places where it suits people to be.*** You don't need to heat/cool the whole structure. It is the people, not the building, who want to be warm, or cool. The building doesn't care.

❧***We can emulate the ways traditional people's buildings blend into the landscape,*** even though they may each be distinctly different. (Ours, by contrast, shout for attention, though they are often identical.)

❧***Build incrementally, as you can afford to, without borrowing money.*** Start small; add only when you need to and can afford to.

❧***Decorate your building as you build.*** Key zones to embellish include projections, entrances, and other foci. Decorate doorways, especially above them; the ends of the roof ridge; the highest point, visually; the hearth; walls at eye level; and corners.

❧***Involve the whole family in house construction.*** In conserver cultures, it is common to see tiny children who can barely walk carrying building materials, grinning ear to ear with involvement, accomplishment, and acceptance.

❧***Don't try out complex new ideas on a building you will live in.*** Practice on animal shelters, courtyard walls, and so on. Be conservative in how you design and build, and do what you know how to do, well.

❧***Build to take advantage of Nature's basic rhythms, shapes, processes, and time-***

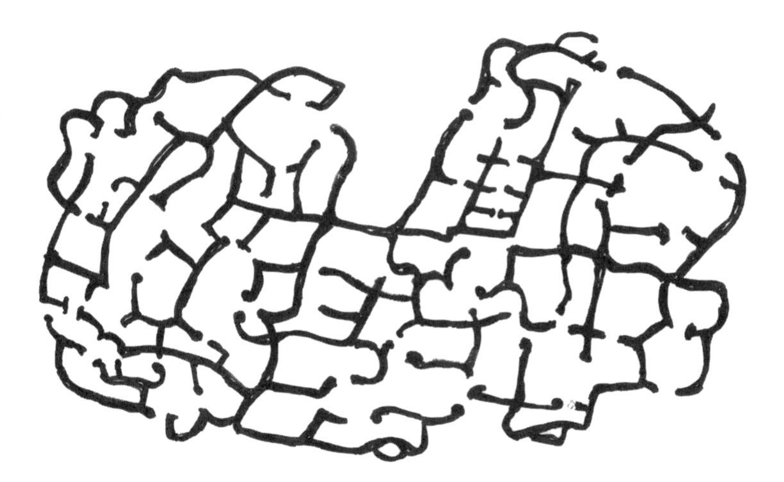

scale. Traditional people observe the natural patterns daily in everything they do.

❧ ***Never buy anything new if you can reuse an old one, borrow it from somebody, make your own, or as a last resort, buy it secondhand.***

Preindustrial people build enjoying every part of the work; it's not a rush to beat the deadline. There's no distinction between art and work and meditation and education. Traditional people tend to do everything mindfully and carefully; they do everything as well as they can. *In every action is sacred process.*

WHAT ARE NATURAL MATERIALS?

At the heart of natural building are natural materials. A convenient definition of *natural materials*" would be "materials that are not industrially processed." But unless you find a hole in the ground or a hollow tree and live in that, your home will comprise materials that are to some degree "processed." The beaver strips branches, then cements them with clay; bees and wasps synthesize wax for comb and "paper" for shelter; many birds create nests of complex combinations of, for instance, straw, clay, sticks, and feathers.

The difference between animals' processing of natural materials and industrial humans' processing is the key to a good definition of natural materials. When animals process natural materials to create their homes, they treat each component part as a discrete entity. Their work celebrates the diversity of the universe. They assemble, by the beakful or the pawful, heterogeneous materials into complex structures. Humans take those same discrete diverse materials and process them into uniformity. These natural materials become the raw ingredients of manufactured homogeneity.

So we might define *natural materials* as "materials that, even when processed, retain their essential nature." Natural materials are respected, and used, for what they are.

Wood is a natural material even after the tree is cut down, even if cut with a chain saw, even after it's sawn into lengths, even after it's split or squared by a hand tool. Each tree used in this way is respected as an individual organism. If the tree is milled into boards by that same chain saw, it is still to some degree natural in that there's a level of personal involvement, of response to specific circumstance and therefore creative choice. Wood's "naturalness" diminishes drastically when a big truck hauls it away to the lumber mill.

Almost any modern building, however natural overall, will require some materials that have been unnaturally processed. It's hard to build passive solar without using glass. Even cob, composed of completely natural components, is somewhat industrially processed if we use mechanically baled straw. There are no hard-line distinctions, but processing separates into several stages, each progressively further from Nature and with progressively more liabilities.

Unprocessed and Uncombined

The list of raw materials for genuinely natural buildings is quite short, dividing distinctly into those that are biological and those that are geological. The biological we pull out of their cycle of growth, breeding, decay, growth. The geological we borrow from the earth; these materials don't grow and they decay extremely slowly. Geological materials do not suffer from being eaten as do the biological; life in general has little effect on them. Rocks and clay tolerate warmth, low humidity or extreme dampness in ways that wood or straw could never survive. Fungus, bacteria, or insects will eat damp biological materials, and dry heat shrinks and breaks up wood frames and thatch roofs.

Animal architecture is minimally constructed with the most readily available materials best suited to the skills and needs of the builder.

In natural building the major materials we use are these: rock, gravel, sand, clay, water, grasses, bamboo, palms, and trees. In the diagram on page 7, they are shown in a sequence linked by water, the geological separated by decreasing scale, the biological by increasing. Water connects the two. To this list add minor materials: natural tar, gums, saps, shells, resins, beeswax, plant and animal oils, wool, skins, and others.

This palette of primaries, as with color, is finite; yet the possible combinations of these basic materials are nearly infinite. We haven't even begun to explore. We are very far from exhausting the possible combinations even of cob's three elements—sand, clay, and straw. In its rush toward industrial processing of everything for profit, our society has neglected even fundamental experiments with unprocessed materials.

Primary and Secondary Processing

Primary processing has been part of folk tradition for millennia. After primary processing, the material remains a separate element: shaped and squared rock, fired-clay bricks and tiles, lime from limestone, milled boards, sand fused into glass, straw bales, iron melted into nails, linseed oil.

A huge conceptual jump takes us to secondary processing, where elements are combined into synthetic amalgams that don't exist in nature (and which are relatively slow to degrade or break down into toxic byproducts): aluminum alloy, stainless steel, plastics, most preservatives, paints, varnishes, particle board, drywall, and above all, cement.

Those materials that have been unnaturally recombined are the source of our deepest misgivings because, like any material with which we did not co-evolve, we have no proven genetic resistance to its damaging us, whereas we have evolved genetic responses to every natural chemical and physical combination in our habitat. When in the space of one or two generations we are suddenly faced with pentachlorophenol or formaldehyde, PCBs or dioxin, our bodies have no ready protections, and we are easily poisoned. We should expect that any synthetic material is likely to be toxic to all life-forms, to differing degrees.

Component Assemblies

Preassembled units are a quantum leap, not chemically but socially, even from secondary processing. Natural materials offer us opportunities—to respect and work with their diverse traits, to see each material's texture and scale and color and hardness, its uniqueness. In the case of manufactured components, all the essential choices have been made for us. We purchase ready-made windows, preassembled roof trusses, vacuum-formed plastic kitchens. Factory-built houses ("mobile homes") are the current extreme.

In lacking the challenge and stimulus of constantly having to adjust to materials' diversities, we are dumbed down. Like the machines these assemblies are made by and for, we ultimately become listless and uncaring, stupefied by boredom, wasting that priceless acuity that comes with active creation.

Living with, and in, natural materials, we can value their visible exposure. They are at their best uncovered, glorious in their God-given nudity. Naked materials reveal the structure of a building, demonstrating the wonder of how it defies gravity, and celebrate the glory of every separate component.

WHY BUILD WITH EARTH AND OTHER NATURAL MATERIALS?

Health

Modern buildings are usually toxic to both builders and inhabitants. Many of the most fervent supporters of natural building are people with acute chemical sensitivities and other environmental illnesses. "Chemically sensitive" people are particularly aware of how modern buildings make us sick, but it is likely that we all suffer from "chemical-sensitivity syndrome," to different degrees. Even the mainstream press often carries stories of cancers and respiratory problems linked to formaldehyde-based glues, plastics, paints, asbestos, and fiberglass. The toxicity of these materials impacts everyone associated with them: workers in the factories and warehouses, builders on the construction site, and inhabitants of the poisonous end products.

Natural materials such as stone, wood, straw, and earth are not only nontoxic, they are life enhancing. The chemical stability of earth, coupled with our having evolved with it as a part of daily life, suggests how safe this material is. Earth seems least likely to stimulate bronchial infections, allergy problems, skin irritations, or any chemical sensitivity. Clay is known in fact to be curative, a healing material that has long been valued for its ability to absorb toxins. During cob construction, skin contact with wet clay is noticeably therapeutic. It can dry up poison ivy rashes, heal cuts and abrasions, and appears to help people feel generally healthy.

Earthen houses, without sealants or cement stucco, breathe gently and slowly through the entire wall surface. The walls may also have abilities to absorb airborne irritants within the building and to soak up and level out excess humidity or dryness.

Psychological Well-Being

There is increasing evidence that modern buildings compromise our psychological and emotional health as well as our physical well-being. As we've noted, right angles, flat surfaces that are all one color, and constant uniformity don't exist in the natural world in which our ancestors evolved. These traits may trigger a subconscious reaction that tells us, "There's something wrong here," keeping us nervous and stressed. Most modern homes don't successfully stimulate our senses with the variety of patterns, shapes, textures, smells, and sounds that our preindustrial ancestors experienced. The monotony of our built environments probably contributes to our addic-

Cob construction avoids predictable uniformity.

tion to sensory stimulation through drugs and electronic media.

From Toronto to Tapachula, in Australia and Algeria, almost every individual we have met who lives in earth loves it. They say they feel healthier, more alive, more productive, yet relaxed and connected to Earth. They sometimes describe their houses as "growing out of the Earth." Even though modern-day people are conditioned to prefer the new, the shiny, the predictable, we respond at a deep level to unprocessed materials, to idiosyncrasy, and to the personal thought and care expressed in craftsmanship. Nearly all natural buildings, regardless of the level of expertise of the builders, are remarkably beautiful. All three of the authors of this book—Linda, Michael, and me—live in handcrafted natural houses. We have grown to expect looks of mesmerized awe on the faces of first-time visitors, and we've witnessed difficulty visitors have in prying themselves from the warm earthen benches when it's time to leave.

Financial Empowerment

Most new houses cost at least $100,000 and take a lifetime to pay for. Real earnings are declining, and housing costs continue to rise, trapping people in lifelong mortgages. Many homeowners take jobs they dislike to pay for houses they do not love. They hand over control of their personal finances to banks, which are some of the most ecologically damaging institutions on the planet.

But it doesn't have to be that way. By using local, unprocessed materials such as earth and straw, by building smaller and smarter, and by providing much of the labor yourself, you can create a home that is almost unbelievably affordable. With earthen building especially, the raw material is almost free, and the skills needed are very basic. We know many people who have built cob homes for under $5,000, and it is possible to construct a small but lovely cottage for $500.

A house you build yourself can be constructed slowly, in phases, as you can afford to buy the components. You can also save yourself money in the long run with a smaller, more efficient house that uses simple passive solar technology for heating and cooling. As the price tag drops from hundreds to tens of thousands or even to a few thousand dollars, it becomes easier to shrug off the yoke of loans and mortgages. As a result you may find your cash needs dropping. Your options for work broaden, and you lose less in taxes and transport. You can cut down the hours you work away from home and spend more time with the kids or grow a big vegetable garden that will reduce your needs and increase your satisfaction even further.

Comfort

In areas where wooden buildings need air conditioners, earthen buildings right next door are cool and fresh all summer long. While neighbors struggle to pay utility bills, cob houses stay snug and warm in winter, soaking up each hour of sunlight to slowly let out heat later.

Comfort also means acoustic privacy, both from exterior distractions and from noise generated in the same building. Row housing, schools, and homes with children can all benefit from the sound-absorbing properties of earthen walls.

Democracy and Empowerment

Industrial cartels, the building industry, and government have all conspired to prevent most people from building their own housing. We grow up being told, "You can't build a house unless you're a professional builder." We're convinced that we need to spend $150,000 and buy a 2,000-square-foot house. And as components for such houses become more technical, heavy, and dangerous to handle, building indeed becomes the province of lusty young men with expensive noisy machines (most of whom probably don't like their jobs any better than does the new house's thirty-year-mortgage holder).

Natural building is democratic. Most natural building techniques, particularly cob, are accessible to old women and little boys, the impractical, the handicapped, the impoverished. Cob is so safe that even tiny kids can do it; there's nothing heavy to fall, no dangerous machinery, no tools so expensive that we can't make or buy them. Natural building empowers those who have been all their lives persuaded they should leave building to "professionals."

House construction is a big-money business, with all the problems associated with other high-stakes industries. In the race to maximize immediate profit, concerns such as health of the environment and of the home's inhabitants often become secondary, as do the effects on the ecology and people in regions where resources are extracted. Yet the rich and powerful expect to make their own homes and lives pleasant (at the expense of less privileged people, who often live, invisible, in distant countries). The building industry and government regulation concentrate power in the hands of government and selected corporations by enforcing compliance with a limited set of options. If the code says we have to use concrete foundations with every building, just think how much profit the cement manufacturers will make!

Huguenot immigrants built these houses of cob and adobe block in Pennsylvania around 1830.

BOTH PHOTOS: BENGI ALLEN

Natural building encourages resourcefulness through the use of found and reclaimed materials and building elements, not only saving money and resources but adding to the life and spirit of the building. Techniques that use human labor and creativity produce a different social dynamic than those that depend on heavily processed materials, expensive machines, and specialized skills—a dynamic where people depend on each other and on personal creativity to get their basic needs met, instead of handing over their power to governments, corporations, and professionals.

Tradition and Heritage

The construction industry would prefer it if we forget that natural building is a strong tradition in many places. Continuing to build with earth maintains that tradition. In England, with forty thousand cob buildings in the county of Devon alone; in Australia with its heritage of cob and adobe; in China, with ninety million earthen buildings; even in the southwestern United States, we still have a strong tradition. There are cob buildings from the 1830s in New York and Pennsylvania, pre-Columbian towns of earth in Arizona and New Mexico. California in 1980 had 200,000 earthen homes. There are century-old adobes in Toronto and Idaho, in fearfully cold conditions, and "tabby" houses in subtropical Florida. These houses are loved and revered by their occupants.

Almost all of us have millennia-long connections with earthen buildings. Hardly anyone is without an ancestor who lived in an earthen home, in many cases quite recently. Small wonder we feel good living in them—this is our heritage.

Durability

All biological materials have a predictably short life. Life made them, so it will take them away. Wooden structures generally survive only for a few generations. As we currently build, using fast-grown softwood, a house is in poor shape in half a century.

Earth by contrast, being geological, lasts indefinitely. Inhabited earthen buildings in the Middle East and India are often more than a thousand years old. In that time forty human generations have come and gone. If you build an earthen home now, think of the wonder on the face of your great-great grandchild!

Environmental Impact

Building with natural materials reduces the push for resource extraction and for industrial processing. It decreases pollution, deforestation, and energy use.

The construction industry is a major cause of mining and industrial processing, with all their attendant pollution, ecological havoc, and social disruption. Modern building materials depend on mining: gypsum for drywall; iron for rebar, hardware, and roofing; lime for cement. And every material used in a typical modern building is the product of energy-intensive processing. The lumber mills, the steel foundries, the factories making plywood and chipboard, the industrial plants using tremendous heat to turn minerals into cement —all consume vast quantities of power, supplied by the combustion of coal and oil, the damming of rivers, or the splitting of atoms.

These manufacturing processes release toxic effluent into the water and hazardous chemicals into the air. The manufacture of Portland cement, for example, accounts for an estimated 4 to 8 percent of greenhouse gasses. And even after our building materials are made, modern construction depends on a stream of polluting trucks to deliver them to us, usually from hundreds of miles away.

It's no accident, either, that the dumpster is a prominent feature of most construction sites. A major byproduct of industrial construction and destruction is landfill waste, which comprises up to 25 percent of landfill volume.

Seventy-five percent of all trees cut in North America are used in construction. Here in the Pacific Northwest, the trail from clear-cut to sawmill to building site is easy to follow. Loaded log trucks roar down every country road, and on almost any day Linda and I hear from our house the crashing of trees, the whine of chain saws. When we built our house here only seven years ago, primeval forest, miles of it, still stood within sight of our windows. Now a few sorry remnants remain, a long drive away.

The United States has lost 15⁄16ths of its original forest, and the lumber industry is trying hard to cut down the last sixteenth. Now Siberia is under attack, and the last great tropical forests, to make cheap plywood for million-dollar wooden houses that will be defunct in half a century.

By substituting earth for wood in the walls, floors, and finishes in a standard 2,000-square-foot house, we can save 60 to 80 percent of the lumber. If as a whole society we switched to earthen buildings, six hundred to eight hundred of every thousand trees currently marked for cutting would be spared. The United States could be completely self-sufficient in timber without ever making another clear-cut.

Earthen building also dramatically reduces the need for extraction and processing of other materials. In cob construction, earth for building is dug locally; generally surplus earth from the building's own footprint is used. Processing is minimal, involving no machinery or chemicals, and can be almost silent. No mile-deep open pit mines, no toxic tailings or effluent ponds, fewer trucks on the road, no company towns. Earthen construction generates no air or water pollution. Excess material goes right back into the ground, effectively

STEEL ROOFS IN KENYA

In Kenya, right on the equator, we encountered an entire village newly rebuilt, completely of corrugated steel: roofs, walls, room dividers. This, in a region noted for its exquisitely built circular earthen homes. We stopped the truck, seeking an explanation, and as always a little crowd gathered under the vertical sun. Why, we asked, were they opting for a material that in the tropics lasts only a few years, is freezing cold at night and is an oven in daytime?

The answer surprised us all. They wanted to be modern and knew how people like ourselves lived in other countries. "In America where you live, is it not correct that everybody can afford a metal house?"

Somehow the word had gotten out that U.S. houses are made of corrugated tin. Maybe a villager saw a movie about Alabama in the 1930s; who knows? Not wanting to look as if they were backward, these people had been misled into abandoning their traditions for an imported material they will spend a lifetime to pay for. "Yes," one woman agreed, in that flawless pedantic British School English only Africans can speak, "it is utterly true that the new houses are too hot, and additionally we have considerable difficulty conducting conversations when the rain comes, but this is the price of Progress."

It is common to see entire communities built of corrugated steel. This one is in central Kenya.

unaltered. When an earthen building is no longer wanted, it returns to its original components and grows a garden again.

Of course, it's impossible to build a house with no environmental impact, but it's our responsibility to minimize and localize the damage. Digging a hole in your yard for clay to make a cob house may look ugly at first, but it's a lot less ugly than strip mines, giant factories, and superhighways, and you personally can take control of turning problems into assets. That hole in your yard would make an excellent frog pond.

THE NATURAL BUILDING RESURGENCE

There have always been some individuals who have challenged the modern building-industry paradigm, preferring to build for themselves using local materials and traditional techniques. During the Back-to-the-Land movement of the 1960s and '70s, thousands of people found themselves building their own homes from available resources, without professional assistance, and without much training or money. They were inspired and aided by the example and writing of contemporary pioneers such as Helen and Scott Nearing (*Living the Good Life,* etc.) and Ken Kern (*The Owner-Built Home,* etc.). The energy crisis of the mid-1970s focused public attention on our use of natural resources and the energy efficiency of our buildings. At that time a huge amount of research and writing was done on passive solar building, alternative energy systems, and sustainable resource use, support for which was subsequently swept away by government policy and public apathy during the 1980s.

The experimental work of conservation-minded builders continued throughout this period, although it was no longer receiving much popular press. In the late 1980s, a flurry of activity surrounded the rediscovery in the southwestern United States of straw bale building, a technique that had risen to brief popularity in Nebraska in the early 20th century. In Tucson, Matts Myrhman and Judy Knox started Out On Bale, an organization devoted to popularizing this elegant and inexpensive construction system. Around the same time, we—Linda and Ianto—inspired by the centuries-old earthen homes in Britain, built the first cob cottage in western Oregon. The interest generated by this wood-free wall-building technique, which had proven itself well-suited even to cool, rainy climates, led us to found The Cob Cottage Company, and, more recently, the North American School of Natural Building. Meanwhile, Iowa-based Robert Laporte had combined two traditional techniques: timber framing (from Japan and Europe) and light-clay (from Germany), using an insulating infill of clay-coated straw, and was teaching his natural house-building workshops across the United States and beyond. Persian architect Nader Khalili had established Cal-Earth Institute, a center in southern California devoted to developing education about, and gaining code acceptance for, earth building systems. Also in California, David Easton was building and writing about monolithic rammed earth walls, while others experimented with compressed earth blocks.

WHERE'S THE DUMPSTER?

WHEN COB CONSTRUCTION is happening, the word gets out locally that something interesting is going on, and a gradual trickle of curious folks comes by. They have a range of (usually) predictable questions, though some are thunderstruck at so radically different a way of building and are silent in wonderment. Last to come are conventional builders, because they already *know* how to build. Generally when they finally show up, they come in little groups (there's strength in numbers), big knowledgeable-looking fellows, nearly always men, with carpenter's belts and denim pants.

One memorable group sauntered in as we were completing an Oregon cottage. They stood around quietly discussing things among themselves, swigging from their soda cans. Finally, one spoke up.

"Where's the dumpster?" He crushed his returnable soda can, looking around.

"Sorry, we don't have one." They all looked at each other.

"Well, what do you do with your trash?"

"We don't have any. Everything gets used, or goes in the stove, or gets composted."

Silence. Then, "Well, you at least gotta have a trash can."

"Nope, no trash can."

That really did it. They may have learned more at that moment than they could have in a week of looking at a cob building. The fact is, if you build with toxins, you have toxic waste. If all your materials are natural, anything coming onto the building site can stay there. No, we don't have the big essential of every construction site—we don't need a dumpster.

By the early 1990s, there were dozens of individuals and small organizations in the United States researching, adapting, and promoting alternative building systems. These visionaries proceeded with their work independently, largely unaware of the existence of the others. Then as the straw bale boom in the Southwest took off, attracting the interest of national periodicals such as the *National Geographic, New York Times,* and *Fine Homebuilding,* and as increasing numbers of natural building workshops were offered and people were trained, the "experts" began to hear about and meet one another.

In 1994 The Cob Cottage Company organized the first Natural Building Colloquium, inviting natural builders and teachers from around the country to spend a week together in Oregon. The idea was for these leaders to get to know one another, to share the building techniques they knew best, and to begin to tie their various philosophies and experiences into a more cohesive system of knowledge. During that original gathering and the annual Natural Building Colloquia that have followed, workshops have been given on wall-building systems ranging from adobe to wattle and daub, roofing techniques such as sod and thatch, and foundation systems including the rubble trench, dry stone walling, and rammed earthbags. Lectures and slide presentations inspired everyone with information on recycled materials, designing with natural forces, bamboo, graywater systems, co-housing, creating sacred space, structural testing and building codes, composting toilets, architectural-education reform, steam generation, and a hundred other topics. Traditional yurts, timber-framed structures, and straw bale vaults sprang up and were decorated with multicolored clays. Ideas and techniques collided and merged, coalescing into hybrid structures, including a straw bale/cob dome, and a straw bale/cob/

light-clay/wattle and daub cottage on a stone and earthbag foundation.

As a result of these Colloquia and the numerous other gatherings and collaborations of people interested in natural building, a few things have become clear. One, that even though we may have chosen to focus on different techniques or aspects of natural building, we are all motivated by roughly the same concerns, and our personal experience makes up part of a consistent larger body of knowledge. Two, that we are not alone. As word gets out into the greater public, we find enormous interest and support from a growing community of owner-builders, professional builders and designers, activists, educators, writers, and conservationists. And three, that together we hold a great deal of power. The power in our ideas and collective action is capable of influencing the way our society thinks, talks, and acts regarding building and resource use. We are helping to create a society where someday natural building will again be the norm in the United States, as it still is in much of the world, and where a new cob house with a thatched roof in any American town will draw only an appreciative nod.

At the 1996 Natural Building Colloquium in California, Michael Smith and Deanne Bednar share a bale.

2 Oregon Cob

❧ Cob offers answers regarding our role in Nature, family, and society, about why we feel the ways that we do, about what's missing in our lives. Cob comes as a revelation, a key to a saner world. ❧

"Oregon cob" is an approach to earthen building that was developed in Oregon in the 1990s, and which is now in use all over North America. It takes the best not only from English cob, but from systems used in Africa, Europe, and Asia, and by Native Americans before the whites arrived.

Cob has its origins in millennia of traditional building, in the oldest permanent human dwellings; we have made shelters this way so long that we may carry a genetic memory of how to do it. For many of us, even a photograph of a cob building can evoke powerful emotions; seeing our first cottage is an unexpected glimpse into a different world that nevertheless feels oddly familiar.

I came home one winter afternoon to find a big black sedan parked in the driveway. A well-coiffured woman of fifty-odd sat behind the wheel, looking rather nervous.

"Hi, were you looking for someone?"

She rolled the window down a crack. "I heard there was a mud house here, but I don't want to bother anybody, and I need to leave real soon . . ."

"Come on in."

"Oh, I can't bother you . . ." With hesitation, she followed me in, *very* nervous.

"Would you like some tea—I was making some anyway? Sit down, make yourself comfy."

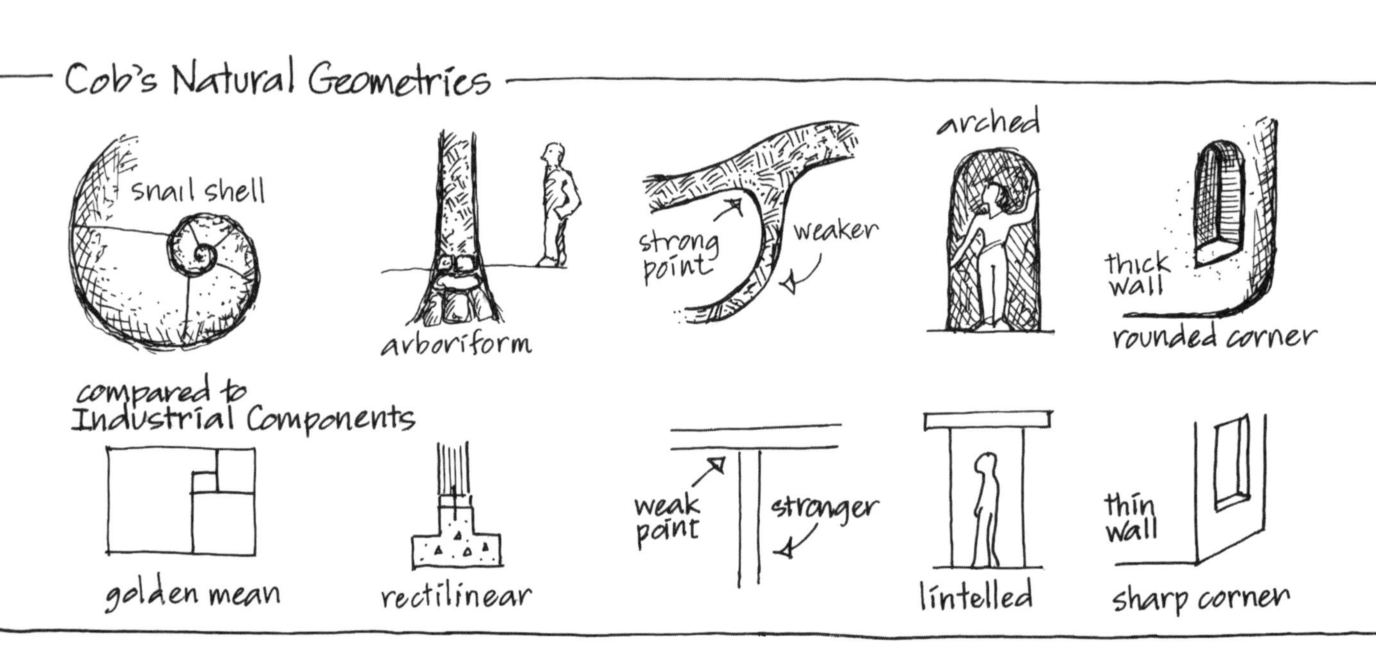

Lighting the gas, I glimpsed around the corner to the stoveside bench. She was sitting rigidly, looking straight ahead, silent, with tears rolling down her cheeks. Without a clue of what was happening, I went over and sat by her.

"All my life," she said, sniffing a little, "all my life I've imagined a house like this. When I was a little girl, I had a book with houses in it just like this one, and I lived in those houses in my mind. But I've never ever seen one for real. I didn't know it was possible."

She drank her tea, but didn't say much more. After a while she got back in the big black car and drove away. She didn't leave her name, and I never saw her again. But the image of her, with blue rinse and a neat gray dress, sitting in my cottage crying, has stayed.

Building with cob offers answers to a wide swath of the questions that seem to plague modern life: questions about our role in Nature, family, and society, about why we feel the ways that we do, about what's missing in our lives. For many people, cob comes as a revelation, a key to the saner world they crave. Some of this is evident before you ever get your feet in the mud.

In this chapter we describe the history of cob building and how Oregon cob came to be. We tell what Oregon cob is, what's different about it, what it does well and not so well. We also tell the story of Linda and Ianto's first cob cottage.

House with 3'–4'-thick cob walls, Dawlish, Devon, England. The exposed dateplate says 1539.

A BRIEF HISTORY OF COB

Unbaked earth is one of the oldest building materials on the planet; it was used to construct the first permanent human settlements around ten thousand years ago. Because of its versatility and widespread availability, earth has been used as a building material on every continent and in every age.

Earth construction can take many forms, including adobe, rammed earth, straw-clay, and wattle and daub. *Cob* is an English term for mud building, using no forms, no bricks, and no wooden structure. Similar forms of mud building are endemic throughout Northern Europe, the Ukraine, the Middle East and the Arabian peninsula, parts of China, the Sahel and sub-Saharan Africa, and the American Southwest (where it is known as "puddled" or "coursed adobe").

Exactly when and how cob building first arose in England remains uncertain, but it is known that cob houses were being built there by the 13th century. Cob became the norm in many parts of Britain by the 15th century,

WON'T BURN DOWN, BUGS CAN'T EAT IT, AND IT'S DIRT CHEAP

BY KIKO DENZER

WHEN I GAVE UP my studio in town, I thought I would soon be working out of my new mud studio at home. Town friends, thinking of me as a capital-A "Artist," would ask me "What are you *working* on?" I would tell them about my fifty-ton mud sculpture, and if they seemed to need a dose of real art-speak, I'd tell them about "four punctuated vertical planes enclosing a large void."

On the other hand, my country neighbors are less interested in art, yet occasionally interested in what and how I'm building. The closest is Tom, a conservative rancher who seemed quite skeptical. But then his father reminded him that his grandmother, after their old homestead burned, had wanted to build a rammed earth house. (I think she had her grandchildren in mind, but Grandfather didn't take to the idea and built with wood.) Well, Tom now lives in the house, and in addition to worrying about fire, he says he can hear the bugs eating it at night.

He also drives a dump truck, and has been very kind about hauling loads of rock and sand when I needed it. But I think it must have been after he heard about Grandma's interest in rammed earth that he got more interested in cob. On his way home from hauling rock for a helicopter-logging operation up the road, he'd stop off to check my progress. When I told him I was getting low on sand, he suggested using the scalpings from the local quarry where he was getting rock. I went to see, and found huge piles of stuff that was near perfect for cob—about a third small gravel, a third sand, and a third silty clay—and free for the hauling. Tom brought me enough to finish my building, and all I had to add was a bit of sticky clay.

One day another neighbor, who drives the big loader at the same quarry, drove past with his wife to see the helicopter at work, flying up and down the valley like an eagle with its talons full of twigs—except that the twigs were alder and fir trees, many two feet in diameter. Tom was visiting, stepped out to greet Jim, and eventually asked if he'd heard about cob. I could hardly keep my jaw from going slack as Tom told him: "Yep, bugs can't eat it, and it won't burn!"

I couldn't help but add, "And it's dirt-cheap."

A week or so later I went up to Tom's to restock my drinking water jugs (my own spring isn't reliable). I stopped to visit before going out to the hose. His wife, Terri, said she had just been down to the county seat to talk to the planning department about building on the old homestead site—a dream they've had for years, not only for history's sake, but also because their son is nearing the end of his teens, and will be needing a place of his own soon. I think it must have been the first I'd ever heard of it, though, because I again had to hold my jaw when Terri said, "We told them we wanted to build with cob, and you were going to be our contractor!"

A few weeks ago yet another house burned to the ground, a mile down the road. The state fire crew arrived in good time to stand and watch and protect the forest. I drove over to see if I could help, but it was already too late. As I walked down the drive with the lady of the house, she said to me, "Now I wish I had a mud house like yours."

She had no insurance, partly because there's no fire department anywhere nearby. Like so many rural working families, she and her husband will take on at least a fifty thousand dollar mortgage for a trailer home—about the only new housing option for working people who don't have both carpentry skills and time to build for themselves.

I don't think barn-raisings were ever part of our culture here, but I hope cob might someday, somehow help bring people together to build beautiful fireproof homes free of the spirit- and creativity-crushing demands of the speculative housing and financial markets. Grandma had the right idea, and maybe her great-grandson will make the dream real.

and stayed that way until industrialization and cheap transportation made brick widely available in the mid-1800s. Cob was particularly common in southwestern England and in Wales, where the subsoil was a sandy clay and other building materials such as stone and wood were scarce. English cob was made of subsoil mixed with straw, water, and sometimes sand or crushed shale or flint. Often native chalk was added, sometimes lime. The percentage of clay in the mix ranged from 3 percent to 20 percent, with an average of only around 5 or 6 percent. It was mixed either by people, shoveling and stomping, or by heavy animals such as oxen trampling it.

The stiff mud mixture was shoveled with a cob fork onto a stone foundation and trodden into place by workmen on the walls. In a single day, a course or "lift" of cob would be placed on the wall between 6 inches and 3 feet high, but usually averaging 18 inches. It would be left to dry as long as two weeks before the next lift was added. Sometimes additional straw was trodden into the top of each lift. As they dried, the walls were trimmed back substantially with a paring iron, leaving them straight and plumb, between 20 and 36 inches thick. In this way walls were built as high as 23 feet, but usually much less. Openings for doors and windows were built in as the walls grew, with lintels of stone or wood set in above.

Many cob cottages were built by poor tenant farmers and laborers, often working cooperatively. A team of a few men, working together one day a week, could complete a house in one season. A cottage begun in the spring would receive its thatch roof and interior whitewash in the fall, and its inhabitants would move inside before winter. Often they waited until the following year to plaster the outside with lime-sand stucco so that the walls would have ample time to dry. Cob barns and other outbuildings were sometimes left unplastered.

Hayes Barton in Devon, England, birthplace of Sir Walter Raleigh.

But cob buildings were not reserved solely for the humble peasants. Many townhouses and large manors, built of cob before fired brick became readily available, survive in perfect condition today. Among them is Hayes Barton, the birthplace of Sir Walter Raleigh, who had so much affection for his childhood home that he offered to buy it from a later owner for "whatsoever in your conscience you shall deme it worth." In the late 1990s an estimated twenty thousand cob homes and another twenty thousand outbuildings remained in use in the county of Devon alone.

And yet by the late 19th century, cob building in England, considered primitive and backward like most handcrafted traditions, was declining in popularity. During the 20th century, however, public opinion slowly evolved until traditional cob cottages with their thatched roofs are now valued as snug, historical, and picturesque. There was virtually no new cob construction in England between World War I and the 1980s, and the traditional builders took much of their specialized knowledge with them to the grave. But enough information survived to allow a cob building revival starting in the 1980s, fueled largely by historical interest and the real estate value of ancient cob homes.

LEFT: *The first new cob construction in England in 70 years, a bus shelter in Devon by Alfred Howard.* RIGHT: *Kevin McCabe's new cob house.*

The first construction project of the English cob revival was a bus shelter built by restorationist Alfred Howard in the 1980s. Since then there have been increasing numbers of new cob structures built in England, particularly in Devon. Kevin McCabe received a lot of press in 1994 for his two-story, four-bedroom cob house, the first new cob residence to be built in England in perhaps seventy years. The 1999 newsletter of the Devon Earth Building Association lists forty examples of new cob construction or significant recent restorations. In 2002, Kevin McCabe will complete a 3000-square-foot, three-story new cob house with 3-foot-thick walls.

The building technique of these revivalists closely resembles that of their ancestors. They mix Devon's sandy clay subsoil with water then straw and fork the mixture onto the wall, treading it in place. Walls are generally 24 inches thick and straight, applied in lifts up to 18 inches high. The machine age has altered the traditional process in only minor ways: McCabe and others use a tractor rather than oxen for mixing cob and often amend the subsoil with sand or "shillet," a fine gravel of crushed shale, to reduce shrinkage and cracking.

In addition to construction and repair, there is a fair amount of research going into English cob. Alfred Howard, for example, has built experimental walls to test a variety of subsoils. Larry Keefe, a former building conservation officer, has catalogued hundreds of old cob buildings and become an expert on why cob walls fail and why they don't. Larry is the cofounder of a unique program at Plymouth University dedicated to furthering earth architecture, which has sponsored earth building workshops as well as several international conferences on earth building. However, most of these developments are quite recent. As late as the mid-1990s, reliable information on cob building was nearly impossible to obtain.

THE BIRTH OF OREGON COB

Ten years ago, earthen building was almost completely unknown in America's Pacific Northwest. In a region known chiefly for the size, longevity, and diversity of its coniferous forests—and for its rain—it seems only natural to build of wood. The weather is cool and damp, and wood is warm and dry. Why would anyone want to build with mud? Surely come spring they'd find themselves living in a big dirty puddle.

In 1985 Linda and I made a research trip to Wales and Ireland, where we found two earthen buildings in the most extreme of cool, damp climates. One, in County Cork, Ireland, was a cob structure that the owner told us had once been a farmhouse, and which was now being used to house animals. Coastal Cork is perpetually damp, misty, cool, a natural rain forest without the trees—a land of fern, waterfall, moss, and lichen.

The other building, in West Wales, was an earthen cottage, long abandoned but with the roof still intact, situated on a wild, wet, windy headland jutting out into the ferocity of the North Atlantic. On these Welsh promontories, trees are reduced by the wind's pruning to squat bushes leaning to leeward. Nothing grows more than twenty feet high. In contrast to the damp seclusion of the Irish farm, this was a site of horizontally driven rain, drenched in salt spray by hurricane-force winds.

We probed each building's walls, now exposed by weather, and marveled at the cob's durability, solidity, and dryness. If cob could tolerate these conditions, might it not survive in the Pacific Northwest, where the inhabited areas are comparatively calm and have several months of dry warm weather each summer and fall? Cottage Grove, Oregon, where we live, is 10 degrees latitude south of Wales. *All* of the continental United States is south of any point in Great Britain, where earthen buildings exist by the thousands to the latitude of the southern Aleutians and in climates little better.

The power of myth over common sense is remarkable. In confidence that three million owners of wooden, brick, glass, steel, aluminum, and concrete homes throughout the Northwest were of course correct, and that building with earth was just a silly notion, we procrastinated any trial for another four years. Somehow, the pressure of public opinion was too great. Every time earthen building was mentioned, someone would ridicule the idea or calmly talk sense into us.

Mercifully, in 1989 we took leave of our senses, abandoned caution, and went to work. Our first cob project was a 10- by 12-foot boxy little cabin with 18 feet of cob wall, 16 inches thick and 7 feet high.

Because we had until then been unable to search out either good literature on cob or anyone who had actually built with it, we inferred a process from limited observations and tempered it with my experience of using earth as a building medium—for cookstoves.

Cookstoves As Inspiration

In 1976 in Guatemala, I had begun work on fuel-saving cookstoves. At that time, more than half the world's women cooked on open fires, usually in their kitchens, inhaling smoke measured as equaling an average of four hundred cigarettes a day. In addition, procuring fuel for cooking was devastating forests and impoverishing families. The obstacle to improvement had been a shortage of capital to buy stoves and a shortage of ingenuity in providing models of owner-built stoves made from local materials. The most promising direction lay in stoves made of earth. Searching the literature provided few clues of how to go about it, but there were mentions of "clay stoves" in use in West Africa and Southeast Asia.

A stove built of clay has one major disadvantage—it cracks horribly as it dries. Cooks would add to their crowded tasks the daily

Mayan women cooking on a huge Lorena stove, Guatemala.

THE COB COTTAGE COMPANY

need to repair those cracks, stuffing more clay into them. The new clay in turn would shrink and crack, extending the process indefinitely. Worse, a cracked stove burns badly, as the air flow through it is uncontrolled. The fuel smokes, energy is lost, cold air dilutes the chimney draw, and smoke ends up in the kitchen.

Twenty-three cracked test models into the project came a major breakthrough—use a *lot* of sand with the clay, much more than any of us ever expected. Mixing as much as 85 percent sand with 15 percent high-quality clay would stabilize the whole mass, preventing cracking. The stove system that resulted was named Lorena (from *LOdo,* "mud," *aRENA,* "sand," in Spanish). This was the first really successful attempt to improve the efficiency of Third World cookstoves. With modifications it has since been used in dozens of countries worldwide, enabling cooks to construct simple cookstoves entirely from local materials, saving fuel and taking smoke out of the house. I worked later on sand/clay stoves in Kenya, Lesotho, Mexico, Nepal, Senegal, and Burkina Faso—always having to first understand local clays and local sands, never forgetting that to prevent cracks and create stability, all you have to do is add sufficient sand. The book I wrote, *Lorena Stoves*, has found its way into the most remote corners of the globe.

Loaves this shape in Wales are sometimes called "cottage loaves." Wrapper reads "400 gram COB HARVESTER."

Our First Cob House

So, building our own house in the Oregon forest, Linda and I did what we knew well. We used a lot of sand, with the clay soil from the building site to hold it together. We trod it thoroughly to mix it and added a lot of straw. As for "cob," cobs are little lumps, and in Wales a cob is both a lumpy little horse and a loaf of bread. So we made lumps, just like forming loaves, and built with them.

Due to the delays inevitable in construction, our cob walls were not started till September 3rd, a foolishly late time of year to begin building. Yet the first wall was complete in three weeks. Every day we two would be up before dawn, digging, shoveling, and treading. The heavy work was done by noon, before the heat of September afternoons. By enclosing, heating, and ventilating the new building, we were able to move in by November, though damp patches were visible until Christmas.

That first cottage felt too prissy for us in its rigid boxiness. So, directly from my cookstove experience came a full-length daybed heated by the stove flue. Necessity forced us to round its corners, then oh, what a difference in feel! The stove was built of two barrels, and their rounded forms forced us to round the shapes of the bench. We ended up rounding out also the room corner where the bench turned. Surprise! The room felt *bigger* with the corner rounded! That led to two rounded windows, one circular, the other arched and churchy-looking, both of fixed glass buried directly into the wall.

Because we wanted a circular window, we searched around for a ship's porthole. But as we live inland, they are hard to come by. Then we priced out commercial round windows—expensive! The advice of carpenter friends was daunting, lots of technical talk about routers and complex joinery, how to cut circular glass, mechanical jigsaws. We almost gave up. It

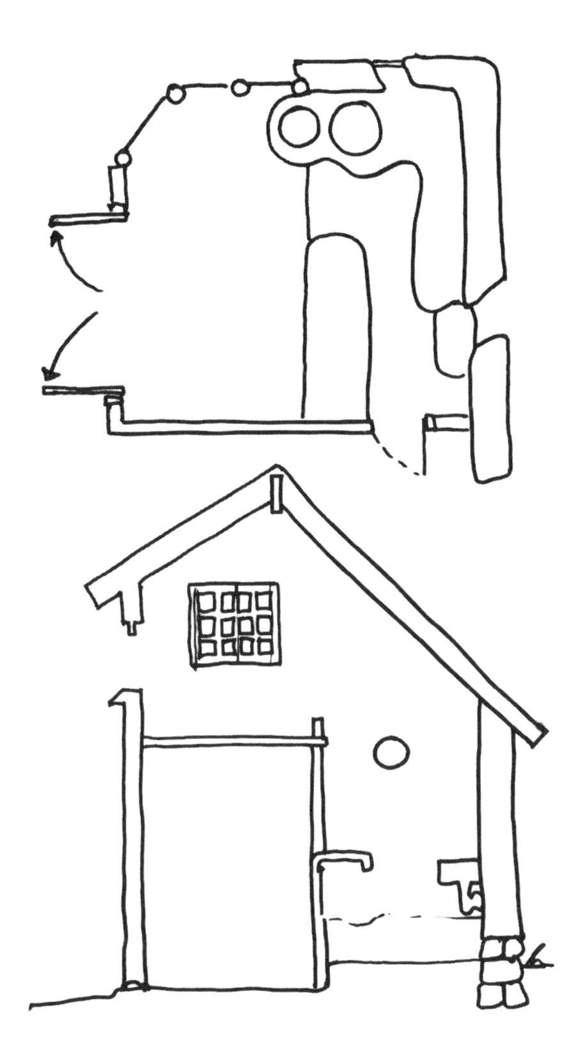

took *two whole days* before the answer came to us: just build in the glass directly, don't frame it at all. Another day or two and it came clear that it's not necessary to cut the glass. It can be left square or made *roughly* round because as all those jagged edges get built into the wall, the mud will absorb them so they are completely buried and safe.

All of this was a revelation: The simplest answer is usually the best. Yet we are so persistent in being unnecessarily complex. We make a big job of everything. Cob helps one see that.

That first cottage changed many lives, not just our own. Almost universally, visitors

LEFT: *Design of the first Oregon cob cottage, 1989.* ABOVE: *The first cottage of the Cob Renaissance. Sadly, the new owners see no value in this historical building and plan to demolish it!*

would marvel at it. But snug and happy as we were, doubts lingered. If it was really such a good idea, how come this was the first attempt in the whole region? Surely there must be a hidden snag, something we had overlooked? It took years to overcome the suspicion that everyone else was right and it was we who were out of step.

We lived in our cottage through four winters and three summers, during which we monitored first its thermal performance and then, increasingly, its social and emotional comfort. We also had use of an adjoining, similar-sized stud-framed wooden cabin with fiberglass insulation and a wooden floor. The two structures had comparable solar access and glass-wall orientation. We recorded temperatures regularly during cold, clear weather and cloudy winter weather, both with and without wood heating. We noticed where we spent time and when.

Thermal comfort was significantly better in the cob cottage, with cold winter mornings dropping to 50°F in the cob, and to the low 20s in the wooden cabin. We stayed snug and dry throughout the eight months of the rainy Oregon Coast Range winter, despite temperatures that dropped several times to 0°F, and snowfalls of up to eighteen inches. The

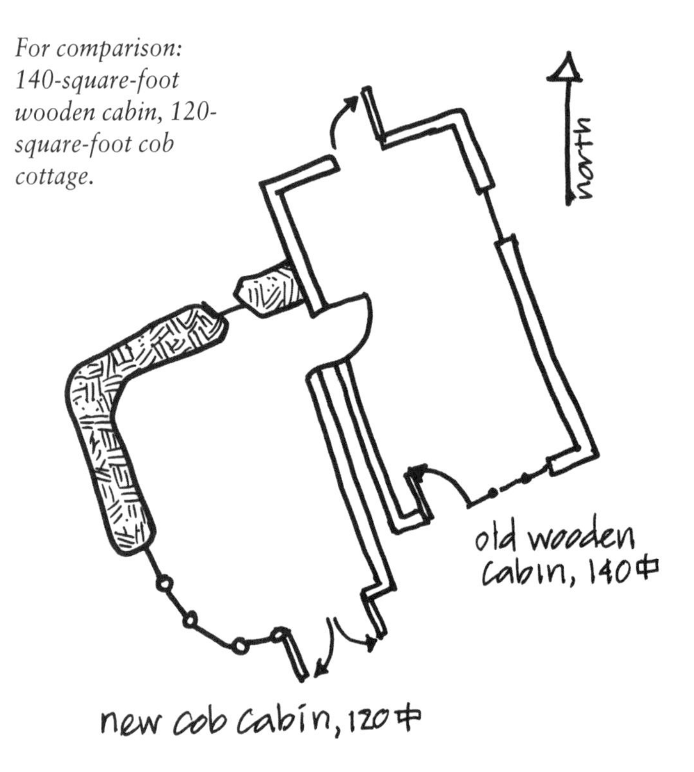

For comparison: 140-square-foot wooden cabin, 120-square-foot cob cottage.

exposed walls showed no deterioration from either rain or snow.

Gradually we recognized our advantages over our neighbors. The cottage kept snug on only about a cord of fir for the entire winter, partly due to its small size, and partly because it was kept naturally warm by the stored solar heat in the cob. Many spring and fall mornings when nearby houses showed woodsmoke from chimneys, we felt no need to light the stove.

Benefits carried through summers. Though we sometimes overheated that first August, before the shade trees grew to shelter our southwest-facing glass, we generally stayed delightfully cool even when temperatures outside rose into the high 90s in the shade. It was rare for the indoor temperature in the cob to go above 75°F, though in our wooden cabin next door it would be over 95°F.

Thermal statistics aside, we were impressed more than anything by the *quality* of the cob building. It *felt good*—not just to us but to every one of the hundreds of visitors who came, as news of a "mud hut" spread. Nearly invariably their response was amazement and broad smiles, then an almost embarrassing enthusiasm. I remember in all that time only one person who objected, a severe-looking lady who said flatly, "I don't like it; it's unnatural."

After four years of research, with the cob still holding up perfectly despite rain and snow, we were finally ready to state categorically that this was a good idea.

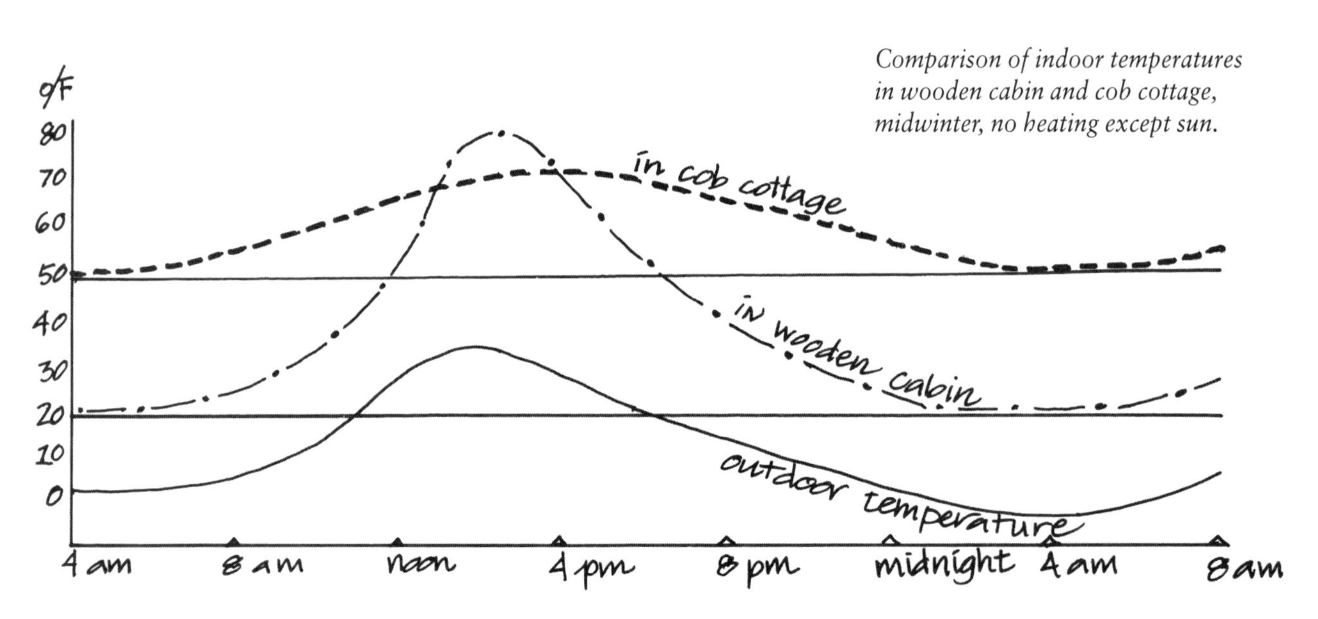

Comparison of indoor temperatures in wooden cabin and cob cottage, midwinter, no heating except sun.

CHARACTERISTICS OF OREGON COB

It was not until we were working on our fifth or sixth cottage several years later that we realized how differently we were building from either English or African cob and that our method offered significant advantages over what we know of traditional techniques. Our system, which we named "Oregon cob," has evolved rapidly toward an even greater range of benefits and continues to improve as its popularity spreads and as experience builds on experience.

Historically, most cob contained very little or no straw, and any straw used was generally short, and of poor quality, with good reason. Straw was expensive a hundred years ago. In contrast, the Oregon cob technique uses the longest, strongest straw available. And unlike traditional cob, Oregon cob uses a carefully adjusted proportion of sand, with just enough clay to bond the mix together. The cob therefore shrinks very little as it dries. This emphasis on precisely adjusting the mix according to the quality of the ingredients produces surprising strength. The builder can sculpt strong but thin earthen shelves, and interior partitions as thin as two inches.

With Oregon cob, structural strength is added also by (a) building walls selectively thicker, as needed, and by (b) curving walls whenever possible. The process feels like *growing* a building instead of violently forcing parts together. We pay particular attention to the elegant geometries of Nature—how strength is attained with a minimum of material—and we use those shapes. A common pattern that results is a series of walls curving in long, loose spirals connected by short, tight curves.

In Oregon cob, each thin layer of earthen mix is systematically bonded to the previous one by "sewing" the straw into a three-dimensional internal textile and by leaving each layer or course rough and lumpy, with pocks and divots to grasp the next course. Structurally critical parts are built with loaf-sized, hand-formed "cobs" that are tossed to the builder, who presses each one into place, bonding it to those beneath and adjoining.

Oregon cob allows ad hoc design changes. Because building is incremental, by the handful, continuous adjustments can be made. Changes suggested by site, materials, the builders' skills, or the evolving building itself nearly always raise quality and are sometimes inspired and magical. Last-minute improvements can be tried, assessed, rejected, or adjusted. The medium melds the edges of sculpture and construction, using responsive, adaptable mixes of the basic ingredients to create the furniture, fixtures, and features with which we live.

What Cob Does Best

As with any building technique, cob is well suited to some circumstances, poorly to others. You will clearly want to be able to select materials and building systems best suited to your particular circumstances. First then, here are some of cob's best features, followed by some cautions, though be aware that this is largely a new technique in North America and that many assumptions are as yet unproven.

Cob Insulates from Temperature Change: The rule of thumb is that heat flows through cob at an inch per hour, so a two-foot thick wall takes about twenty-four hours to transmit the effect of heat or cold all the way through. In reliably sunny conditions, a foot-thick wall can keep a building coolest in daytime, warmest at night.

Where daytime temperatures are very high, as in dry continental climates, cob's massive walls moderate that heat, absorbing it during the day and then releasing it each night as the temperature drops.

Where days are often sunny but nights are cold (even if the sunny days are cold), the great thermal mass of cob will soak up and store the sun's energy, then release it over hours or days.

Cob is a good thermal choice, also, where the air is warm while the ground stays cold, as on hot spring and summer days in most of Canada and in the United States west of the Rockies above the 38th parallel. This is true even if there are hot nights for short periods, provided the constant ground temperature at the site is below about 50°F. Cob is particularly useful here for floors, which gradually pull heat out of the room and store it for days or weeks.

MASS COOLER

OUR HOUSE has no electric refrigerator. In our Oregon rain forest climate, in-the-wall refrigeration works well. A three-shelf closet cut right through the north wall at eye level keeps perishable foods in good condition. It opens by a small wooden door above the kitchen counter, and the outdoor face is fly-screened. Its interior is gypsum plastered directly onto the cob, and it has a ceramic tile floor. Even in summer, with temperatures in the 90s, milk covered with a wet cloth will keep fresh for three days in this "mass cooler." Fungus and bacteria are inhibited by dry air and drafts so fruit and vegetables stay fresh, as higher summer temperatures are counterbalanced by much better airflow than in sealed electric refrigerators.

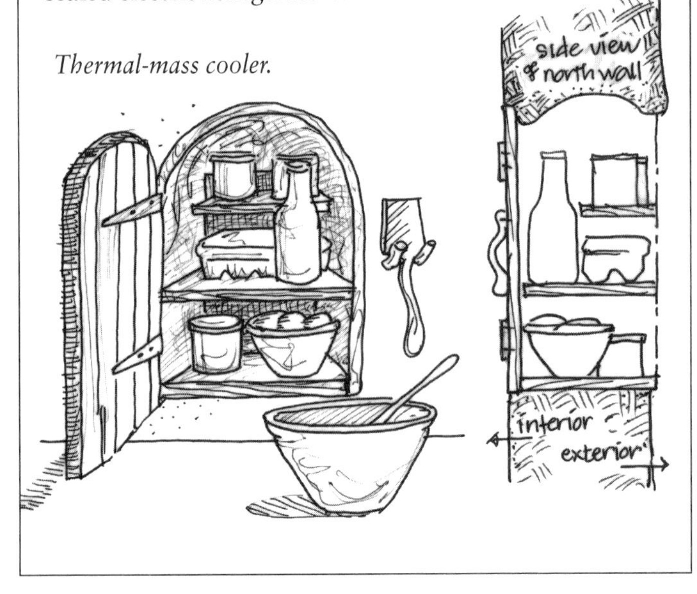

Thermal-mass cooler.

Cob Complements Passive Solar: Cob is useful in passive solar buildings as storage for sun-heat, especially when used for floors and interior walls, and with heavy, natural plasters. Cob is of moderate density, with a specific gravity of 1.2 to 1.9, and provides much better insulation than rock, brick, or concrete (for more details, see chapter 6).

Organic Shapes Are Easy: For creating buildings with irregular, curved, organic shapes, cob is ideal. Curvilinear, thick-walled spaces feel larger than rectilinear ones of the same measured area. Spaces can thus be designed tiny, fitting snugly around the uses they protect and responding to personal spatial needs rather than creating use-neutral containers. Curvilinear buildings require less space and thus less heating and cooling, less maintenance, and fewer resources.

Corners Aren't Necessary: Where cold winds are a problem, as in the Great Plains, cob's rounded aerodynamic shapes reduce infiltration of cold air. Whereas boxy shapes create areas of extra-high and extra-low pressure surrounding the building—sucking warm air out and forcing cold air in through the tiniest cracks—more rounded buildings lack corners to protrude into the cold wind. Interior corners are the coldest parts of a building, as outdoor winds speed up there, moving heat away faster.

Cob Rewards Time and Creativity, Not Money: Cob may be a good choice if you are short of money but rich in time. Cob is simple enough to learn mostly by practicing. Very little can go dangerously wrong; almost nothing is irreversible. At worst, even if time is wasted, money is not lost.

Cob Insulates from Sound: Cob is a poor transmitter of sound and is useful where outdoor noise—from highways, railroads, flight paths, factories—is a problem. An earthen house in Toronto, built in 1827 with 20-inch walls, is still lived in. The current owner states that one great advantage is sound protection from the railroad that passes only fifty feet away. Not only can cob keep noise out, it can keep noise in, for example from a machine shop or music practice studio. Additionally, cob is well suited to surface modeling for sound absorption.

Cob Works Where Other Earthen Techniques Don't: Cob is often useful where other earthen techniques are undesirable, for instance, in place of adobe in cool damp regions, instead of rammed earth if machinery is too expensive or unpleasant to work with, or where materials for wattle are scarce.

Cob Is Safe for Inexperienced Builders: There is no need for mechanical tools or power equipment on the site. The building process is safe because the building has no heavy components, no sharp parts, no toxic chemicals. Mud pies are not at all intimidating to people new to construction. Materials are familiar and almost impossible to misuse or waste.

Cob Is Democratic: Because cob is built by the handful (and it fits any size of hand), it is making construction accessible to women, children, the elderly, and the frail. Other building techniques, it seems, were designed to require the strength of energetic young men. Concrete blocks, sheets of plywood, bags of cement all come with the assumption that you are active, healthy, and muscular, with a good back. By contrast, cob fits almost anyone's physical strength. You can use any size of unit, any weight of loaf to build with. A mix can be any amount of material that you can roll. There are no heavy loads to lift.

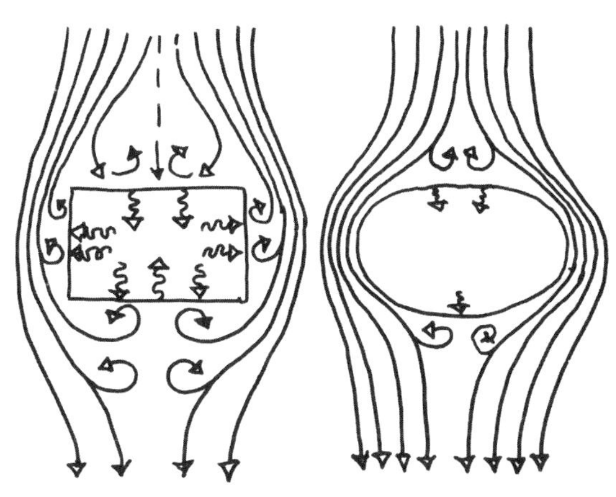

Streamlining. Air flows easily around curved buildings, so there is less air infiltration due to wind and no heat loss caused by projecting corners.

Cob Encourages Site Respect: The process of building with cob can be slow, deliberate, and accretive, which almost *forces* builders to pay attention to their site. Digging your own building material by hand, often right alongside your building, encourages careful observation. Buildings created this way stand a much better chance of responding to the site's aesthetics, ecology, and microclimate.

Cob Is Fire-Resistant: In forest fire zones and arid areas, local governments often mandate nonflammable roof materials. With cob and an earthen roof, the whole structure is nonflammable. In the tremendous bush fires that in 1994 almost surrounded Sydney, Australia, the only surviving building in one entire neighborhood had an earthen roof.

Cob Is Wind-Resistant: In hurricane, cyclone, and tornado regions, the solidity of cob is protective—like hiding in a basement. Windows should be shuttered and their sizes kept smaller in such conditions, and you should either use a heavyweight earthen roof or anchor your roof very securely in the walls.

Remote Building Sites Aren't a Problem: Cob is a good solution on remote sites where it is difficult to import building materials or where lumber and other processed building materials are expensive.

Cob is Durable: Well-maintained traditional cob buildings have lasted for centuries, possibly millennia, without major repairs. Wood-frame houses built today often need major repairs in ten to fifteen years. (There may be good reasons you can't get a forty-year mortgage: lenders don't want to risk forty-year money on a decomposing security that could fall apart in thirty.)

Where Oregon Cob May Be Inappropriate

No building system is appropriate under all conditions, and there are situations in which cob should be used cautiously, or not at all. The process of natural building starts with careful observation of the building site, including consideration of its climate and microclimate, its surface and sub-surface geology, and the availability of on-site materials (see chapters 5 and 9).

Some of the circumstances in which we would not generally recommend cob as a major structural part of a permanent building include:

In Some Cold Climates: In climates with very cold, dark winters; on north slopes; or on cold sites that get little winter sun, cob's high thermal mass can work against you. These are places where passive solar design is much less effective. During periods of cold, overcast weather, the "thermal battery" in heavy cob walls will run down, so more energy will be needed to heat an all-cob building than one with more insulative walls. However, cob could still be used if the walls were insulated from outside. Cob interior walls, floors, trombe walls, and built-in furniture are effective at storing heat and regulating temperature inside a well-insulated shell. In Ontario and Denmark we have experimented with exterior straw bale insulation on cob buildings to achieve a similar effect.

In Flood Plains: While cob walls have a remarkable capacity to withstand normal moisture and weather conditions, the most severe threat to their stability is prolonged or repeated soaking. Don't build a major cob structure in a river flood plain, seasonal creek or gully, or below the high storm-tide line at the seashore. You probably don't want to be living in these places anyway, unless your house floats. In areas of concentrated seasonal rainfall, where water occasionally pools or runs over the ground, provide your cob cottage with a good high foundation and excellent drainage (see chapter 10). For a more thorough discussion of cob and water damage, see appendix 3.

In Areas without Soil or without Clay: Where there is no soil or no clay close by, cob may not be the most sensible choice. Although it is standard practice in modern building to import all of the materials from hundreds of miles away, it makes little ecological sense. In natural building, we strive to use whatever materials are most available on and near the site.

In Extreme Seismic Zones: All buildings are threatened by earthquakes, and cob is no exception. While cob seems to be considerably more resistant to earthquake damage than adobe and other unreinforced masonry (see appendix 4), it is still probably less safe than very lightweight, flexible building materials such as bamboo and steel. Good design strategies in earthquake areas include keeping the overall height of the cob walls down to one or two stories, using a post-and-beam structure to support the roof, and making the cob walls extra thick at the bottom, emphatically tapered, curved in plan, with occasional buttresses. Also, use a lightweight roof structure. We have employed all of these strategies in California, including at a site practically on top of the San Andreas fault, and have visited cob buildings in New Zealand that have survived multiple major earthquakes unscathed.

In Buildings Where Heating Is Needed Only Irregularly: When a high-mass building is allowed to cool down, it takes quite a while to heat up again. In colder climates cob is much better suited to structures such as greenhouses and houses that will be continuously kept warm rather than large, occasional-use buildings such as classrooms, churches, and meeting halls.

COB IN HYBRID NATURAL BUILDINGS

Cob combines easily with a wide range of other natural building materials, and there are many reasons why you might consider a "hybrid" structure rather than one made entirely of cob.

One common reason is thermal performance. Especially in very cold, cloudy, winter climates, it's often desirable to have better insulation in the exterior walls than you can get with cob. A good solution might be a straw bale shell, with cob floors, interior partitions and built-in furniture, bookshelves, niches and arches. The cob provides heat storage and temperature regulation.

Even in places with mild winters, cob/straw bale hybrids can be a good idea. We've worked on several cottages with straw bale walls to the north, and cob to the south or wherever sun hits directly. That way you avoid a large mass of unheated, north-facing cob that might otherwise bleed heat out of your building.

Poor location for cob: flood plain, steep north slope, and dense trees blocking sun on south side.

Another reason for combining cob and straw bales is speed. In general, straw bale walls go up much faster than cob. This is particularly notable with large structures. Straw bale walls might make the difference between getting your building finished in one season or not. Cob should be used where its special characteristics are most desirable: for thermal mass and compressive strength, and for its sculptural qualities. For example, rounded cob windows, niches or shelves can be sculpted into gaps left in a wall of straw bales or nearly any other material.

The availability of local building materials should certainly influence building design. In a location where clay or sand is scarce, an all-cob building might not make sense. But you can easily import enough raw materials for functional decorative elements such as a cob fireplace or trombe wall. Conversely, the amount of cob you have to mix can be reduced by embedding other locally available materials into the wall. To save time mixing and make cob walls dry faster, we have built rocks, broken concrete chunks, and cordwood into the wall, sometimes as much as 70 percent rock by volume in interior walls and benches.

An all-cob building is certainly beautiful, but using multiple building systems creates a richness of different shapes, textures, and materials. Hybrid buildings are a good way to learn more about and to compare different options. Michael recently designed and helped build a small cabin with a rammed earthbag foundation, walls of cob, straw bale, wattle and daub, and light straw-clay, bamboo roof trusses, a poured adobe floor, and multiple earth- and lime-based plasters. The structure has provided almost endless training opportunities and now serves as a compact demonstration of diverse natural building alternatives.

Combining multiple building systems usually requires extra care in the structural aspects of design. Different materials react differently to changes in moisture and temperature and to earthquakes. They also behave differently over time. For example, cob shrinks a little vertically as it dries, and straw bales can settle substantially under compression, but wooden posts are quite dimensionally stable. So before putting a roof on a building that has walls of load-bearing cob, a wall of load-bearing strawbale, and a post-and-beam solarium, you need to make sure that the cob is fully dry and the bales are precompressed. Similar considerations apply when attaching a cob addition to an existing wooden structure.

Give extra attention to the places where different wall systems come together. Some sort of mechanical connection or key is generally necessary. For ideas about attaching cob to wood, see the section in chapter 14 on connecting cob to door and window frames. We've also developed several systems for connecting cob to straw bale walls.

If multiple wall systems will interrupt the structural continuity of the building, the integrity of the foundation is extra important. You may want to use a reinforced foundation system such as poured concrete rather than one built up of multiple units like stones or concrete chunks. Consider also a continuous bond beam of sturdy wood or concrete to hold the top of the walls together beneath the roof, or beneath the stemwall, on top of a rubble trench foundation. These considerations are particularly important in earthquake zones.

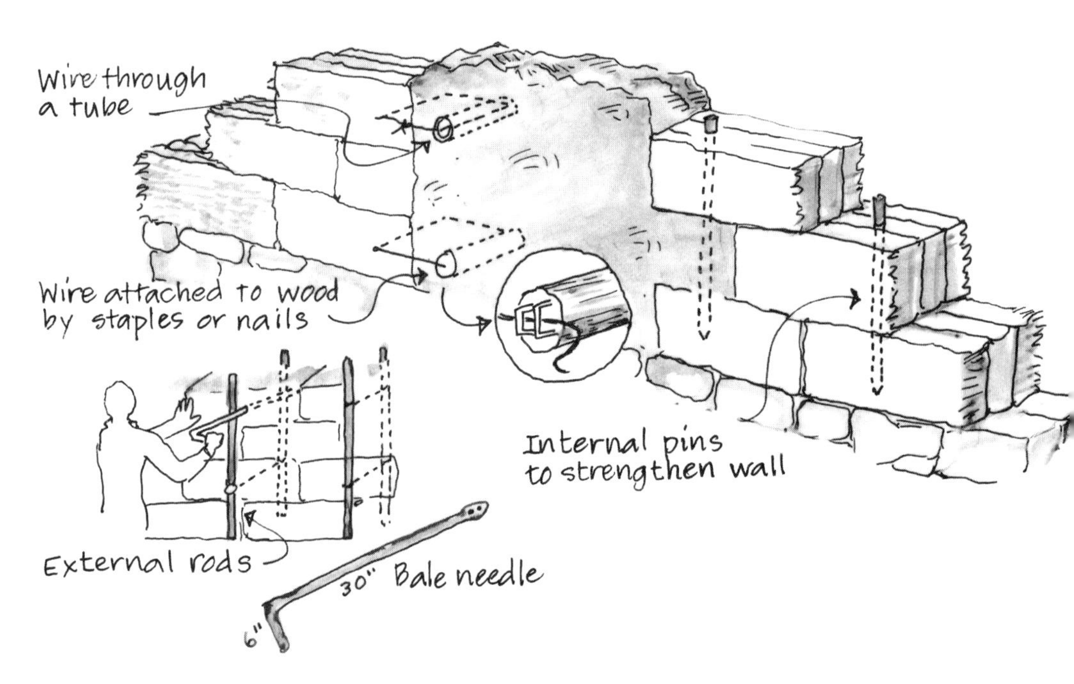

There are many ways to connect cob and straw bale walls. If cob wall and bale walls are built at different times, they can be cinched together with wire (left). For the strongest connection (right) build cob and bale walls simultaneously and drive a firm stake through the overlap. Instead of interior pins, straw bale walls can be held together with external rods or sticks, using wire or baling twine and a needle long enough to pierce the bales.

COB AND STRAW BALE COMPARISON CHART

	BALES	COB
Shape and Size	Preassembled bales are a modular industrial product, relatively homogenous. They usually come in only two sizes, both very large, both rectilinear. Deliveries can be of any quantity. They store well, if kept dry. 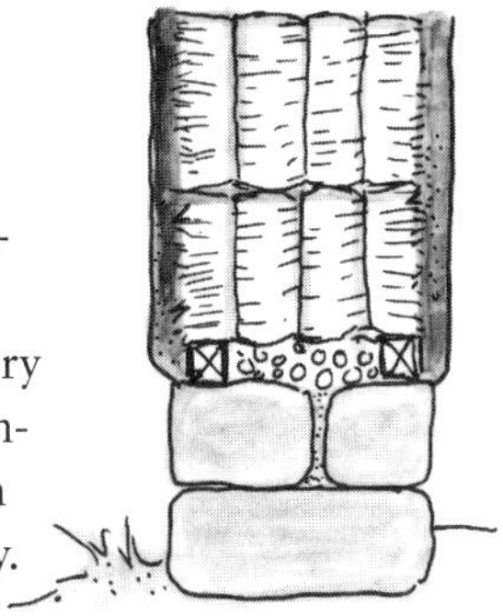	Cob is made on-site to order, in tiny to large amounts. Amorphous, it can be formed to almost any shape. It can be applied by the spoonful to the bucketload. It can be custom mixed to a wide range of densities, strengths, etc. Intensely sculptural at a small scale; can be built thin. 
Thermal Properties	Excellent insulation, poor thermal mass.	Good thermal mass, poor insulation. Stores heat or "coolth."
Load-Bearing Properties	Bales compress under vertical loading. Bale buildings should resist earthquakes well. Long-term stability in high humidity not yet known. Water penetration is potentially disastrous, at any time from manufacture on. Multistory buildings need special care.	Needs no additional structure, roof-bearing. Cob settles as it dries, then is very stable. In earthquakes, will perform better than masonry, worse than bales. Tolerates liquid water until soaked through, then possible failure rapidly without warning.
Construction Conditions	Any temperature, not in rain. Rain protection and dry storage are essential.	Not in frost or very humid, still conditions. Protect from heavy frost and rain.
Protection Needed	Protection essential from moisture, rodents, and fire.	Excellent weather resistance. Needs interior plaster and driven rain protection. Fire- and rodent-resistant.
Durability	Proven one century in dry conditions.	Proven many centuries in various weather conditions, including windy, rainy coastlines.
Rate of Building	Slower than you think. Fast assembly, but needs extensive finishing.	Faster than you think. Slow building rate, but little finish needed.
Most Appropriate Use	Large, simple buildings; in earthquake zones, single story. For walls, roof insulation; for exterior walls in extreme climates; fast or temporary structures; buildings that need fast heat for occasional use: classrooms, meeting halls, etc.	Small, complex buildings; sculptural, artistic, curved buildings; on stable ground multistory; for floors; for interior walls; for exterior walls in mild areas; long-lasting buildings; cabins, greenhouses, buildings that are solar heated and need thermal stability.

3 Creative Economics

❧ BY BUILDING YOUR OWN EARTHEN HOME, YOU CAN CREATE A PERMANENT DWELLING THAT COULD LAST PERHAPS A THOUSAND YEARS, AT A CASH COST OF ONLY A FEW MONTHS' RENT . . . ALLOWING YOU TO SPEND LESS TIME WORKING FOR MONEY AND MORE IN PERSONAL SATISFACTION. ❧

A HUNDRED AND FIFTY YEARS AGO, Henry David Thoreau wrote, "The mass of men lead lives of quiet desperation." Building your own cob cottage can be your escape from economic slavery, empowering you to make conservative and careful choices about income and how you spend it. This chapter offers you help in getting free. We show how to trade time and skill for the money a house usually costs. We look at ways to find cheap or free building materials, discuss finding inexpensive places to build, and offer a checklist for keeping costs down.

BUILDING FOR YOURSELF

Most of us struggle to earn the money to pay for everything we are accustomed to buying—food, housing, transport, clothes, meals out, luxuries. As world resources decline and population grows, we spend more time earning that money, and it buys less. We feel trapped into working at jobs we may not particularly like, without inspiration, for the security of having a home.

The average new house now costs well over $100,000. Even when owner-built, it is difficult to build a wood-frame house of industrially processed new materials for less than about $30,000, plus the cost of a site, which can be another $30,000. Rental of a modest older house in 2002 on the West

THOREAU ON SHELTER

In the savage state every family owns a shelter as good as the best, and sufficient for its coarser and simpler wants; but I think that I speak within bounds when I say that, though the birds of the air have their nests, and the foxes their holes, and the savages their wigwams, in modern civilized society not more than one half the families own a shelter. In the large towns and cities, where civilization especially prevails, the number of those who own a shelter is a very small fraction of the whole. The rest pay an annual tax for this outside garment of all, become indispensable summer and winter, which would buy a village of Indian wigwams, but now helps to keep them poor as long as they live.

I do not mean to insist here on the disadvantage of hiring compared with owning, but it is evident that the savage owns his shelter because it costs so little, while the civilized man hires his commonly because he cannot afford to own it, nor can he, in the long run, any better afford to hire. But, answers one, by merely paying this tax the poor civilized man secures an abode which is a palace compared with the savage's. An annual rent of from twenty-five to a hundred dollars (these are the country rates) entitles him to the benefit of the improvements of centuries, spacious apartments, clean paint and paper, Rumford fireplace, back plastering, Venetian blinds, copper pump, spring lock, a commodious cellar, and many other things. But how happens it that he who is said to enjoy these things is so commonly a poor civilized man, while the savage, who has them not, is rich as a savage?

If it is asserted that civilization is a real advance in the condition of man—and I think that it is, though only the wise improve their advantages—it must be shown that it has produced better dwellings without making them more costly; and the cost of a thing is the amount of what I will call life which is required to be exchanged for it, immediately or in the long run.

Thoreau's house, built in 1845 for \$28.12½. The annual rent for a Harvard dorm room then was \$30.

An average house in this neighborhood costs perhaps eight hundred dollars, and to lay up this sum will take from ten to fifteen years of the laborer's life, even if he is not encumbered with a family . . . so that he must have spent more than half his life commonly before *his* wigwam will be earned. If we suppose him to pay a rent instead, this is but a doubtful choice of evils. Would the savage have been wise to exchange his wigwam for a palace on these terms?

Henry David Thoreau, from *Walden,* 1854

Coast ran \$500 to \$900 a month in cheaper areas, in higher-income zones a lot more. The monthly mortgage rate for a similar property would be about the same. Many people in the United States now spend 30 to 40 percent of their entire income on housing, one of the highest percentages in the world. (Interestingly, I am told that in Cuba housing costs have been pegged at 10 percent of income, and for a long time in the former U.S.S.R., 3 percent was the most you could be charged.)

Housing is for many of us our biggest expense. If we could stop paying rent and mortgages, everything would get a lot easier, the pressure to earn would be less, and the covert threat of losing our house to the bank, gone.

Who hasn't at some time dreamed of building their own house? Yet the prospect is

Many people in the United States now spend 30–40% of their entire income on housing.

daunting. Modern building, using industrially made components, requires a big range of expensive tools and frightening machines. We need to learn dozens of technical skills we probably don't have. We will need to buy a site. So "building themselves a house" has largely come to imply well-heeled persons employing a contractor to custom-build for them. Few of us have the courage to actually *build for ourselves,* with our own bodies, skills, and time.

Cob cottages are a possible way out of this trap. The skills needed are minimal—anyone can learn them in a short workshop, or even from this book, with a little practice. Cost of materials is very much less than with most other building techniques; you will manufacture your own building materials from free, or very cheap, components. Tools are rudimentary, inexpensive, and easy to handle. Total cost can be low enough that most people can save enough in a year or two to build phase 1 of a small but comfortable home.

Perhaps most encouraging is that in working with earth we develop the confidence to abandon old ways of being that are dragging us down. Building for ourselves, we can prioritize creating a workspace in which we can work from home, saving the time and cost of commuting and choosing our own hours. Not needing total reliability in a car to get to work, we can spend less on a vehicle, perhaps invest in an older and more utilitarian truck in place of a late model sedan. The home worker has little need to drive anywhere, so families might dispense with a second car. With overheads down, we can contemplate spending less time working for money and more in personal satisfaction. We may put time into growing our own food, after which eating out has very much less allure, as the quality and flavor of homegrown food is so obviously superior. A cob house is satisfying and intriguing, so friends visit more and we spend less on purchased entertainment. By building your own earthen home, you can create a permanent dwelling that could last perhaps a thousand years, at a cash cost of only a few months' rent.

The lifetime cost in materials and labor for building, stretched out over two or three centuries, reduces both the amortized cost to users and the amount of annual effort put into upkeep. There is on record an English cob house built in 1585 that required no maintenance at all until 1810. This is 225 years, three or four times the average total lifetime of recently built stud-frame houses. A typical frame house, good for maybe fifty years, takes thirty years to pay for, 60 percent of its lifetime. A well-built cob home should last at least five hundred years. If it takes even five years to pay for, that's *1 percent* of its lifetime, a completely different view of household economics.

Trading Money, Skill, and Time

A typical phone call to our office:

"How long will it take me to build a cob house?"

"With an organized, industrious helper, six months to two years, full time."

"Oh, but that's such a long time!"

"Yes, it takes a long time to build a house. Any kind of house, a cob house is no different; it takes a long time."

Audible disillusion down the phone line. She imagined that because it's easy it would be fast. Let's be clear: Creating a house is probably going to be the biggest single investment of your life, whether of money or time.

Paradoxically, building with wood looks really fast, yet it isn't. A team of good framers can throw up a large structure in a few days, but it can be months before the details are all in place. By contrast, cob building looks slow, a real tortoise of a method, yet in many ways it's much faster than it looks.

Whatever the technique, it takes time and skill and money to build a house. *In an industrial consumer society, money, skill, and personal time are all somewhat interchangeable.* A person with unlimited money need supply neither time nor skill. A skilled builder with plenty of time can build a house with very little money; and given enough time even an unskilled person with enough determination can build a competent home at very little expense. Only skill can't stand alone; that is, you need to multiply it by time or it has little value.

As time, skill, and money are largely interchangeable, we have control of what proportions of each we supply. Take the trouble to analyze how much money, skill, and time, respectively, you want to invest in your new house. Get advice from people who know you well, particularly those who know how *determined* you are. Determination is a fourth ingredient, without which an owner-builder can't succeed. If you choose to contribute time and skill instead of some proportion of

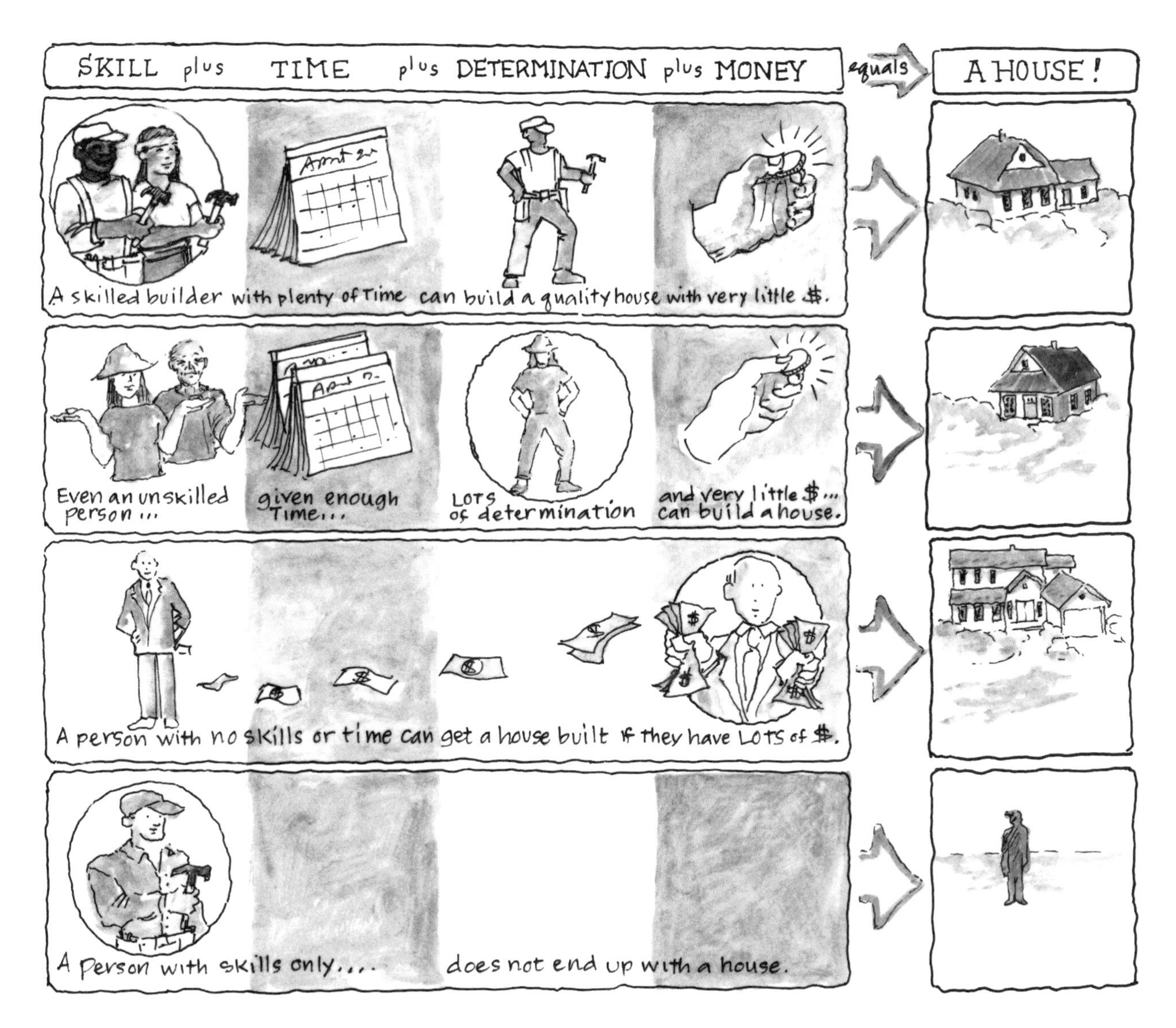

The 30-year mortgage.

the otherwise necessary money, you will need lots of determination.

What do we mean by "skill"? Partly, the manual dexterity to build, in whatever medium you use. Driving nails straight, cutting wood exactly where you wanted to, attaching cob so it is structurally sound. Skill is also organizing the job so that your supplies are delivered on time and you have the right tools when you need them, and so that all your helpers arrive at the right stage. Skill can mean the speed with which you work, how satisfying you make your work rhythm, whether you can keep everybody happy on the job (chocolate, beer, trips to the swimming hole, all at crucial times). It also means, most importantly, not making basic planning errors that can take extra time, skill, or money to remedy, or discourage you so you give up (see appendix 1). This book can help improve your skills in all of these areas. It should pay for itself several hundred times.

Many people abdicate their skill and time to a builder or a landlord or a developer and put in only money. They invest their time and skill elsewhere earning that money. The process is indirect, so they suffer from the inefficiencies of oblique procedures. For instance, the contractor wants not just to be paid for the work, but to make a profit too. *You* pay for this profit. The bank will lend you money but wants to be repaid an extra bonus. On a thirty-year mortgage, you pay for the house three times over.

The money you pay (to the bank, the developer, the builder) is devalued even by the time you get it. There are hidden inefficiencies in trying to earn the money to pay other people to do work on your behalf. Suppose a carpenter charges $25 an hour to work on your house. It's easy to think, "Well, that's a good deal because he's a faster builder than I would be, and besides, I can make $25 an hour working at my own job." Be careful with this logic. Your salary is diminished by taxes, the cost of getting to work, the time spent traveling to work, the cost of special work tools, clothes, appearances, by lunches out, time you're without work, and the cost of finding a job. A specialized education for the work you do may already have cost a small fortune. Unless you really enjoy your job, you pay for recreation to recover from it, and toys and treats to compensate yourself for having to work at all. You may pay in health and just possibly with your life itself. And you could spend your most productive years driving to work to pay for the gas to get to work to pay the taxman and the bank for a home that may never become yours. In 1995 only 10 percent of Americans owned the houses they lived in, while 50 percent were renters. The other 40 percent? Mortgage-holders. According to *Webster's,* the etymology of *mortgage* is "death pledge."

Owner-builders choose to expend time and to learn skills. In consequence, they need much less money to house themselves and can normally avoid going into debt. Cob offers three additional advantages: (a) the raw ingredients are cheap, (b) you manufacture your own building material, so effectively pay yourself for it in savings, rather than pay a commercial source, and (c) the technique is easy to learn, so skill comes quickly. The Cob Cottage Company has alumni who without previous construction skill have built beautiful homes for less that $5,000. Michael and I built Linda's and my cottage for under $500, as a deliberate demonstration of how little cash is necessary (although it required extensive planning time and a lot of creativity, and it's tiny). Even cheaper cob cottages have been built.

CHEAP HOUSING FOR THOSE WHO NEED IT MOST

As more and more of us spend more and more of our incomes on housing, more and more of us will end up living in our cars, sleeping under bridges and on hot-air vents, rotting in prisons, or sharing ill-conceived houses with unsuitable housemates. Our society has failed to define housing as a basic constitutional right, and we have been brainwashed to believe that a justifiable price for making profits is to have a large population without homes to go to.

For the snug middle class, to which most of us reading (and writing!) this belong, the problem has not yet become completely real. This couldn't possibly apply to us. Or could it? The homeless up until now have generally not been individuals we personally know. But this can suddenly change. For the first time it is now increasingly likely that a member of your own family will be unable to pay the rent or will lose his or her job; then watch the bank seize their home. They're relatively lucky if they at least have a family; there's a chance someone will give them temporary shelter. Knowing all this, we live in perpetual low-key fear, compromising our beliefs, our time, and our human dignity by working longer hours, wasting our days, just to maintain a sense of security and keep a roof over our heads.

Many are also tempted to engage in any of a number of activities that, though they harm nobody, are illegal and can thus land them in prison. The prison-industrial complex, a frightening new growth industry, depends on our desperation and carelessness to keep up its profits. With almost *two million* people in U.S. prisons (by far the highest proportion in the world, and six times the number of fifteen years ago), we watch helplessly as more jails are built, knowing that somebody will have to fill them. Not long ago the prison population was largely composed of people we could easily see as a "criminal class." Now, that class has been redefined to include most of us. If we don't already know someone incarcerated, we will soon, as nice honest respectable family people get caught in the wrong place at the wrong time. For instance, we knew someone who ended up in jail for not having car insurance.

Relieving worries about housing, and reducing actual homelessness, are among the salient possibilities of natural building. The solutions are not simple, but we now have at least a glimpse of the potentials of these materials and techniques. What, solve the homeless problem with three million mud huts?! Perhaps not (but perhaps so); in any case we are well on the way to providing affordable alternatives. There have been several obstructions in the path. The natural building movement has been able to tackle and demolish some of them; others remain standing, yet already the path is clear enough that we can see light at the horizon.

Getting Free from the Trap

To many intelligent people caught in the rent or mortgage system, the difficulties of building their own home might look like this. "I make minimum wage and live in a dreadful rental. I know the landlord is ripping me off, but what can I do? I have to house my kids. I would build my own house, but the cost of materials is too high. To pay the $30,000 in materials I need for a frame house would take ten years' savings, but if I begin saving now, in ten years' time costs will have inflated so much I'll have to start saving again. The bank won't lend me the money because I don't have a secure enough job or a good credit rating. And they won't lend at all on a house under 1,500 square feet. So I have to go on spending half my income on rent. I'm too exhausted working two jobs to have the energy to get training in construction, so it's an overwhelming prospect to consider doing my own building, and even if I could, a building site costs another twenty grand that I don't have."

Workshop graduates after a week's training. Some of these people felt ready to go right out and build a house.

Let's play "what if." What if the building materials were almost free? What if you could quickly and easily learn how to design a spacious little house and build it? What if you could come up with a few thousand dollars over the next year or so, meanwhile learning skills and assembling materials? What if you had a free building site? Could you make yourself a house under these circumstances?

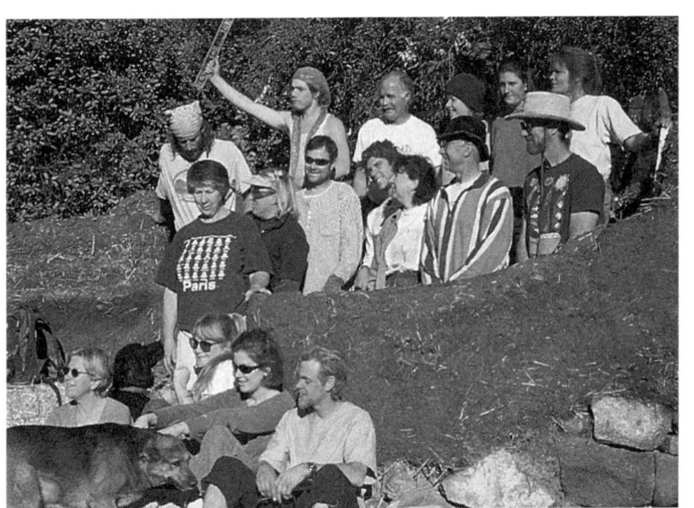

First, check off the easiest part, *plan your project around the least expensive building materials*. Cob is free, or almost free. You may have to buy straw; you might need sand or clay to be delivered. Check out all the sources for materials (see chapter 8). A small used pickup truck is a good investment—you could get to work in it *and* haul free building materials. Even better if you share it with neighbors and trade in your car.

Next, *learn some skills in order to gain confidence as well as experience*. Start reading up on natural building. Begin with this book, then use the bibliography section in the back to order other literature through your public library. Get a good feeling for what you will need to know. If you can, get to a practical workshop on cob building; it will save you time and money in the long run. Training centers are listed in the resources section. If you're a first-time builder or have never worked with natural materials, take a general practical construction workshop from a skilled teacher. One of the best ways of learning not only construction techniques but also planning and project management methods is to help someone else build. Ask lots of questions. Natural building does not require long technical training; much of what is essential is basic common sense. The skills you can acquire in a couple of weeks of workshop or helping on other builders' projects can save huge amounts of time and money.

Now, *scale down your plans to the smallest possible building your whole family could overwinter in* (see chapter 7). Make sure you take on only as much as you can complete in one spring and summer, and move in before the weather gets bad. Building small, the materials you actually have to buy are minimized, as is the drain on scarce cash, and you'll make mistakes on a small part of your

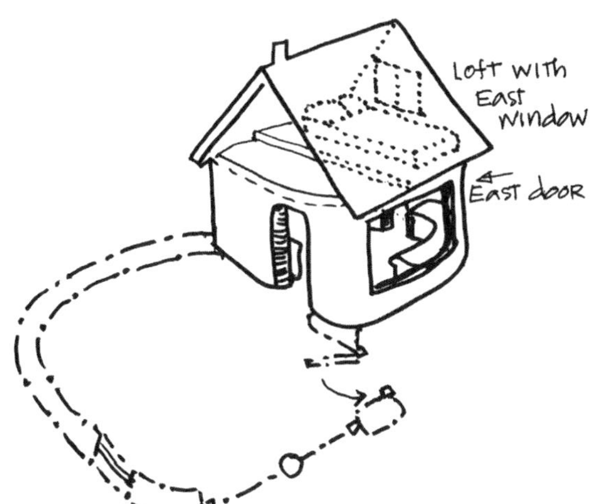

Phase 1: Build small, but plan for phase 2. Begin with a sleeping loft over a small kitchen.

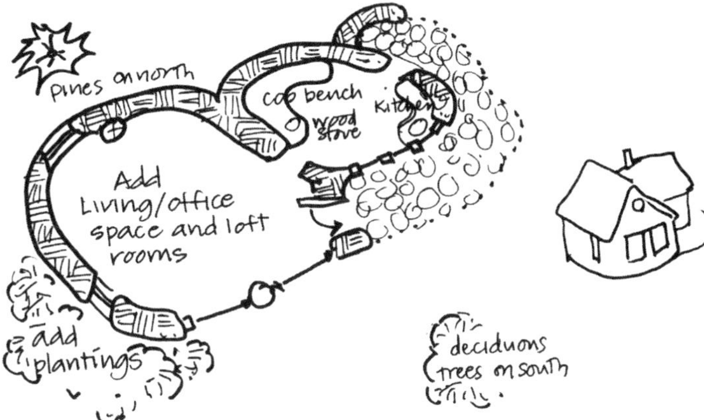

Phase 2: Add cob-walled courtyard space on east and living/office space on west.

house instead of all of it. You can add to the basic core of the house next year, incorporating the new tricks you've learned.

Save enough money to buy the true essentials—a few tools, perhaps a few bags of lime or cement, a load of sand, straw bales, some secondhand hardware, doors, roofing materials, and so forth. Start saving *now!* Put aside as much as you possibly can, every month, every payday. Go without that new VCR, delay buying clothes, don't buy anything you don't absolutely need. A one-year moratorium on eating out means an enormous savings for most of us; go to your natural foods store and get a 50-pound bag of rice instead.

Borrow from friends and family. When you're ready to build, don't let a cash shortage stop you. Ask friends and family if they could make fixed-term loans. Once you have free rent it will be possible to repay them.

Finding Inexpensive Places to Build

To the people who most need it, there is little point to a thousand-dollar cottage if the site to build it on costs thirty times that much. Rather than spending half a lifetime paying for the ground under your feet, you have some attractive options. Among them are these:

Lease Rural Land: Often undeveloped rural land is cheap. Provided the owner has no objection to you living there, even if possibly quasi-legally, you could take a long lease (preferably five to ten years or longer), build a starter cottage to learn the skills, then be

EDWARD'S ALMOST FREE HOUSE

We were working on a simple structure by a rushing mountain creek, up in the Idaho Rockies—cob walls on a rock foundation four feet deep. The door frame was up, the walls almost completed, windows were in, the floor leveled, and plasterwork was started. We'd just completed a cob archway five feet across leading to the spiral staircase in a three-story cob tower. It was time to go out to the forest to select a curved ridge beam, to get the roof going. Edward was reflective: "We made a good job of that foundation; it's quality rockwork. You know, except for the cement for the mortar, this building hasn't cost a penny so far."

prepared to move on when the lease expires, having thus saved enough cash to find a more permanent prospect. Make sure you have some security of tenure; you wouldn't want to be forced to leave having just completed the cottage.

Lease Land with a House on It: Move out of the house as soon as possible. Sublet it to pay the lease, and begin building. If you move out in early spring, you could camp or live in a tent or travel trailer through the warm months, until the first part of your new cottage is livable in the fall.

Share Rented Housing on Rural Land: Then gradually move out as parts of your cottage (on the same property) are completed. For reduced rent, you could sleep in a temporary structure and share only storage, kitchen, and utilities until your cottage (on the same property) is complete. Linda and I have lived under this agreement for nine years.

In each of these first three scenarios, the land-owner eventually ends up with your house. You end up with the skills and experience you gained while building. If you do a nice job, the owner may be willing to reimburse your expenses when you leave.

Borrow Land from Friends: Make a tight agreement with a friend or family member who already owns land; if necessary advertise for a place. Make a deal that you will pay for and build a cottage in exchange for its use for a set length of time—say, four or five years. After that it's theirs, a charming, cozy work of art they could rent out, or use as a guest cottage or weekend place. After your agreement expires, you might rent from them or move on, having practiced on their building, then repeat the process. Or you could stay on as caretaker in exchange for living there.

Get the owners enthusiastic about your project—it *is* exciting, after all. Show them this book, the color photos. You can guarantee a lovely small cottage they will enjoy. Make sure they really get a good deal out of the arrangement, and try to cover all eventualities (for instance, specify your security of tenure if they sell the property). *Write up the agreement you make.* It's not a legal contract; it's a record of what you've both agreed upon, and the record may be invaluable if misunderstandings occur in years to come when memories grow fuzzy. This arrangement seems to have real potential. Linda and I have built on borrowed land, as have Eric Hoel (see chapter 5) and Brigitte Miner (interviewed at the end of this chapter).

Borrowing Money

If you save until you can afford to build a modest house, or, better, build parts as you have the ability to pay outright, you never get in debt. The title or lease is yours free and clear, you have no worries about repossession or losing your job while you're building and you can design exactly what will best suit you.

Admittedly, if your best option is to buy a building lot, there's a real possibility you'll need to borrow money. If you have to deal with a bank or mortgage company, they will put constraints on how you build. Lending institutions exist for the benefit of their shareholders. Your personal health, happiness, even your safety are not their concern. They don't want you to build a magical house that is custom-fitted to your family; they want to be sure of easy resale if you can't make the payments. It will be difficult to build with natural materials; instead everything will need to conform to a predrawn set of plans (perhaps even plans from their own catalog), and you'll suffer some extra large costs, including building permits, sewer connections, and inspection fees. You may need to put in unnecessary and expensive devices such as baseboard heaters or a furnace as well as a woodstove.

Many lenders demand that you build a much bigger and more expensive house than you actually want; in fact it is difficult to get an institutional loan at all on a house under 1,000 square feet. If you ever want to alter the building, you will have to get their agreement, and there may be constraints on your freedom to sell, lease, or rent.

If you absolutely have to borrow, do so privately; if at all possible, borrow from family, friends, neighbors, or colleagues. The money stays in your community. Interest terms should be cheaper, and friends or family are more likely to be understanding if you make a late payment. And this way, if you want to build only 600 square feet, or build with cob, or have a five-story lookout tower, nobody can stop you.

A private loan doesn't need to be one lump sum; in fact a series of small loans gives you more repayment options. Write a letter, explaining your project, and send it to everyone you can think of who might conceivably be useful. Tell them that here is an opportunity to reinvest a little of that money they are currently lending to the banks. Explain that your first choice would be loans for, say, five years, interest-free. You would next consider one-year loans at no interest. Then interest-producing loans, at a low rate. Finally, if you can't raise enough by these methods, you could pay a higher rate of interest if your lender really needs such terms. You can offer your friends a better rate than commercial banks would pay them for a savings account, and the loan will still be cheaper to you than the borrowing rate you would have paid to a bank.

CHECKLIST: HOW TO KEEP COSTS DOWN

Building your own home tends to develop your self-confidence, ingenuity, and community. Not spending money is an effective political action, even when frugality is not an economic necessity. It takes power away from the manufacturers, retailers, and corporations that thrive on people's dependency. Following is a checklist reiterating sensible economics when building your own house:

❧ ***Save enough in advance.*** Stockpile money to pay for all the materials you will need for tools, specialist help, fees, bribes, and permits. If you can, save enough also to cover basic living costs while you build so you that you can concentrate your full attentions on the project.

❧ ***Practice extreme economy.*** Be fanatically frugal until after you move in—it will only be a matter of months. Don't eat out; don't buy anything unnecessary. The whole idea is to finish phase 1 of your cottage without incurring debt.

❧ ***Move your home onto the building site.*** Live in a temporary shelter on-site, so you're not having to pay housing costs while you build. Borrow a big tent or buy a $200 travel trailer. Make sure it's unsavory enough that you have incentive to complete at least a part of your building before cold weather comes.

❧ ***Don't involve government.*** Steer clear of officials, with their permits and paperwork, if you possibly can. Their attention can be expensive and is not usually helpful.

❧ ***Start small.*** Make the building tiny at first to keep within budget and on schedule. Plan to add on to it only as you really need to, as time and resources permit.

❧ ***Plan ahead.*** There's a Permaculture principle that states, "You always pay more for emergencies." This is doubly true in building. When you begin construction, make sure that you have everything you need on site. Last-minute trips to the store waste time, energy, and money. Start collecting materials as soon as you can, even if you don't know when or where you will build. That way you can take advantage of what comes your way for free or cheap.

❧ *Use what's available or do without.* Adjust your design according to the materials you have on hand. As much as possible, use natural materials from the site: soil, rocks, standing dead trees. If you can't find or afford something you think you need, consider the possibility that you may not really need it. What else would work? If you simply must have those French doors or that stained-glass window, but can't afford them, go ahead and frame them in. Then cover the spaces temporarily with something else until cash for the element you want comes along.

❧ *Develop creative problem solving.* Learn to ask yourself, in this order: What am I trying to solve? What's the simplest, most creative way? How can I use what I already have, use what I can get for free? Can I trade for it from a family member, a friend, or a neighbor? Spending money should be a last resort, reserved only for condiments; select main staples that are free.

❧ *Do it yourself, with friends and family.* Design the building according to the skills you have available—your own or volunteer helpers. Labor is expensive, particularly specialized labor such as that of a skilled mason, carpenter, or roofer.

❧ *Avoid machines.* You can get by without them. It may take a little longer, but unless you are a contract builder, doing hard work will save you money. Machines always cost more than you expect, in fuel, maintenance, accidents, mistakes, delays from downtime, environmental impact, and fixing up by hand the damage they cause.

❧ *Check demolition sites and construction dumpsters.* Other people are throwing away the stuff you need, especially lumber, windows and doors, plumbing, and wiring supplies. You do them a favor by taking these things off their hands, saving them the dump fees. Many dumps set aside free building materials. Call excavation contractors for free truckloads of soil for cob and concrete rubble for foundations.

❧ *Ask for donated materials.* Tell people about what you're doing. They may get excited and want to help. People cleaning piles of useful stuff out of their garages need someone to give it to. Offer to help them clean up and haul away whatever they don't want. Set yourself up as an educational resource center; people will see you as offering a valuable service. We have been given wool for insulation by sheep farmers who wanted to support our work. Lumber, windows, doors, used roofing have all been given to us, as well.

❧ *Make it!* Make your own windows and doors, skylights, furniture, and cabinets. Make your own tools. They can be crude, functional, and beautiful.

❧ *Fix it!* Replace a broken pane of glass, splint a cracked mullion or frame. Duct tape and wire are indispensable parts of any tool kit.

❧ *Borrow it!* You can often find someone who already has specialized tools that you will need only occasionally, which they would be willing to lend to you. Consider setting up a community cobbers' tool-lending bank.

❧ *Trade it!* Get to know other owner-builders and do-it-yourselfers in your area. Offer to trade labor or surpluses for things they have that you need, or for specialized skills. Offer to teach cob in exchange for labor or materials. Consider teaming up to co-build a number of houses among a group of families.

❧ ***If all else fails, buy it, used.*** Before you hit the building supplies store, try the recycling center. Many towns have businesses that specialize in salvaged materials, including lumber, hardware, and glass. Besides being socially and environmentally responsible, buying secondhand can give you enormous savings. Used materials also add character to your building.

INTERVIEW: BRIGITTE AND ELYSE'S HOUSE

In 1993, in the early days of The Cob Cottage Company, a single mother and her three-year-old daughter attended a cob building workshop. After the course was over, they stayed a few days helping with construction, then left to build their own house. A year later they were moved in.

Brigitte had no building experience, no experience of designing houses, and no stockpile of materials. As a full-time single mother, she had very little income and almost no savings. Worst of all, she had no assets, no property, and no site to build on. A week-long workshop gave both her and her daughter, Elyse, confidence that they could build, at least the earthen parts of a cottage they would design themselves. But in western Oregon where they live, a building site of any kind can cost twenty to fifty thousand dollars. A site for under fifteen thousand dollars is almost unheard of. How could they possibly buy a building site?

Purchase being out of the question, and a quality home an urgent necessity, Brigitte chose to borrow land. When she let it be generally known what she was looking for, several opportunities were offered. After weighing the options, she chose to build on the land where her parents live, a densely wooded tract of several acres where an inconspicuous cottage would be unlikely to attract attention.

Heart window in Brigitte and Elyse's house.

Brigitte and Elyse began building in May, when Elyse was four, and were able to move in before winter. Their 450-square-foot cottage has cob walls and a cob floor with a north wall of straw bales. Together they did all the cob work—"5,400 individual cob loaves" says Brigitte. Carpentry was done mostly by Elyse's grandfather and friends. Windows and doors were found secondhand, and most of the cash cost was in new lumber for the loft and roof structure, as well as roofing. Then there was cement for the foundation wall and gypsum plaster, plumbing fittings and straw bales, plus transport of materials and rental of a trencher to bury the water line and create drainage.

Brigitte and Elyse were the second family to occupy a self-built cottage of Oregon cob. Elyse, at nine, built herself her *own* cob cottage and by ten was running her own mail-order business. Ianto talked with Brigitte, nearly four years after they moved in.

IANTO: Why cob? Why did you do this?

BRIGITTE: In a few words, I knew *mud house, simplicity,* nothing else at first. It came to me as a personal revelation that this is what I needed.

IANTO: Were you doubtful this would work? Was there a point of commitment when you decided, "I'm going to go for it"?

Brigitte and Elyse's house. Note the quality of light through the deep window reveals.

BRIGITTE: After the workshop I had no doubts, but I soon discovered others did. My dad had said he would help clear the site, but I don't think he took my project seriously at first. He suggested that I buy a little trailer—a quick and convenient solution. When he didn't help clear the site I started myself with my bare hands, pulling out poison oak. That got him out of his chair! He knew I was serious and the project began to interest him.

IANTO: Were there things you learned during the construction?

BRIGITTE: It's everything you philosophize about, everything I heard you say in that workshop. Anyone can do it, and it maximizes your interaction with the natural world. It releases something intuitive inside.

IANTO: Yes, it was a revelation to me, almost ten years into this project, to realize that I have encountered no opposition, not from anywhere. It's almost uncanny. I think that's why I've stuck with it so long. Have you found the same?

BRIGITTE: At first I encountered the skepticism of neighbors, but it takes so little to show people what it's all about. It bothered me at first the walls aren't straight. It only bothered me because I would show it to other people—I was concerned about what they would think of me, not of the walls. But things can be funny in a cob house and you can be happy with that, to have a fun house, instead of a look-alike house that's not supposed to be fun.

IANTO: Are there ways this experience has changed your life?

BRIGITTE: You can't imagine you can change so much, be so much. I feel a more complete person. I feel I have so much more to offer people, not just building expertise but of who I have become. You get so much respect when you've done this, it's a passport into any circle. This is the only way you can change the world. I'm walking my talk, enjoying my place, getting out of those vicious cycles that society pressures us into.

IANTO: Well, "the proof of the pudding is in the eating thereof." What's it like, living daily in your cob cottage?

BRIGITTE: Living in it is a daily affirmation of who I am and where I'm going. It enhances my growth by nurturing my body, mind, and soul. What other building process can do that? You don't get that from brick or bales.

It's right there in your face when you get up every morning. Every day begins magically. There's no bad side to the bed anymore. My windows are special because each one was placed carefully to frame a special view. Each one glows from the surrounding white-washed walls. We have twenty-two windows in our house, and each one of them came from somebody. It's a quilt within the house. I have a sweeter affinity for what's outside. There's something sensitive that goes with the windows that makes me look out.

Living here is like a marriage in many ways. It's not perfection. I still see the mistakes, but it's a commitment. I used to feel, "dammit why can't I take it with me—couldn't I put it on wheels?" I wanted to be free to move, that's our society. But there's something about settling into where you are, a centering magnetic pull. It makes you put down roots and face life full on. This is the most nurturing, spiritual place I have ever felt. Every person who has ever come to my funny wobbly little house has felt the same.

IANTO: More pragmatically, what about costs? How much money did you put into your home?

BRIGITTE: I'm not totally sure because I didn't add up all the details or subtract the food I bought, or the cost of running my car while I was building. I just know that I had four thousand dollars when I started and we used it all up. So four thousand, certainly not more. Mostly it went to lumber—the beams, roof lumber, paneling for the ceiling, the loft, skylights.

The big economic benefit is in living here. I have a different attitude about not wasting things. When you build your own structure with your own hands, you're more likely to self-decorate. You don't buy Wal-Mart stuff to go in a cob house. You don't get into that redeco-madness cycle. When I was building I felt the freedom to express myself in the moment. It's so much better than something premeditated; as a result I've changed very little in my house because I got to express myself freely the first time. There's a liberating lack of maintenance I don't have to do; I don't need to kill this, repair that.

IANTO: Finally, how does your house behave thermally? Is it comfy?

BRIGITTE: On hot days everybody's got their air conditioner on, and here it's absolutely cool. Even on hot nights it's delicious. Elyse and her cousins last summer would choose to come here rather than be in Grandma's house next door; she had the air conditioner on. I have lots of curtains. If I close the curtains, it's always cool, and with twenty-two windows we're never short of light.

4 Tilt and Spin

❧ IT IS MY DAILY PRACTICE TO PAY ATTENTION TO THE SKY. I FEEL A COMPLETENESS ROUNDING OUT MY LIFE, A COMFORT FILLING THE CORNERS OF MY IGNORANCE. ❧

BEFORE YOU PUT DOWN ROOTS, IT'S GOOD to know where you are. This chapter helps you orient yourself to the yearly tilt and daily spin of Earth, to the cosmic geometry of the planet. It will help you determine where in the Universe you are, which is a precursor to knowing where and how to build your home.

If a house prevents us from participating in cosmic awareness, it is *un*-natural, or worse, anti-natural. A truly natural building must be arranged so that it reveals the recurrent magic of Earth's spinning and tilting to the people who occupy it. We can invoke cosmic magic in small ways such as a Solstice Moon peephole in the wall of a house, in big ways such as Stonehenge or the Great Pyramid, or we can do both.

The Sun doesn't really rise; Earth's horizon dips. Sunset is Earthrise. The Moon and stars don't all rush across our sky every night. They *are* the sky, and we spin past *them*. But somehow Galileo and Kepler never quite convinced us. Our language patterns must have been pretty well fixed even by the early 17th century—we still subconsciously believe in an Earth-centered universe. So little attention does our culture pay to real cosmic movements that most of us don't have a clue where the Moon is at any given moment, still less Venus. When we're inside a building, we don't know where the Sun is. How many of us really understand what an eclipse is, or what the Arctic Circle means?

Schools don't often teach kids how to pay attention to cosmic geometry. Children won't learn it from most religious institutions, gov-

WATCHING EARTH TURN

❧ Once when I was perhaps fifteen, just after sunset, looking east across the English Midland plain I watched a band of deep blue creep quickly up from the eastern skyline, swallowing the pink glow reflected from the setting Sun. Gradually I realized this blue band had a curved top, that it had to be the shadow of Earth cast on the atmosphere, our own shadow! I've watched it hundreds of times since, pointing it out to others. Almost never do they know what it is.

❧ The equinox is the Great Leveler, the day when the Sun both rises and sets at 6 o'clock; that it rises due east, sets due west; in flat country the sunlight shines exactly 12 hours, and this is true *all over the Earth.*

❧ Try as we might, it's hard for us to *see* Earth spinning. Like the clock's hands, you look away and they have moved, but you never catch them moving. But one evening when I was already over forty, I saw for the first time the movement of the horizon against the rising Moon, watching distant trees sliding obliquely across the disc. My metabolism had slowed sufficiently, or else perceptual time had sped up to the point where it was observable and real. Now I am sometimes aware of us moving even at midday.

❧ Finally, not long ago, I watched a tree inch across the full Moon, counting seconds. Curious, the whole transit took almost exactly 4 minutes. Time it again, with the watch: 241 seconds. *Four minutes is a fifteenth of an hour which is a twenty-fourth of a day.* Tossing the figures in my head, then Good Grief! 15 x 24 is 360! The face of the Moon spans exactly one angular degree. What a strange coincidence! Wait a minute . . . of course it's not a coincidence; we divide the circle into 360 *because* there are 360 moons around one complete rotation. I've heard other explanations, but this makes a lot of sense to me. Mostly though, it was a reminder that I really haven't been paying much attention, that there's a whole web of interconnected understanding that goes with watching tilt and spin.

What is probably most significant about these incidents is that I, a compulsive reader with an oversupply of formal education, never learned any of this in school. Nor did I ever encounter in fifty years of reading any explanation of these happenings that are so fundamental to where and when we are. Now in middle age, it is my daily practice to pay attention to the sky. I feel a completeness rounding out my life, a comfortable filling of the corners of my ignorance. Observation and observance connect me directly to the basis of our existence; this is very different from TV. These natural wonders repeat daily, seasonally, and I know that they have done so since before me, before humankind. It's very comforting.

ernment, or the media. What's left to influence a young person? Parents of course. Yet now, ten generations after the first Industrial Revolution, parents are as bewildered as their kids. Few of us really know how to be aware of the spin and tilt of Earth, or how profoundly they affect us.

It's hard to be useful to family or society if you don't first have your own bearings. If you're confused, you don't feel good. We exist at the intersection of two systems—where and when. *When* is the time we reside in, our position in the history of our tribe, and a recurrent cyclic placement that repeats daily and yearly. *Where* is a location defined by cosmic coordinates and terrestrial ecology, connected to these recurrent cycles. *Where* is constantly moving. As is time.

The reasons for seasons.

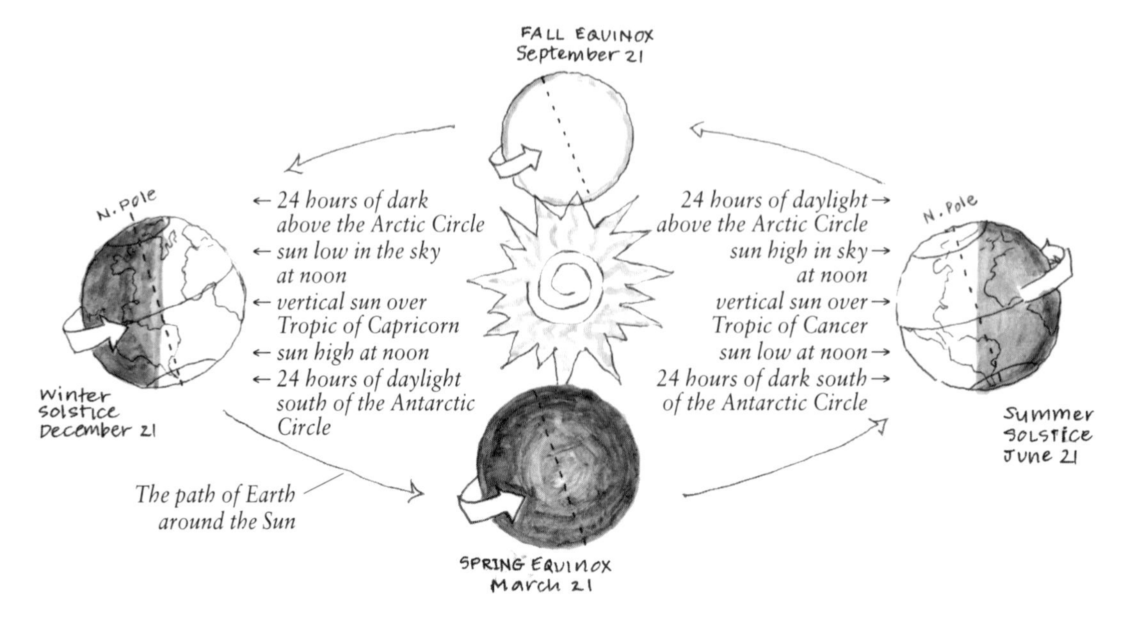

SOME COSMIC FUNDAMENTALS

To unravel some of the confusion, let's run through some cosmic fundamentals. Most of the solar system operates as if the parts were attached to a big invisible wheel. All the planets revolve around the Sun in approximately the same plane, all in the same direction, and most of us spin that way too. Earth spins toward the east at a constant rate, one rotation every 24 hours. This rotation explains why we have day and night.

The Sun, Moon, and visible planets (and their moons too) all appear to slide across our sky along one broad path, called the Plane of the Ecliptic. But Earth *spins* tilted a little away from the plane, which is why different latitudes get variable attention from the Sun, depending upon which way we tilt at different seasons. When our axis of spin is tilted toward the Sun, the Sun looks higher in the sky, and we get more heat and longer days.

Both poles of the Earth tilt toward and away from the Sun, spending six months slowly swinging away from the Sun, then reversing and rolling back for another six months. This accounts for why we have seasons and why they are most pronounced farthest from the equator. This tilting swings us conveniently just a tad over 45 degrees, or half a right angle.

When is *now?* Who broke our day into 24 parts, divided each hour into 60? The clock now says 9:22, but who decided that? Clearly this kind of time is a human construct, based on fairly arbitrary agreements. For convenience, we numbered the years and invented clocks, very recently in our species' history. We stumble over the analog precision of numerically defining each second yet ignore the inevitable cosmic recurrences of Sun, Moon, stars. There's nothing very magical about A.D. 2000 if we see it as a result of Christian chauvinism and decimal arithmetic. Had we had six fingers on each hand, the year 2000 would still be a long way ahead.

There's another kind of time, structured on the movements of the Sun and Moon, the cosmic bodies that ultimately control everything we do. True, there *are* 365 days in a year; it is indisputable that we spin that many times each revolution around the Sun. But the notion of *Tuesday* or *seconds* or *October 14th* is completely fabricated. It distracts us from the reality of watching the Sun rise; Tax Day becomes more significant than the Solstice.

A COSMIC QUIZ

HOW'S YOUR COSMIC AWARENESS? Try answering these questions for yourself. You'll find possible answers to most of them somewhere in this book, perhaps in this chapter, but that's not the point. Sometimes it is worthwhile to answer questions for yourself, without an examiner to pass or fail you. Firsthand knowledge is a key to knowing yourself, so watch the skies for answers.

1. Do we spin toward the west or east?
2. Can anyone actually see the movement of Earth spinning?
3. Point to where the Sun rose this morning. Are you sure? Check it tomorrow.
4. Does the Sun always rise due east, and set due west? If not, why not?
5. From what direction does the sun rise on the Arctic Circle at the equinoxes?
6. Explain three ways you can tell where solar south is.
7. Explain why in midwinter, evenings start getting lighter before mornings do.
8. What exactly is the Summer Solstice?
9. How can the Moon be useful in determining the siting of passive solar buildings?
10. At night, point to the North Star. Where will it be at noon? Point there too.
11. Quite often you'll see the Moon depicted in illustrations like this. What's wrong, apart from that the cheese will run out of the horns?

12. Do you ever see planets in the northern half of the sky? From where? And when?
13. Rainbows are nearly always to the east or north of us and seldom appear in the morning. Why?
14. Which planets can you see in full sunlight with the naked eye?
15. Why are circles divided by 360 degrees?

LET YOUR HOME REFLECT COSMIC AWARENESS

To feel more human and less mechanized, you can choose to live in a home that reflects cosmic cycles. Your house can reveal and display the procession of Sun and Moon across the sky and the movements of the planets. It can offer a snug place to contemplate the Perseid meteor shower at 4 A.M. on the morning of August the 12th. Or it could have a tiny peephole directed precisely at the North Star, Polaris, the only star whose position never moves. You'll need to pay continued attention to the sky and weather to fully exploit the possibilities, but any extra observation will reward you richly in the comfort and security of regular reminders that all is right with the world.

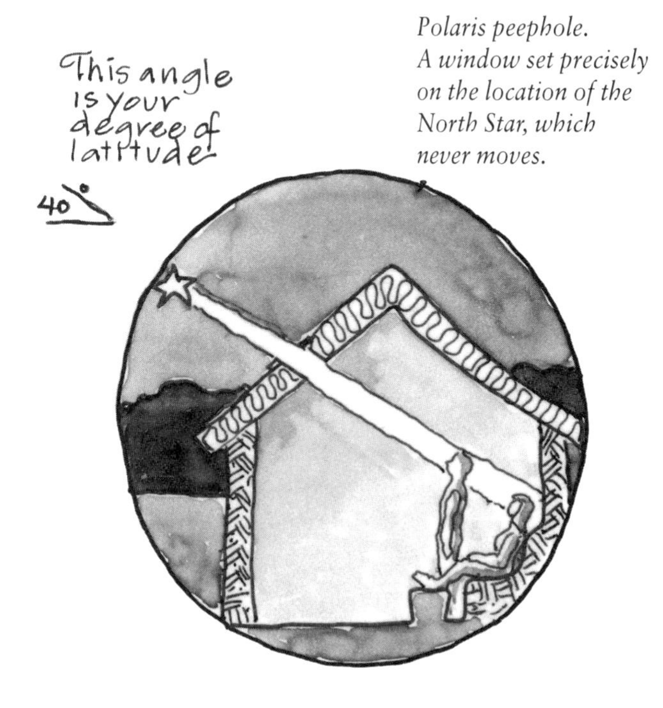

Polaris peephole. A window set precisely on the location of the North Star, which never moves.

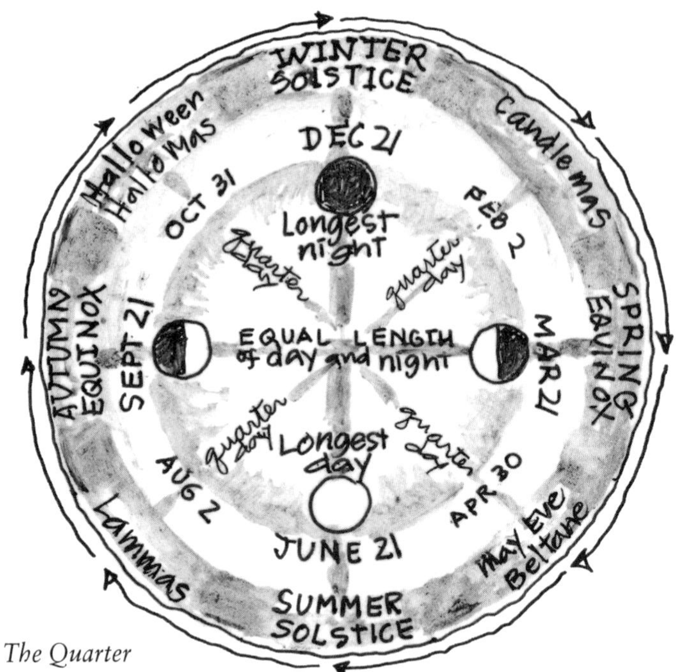

The Quarter Days: Candlemas (Imbok), May Eve (Beltane), Lammas, and Halloween or Hallowmas (Samhain).

In building our Heart House, such an opportunity arose. The cottage is set in a tiny clearing in the rain forest of the western Cascades. To the east is a solid stand of Douglas fir stretching half a mile to the next clearing, tight ranks of heavy, dark trees festooned with moss, ferns, and lichen.

It being the height of summer, Michael and I were working before dawn to glean what little cool was available early in the day. At the beginning of August, no sun reaches the house until after nine o'clock. As the east wall of the cottage grew, it became apparent we should put in a window to throw morning light into the kitchen, but exactly where had not revealed itself. One day, right at dawn, Michael called out, "Look—I can see sunlight!" Sure enough, miraculously, a tiny beam of light was twinkling like a distant flashlight through the forest. While we watched, it twinkled, it winked, then suddenly it was gone. Next morning we were ready with pruners and as the light showed again, we clipped just a few twigs, opening a tiny porthole for the sun. The third morning the same. So we put a window precisely where, in that east wall, the dawn sun shone into our kitchen. We call it the Lammas window. Lammas is the first of August, the old quarter-day, halfway between Summer Solstice to Fall Equinox.

Now every Lammas and every Beltane (the spring quarter-day around May 5), if the skies are clear, we see the rising sun for no more than seven minutes on four days only, as it enters in through the Lammas window to light up the inside wall clear across our little cottage. When I tell the story, visitors are entranced, even those who don't know the meaning of Lammas and have never paid attention to such patterns. Some of them comment that this is like Stonehenge, the only reference most of us have to the ritual importance of observing sunrise at key moments in the seasons.

Our distant ancestors saw clearly a relevance in cosmic motions that we have long lost. Consider how our society positions our houses. Drive the streets of any town in the United States; look closely at the house fronts. What tells you it's the front? For two generations now the front has been where the garage door is, but it also is and has been for centuries where the main (ceremonial) door is and, particularly in the past century, where

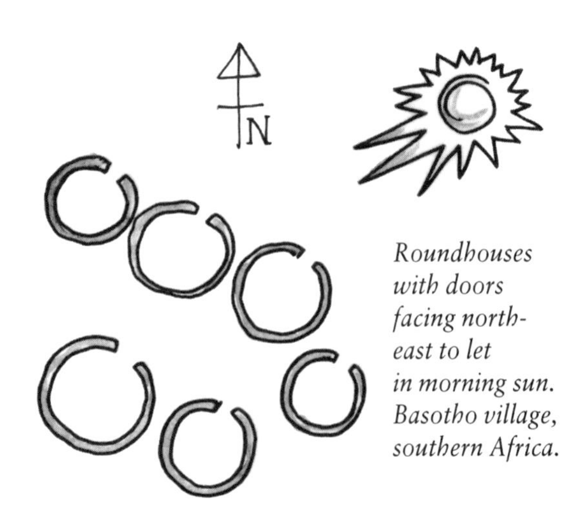

Roundhouses with doors facing northeast to let in morning sun. Basotho village, southern Africa.

the majority of the important windows face. Most of the biggest windows in most American houses face the street, looking onto parked cars and the windows of the house across that street.

By contrast, traditional (preindustrial) houses all over the world are carefully oriented to take best advantage of the sun, and to reflect moods and activities that vary with time and season. In southern Africa there are huge villages of roundhouses, *rondavels*, with every door facing roughly northeast. In cold weather all those doors are thrown wide open at dawn for the morning sun to warm the inside and cheer people out of bed.

Old Welsh farmhouses faced southeast, with almost all their windows on that side, probably for the same reason. The southwestern gable end had no windows but often had a chimney stack built in, to protect and warm the endwall that faced into the teeth of Atlantic gales. Aerial photos of rural Wales show a dominant grain of long farmhouses running southwest/northeast, with the active side, the main door and windows, to the southeast. Curiously, after the middle of the 19th century this pattern was disturbed. More recent houses, often in the same villages, face almost any direction. Nearly all the southeast-facing houses were built before 1850. What happened? Probably the railways were responsible, because they made industrially manufactured bricks available all over the country. Brick houses were more rain resistant, so they could face almost any direction. Often, though, they were oriented to face the street.

An expensive feature of settlement patterns in the 20th century has been our obsession with facing houses onto the streets that serve them. The net result is that in the entire North American housing stock, this random placement puts only a quarter of all homes toward south. Another 25 percent are oriented to the west, such that they see chiefly the afternoon sun and so may overheat easily in hot weather but enjoy little sun in the winter. The third quarter face north, and thus get no sun at any time. And the remainder, east-facing, are comfortable only where summers are cool and winters mild.

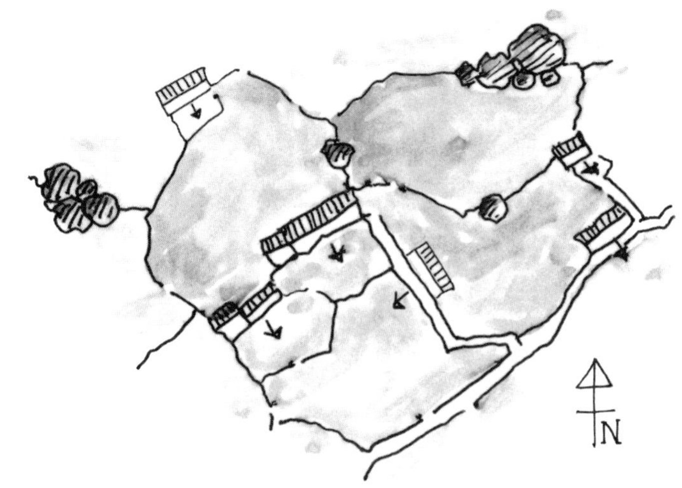

Welsh pre-railroad farms. All face southeast. Newer houses are randomly oriented.

Humans, like almost all life-forms, have evolved over countless generations to respond to the quotidian and sidereal cycles of light and dark: long winter nights, long summer days. The pituitary gland is quite sensitive to these rhythms, and our genetic inheritance demands that we pay good attention, or else experience confusion and despair. Many public buildings—including schools—have no windows at all. Whole populations of workers and children spend months each winter commuting to work or school in the dark, working without natural light, then commuting home again in darkness. Is it surprising that so many of us are disoriented?

Responsible buildings should emphasize the daily spin, the yearly tilt, of this planet we live on. Our homes and schools and workplaces should be magnifiers of the inevitabilities of the cosmos. They should exemplify the basic satisfactions of consciously exposing sidereal time. Without this connection to the Universe, can we truly be guardians of Earth?

5 The Site You Build On

❧ For the Magic to flow, the building and the site need to grow together, each improving the other, like an excellent marriage. ❧

Natural building implies building in harmony with Nature—a challenging goal for those of us who grew up in industrial consumer society. We early-21st-century natural builders are pioneers. It has been so long since humans built without trashing their place that we often don't know how to begin.

There's the danger we will think of a building *as something we put on a site,* rather than as *a creation growing out of a place* that already exists. Ecstatic architecture grows like a tree, responding to the nuances of the place where it develops and to the evolving needs of the dweller.

Selecting a building site is one of the most critical design decisions you will make and should precede the rest of the design process. *You can't have a good building on a bad site.* For the magic to flow, the building and site need to grow together, each improving the other, like an excellent marriage. Your site should empower all your home's potentials, so that you can live a balanced indoor/outdoor life, so that your house itself can enjoy a long life, and so that future inhabitants will love this place as you do. A good site will make a house easier to build and satisfying to inhabit.

The wrong site can have long-lasting negative effects that are difficult or impossible to mitigate, both during construction and throughout the building's lifetime. On a good site, if you screw up during the building process, the worst consequence could be that you will need to rethink and rebuild part of the structure, but if the siting is wrong, you're stuck with problems as long as the building stands.

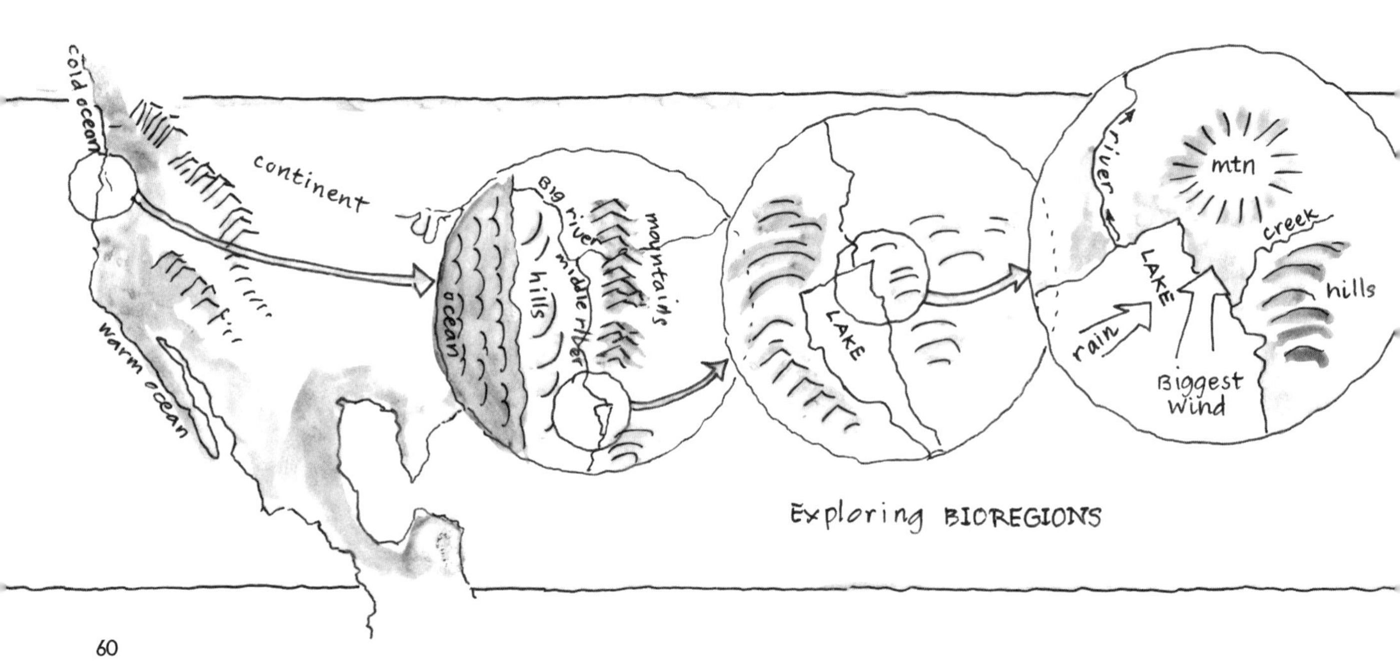

On the other hand, cob is well suited to some building sites that would normally be problematic. You can, for instance, build a cob house in a place that's noisy at night (e.g., commuting shiftworkers on a nearby highway, freight trains at a crossing), and because of cob's sound-absorbing qualities little noise will reach your bed. Since cob is fireproof, you can take advantage of a site priced cheaply because it is in a brush-fire zone.

This chapter aims to help you select land that can really sing to you in the long term and help you avoid places with hidden difficulties. But read on even if you already have land to build on. We will help you decide on the best specific building site, give you pointers on how to deal creatively with unavoidable problems, and discuss how to use your land to best advantage.

CHOOSING A SUITABLE PROPERTY

Here are some guidelines to help make sure the property you decide on will fulfill your hopes and expectations. Think of this as a set of clues for how to reflect upon the process of choosing land to build a home on.

Consider the Local Bioregion

Don't go directly to your potential building site. First seek to understand the influences on it. Look at North America in an atlas or on a globe. *Study the geology, economics, settlement pattern, natural vegetation, transport systems, and climate of your site's bigger environment.* Then look at the local bioregion in more detail: this may be an area of perhaps a few hundred square miles, characterized by similar geography, ecology, and climate. Describe your site's regional context to yourself: "The region is defined by these features: It's on recent sedimentary rocks with an oceanic climate moderated by the Cascades and the Rockies. It was never glaciated. It was invaded by European Americans 150 years ago, and the traditions of the natives have been mostly lost. Settlements are still relatively scarce, but concentrated along a North–South corridor only a few miles from the site. The native vegetation is cool temperate coniferous rain forest with much recent destruction by a declining lumber industry that has left an abundant road system almost everywhere. The region has cool damp winters and warm, sunny, dry summers."

Within the bioregion, your site sits in a smaller watershed, possibly a single creek drainage of a few hundred acres. Go to the library and take out the U.S. Geological Survey topographical maps and aerial photos that locate the site's neighborhood. Check county records, country stores, realtors. Investigate possible changes to the neighborhood over which you have no say—by government (a proposed dam, airport, road widening), developers (a high-rise that will be right in your lake view), or utilities (plans for installing power lines, lopping trees). Do all this research before setting out to look seriously at any specific piece of land.

When you arrive in the vicinity of your potential site, explore the surrounding roads

Get a feel for the ecological and human neighborhood of the land you are considering.

and footpaths (bike if possible, or walk). *Get a taste of the ecological and human neighborhood and look for glimpses of the land you are considering.* You may find a site that is lovely, but it's at the end of an ugly stripmall. Remember that the daily approach to your home will condition how you feel there. You have little control over the road that leads home, so be sure you like that journey.

Still without setting foot on the property, walk its boundaries; feel out influences that may affect it, and get a good idea of how neighbors might react to a new building going up. Find out who owns surrounding land and what they plan to do with it. Might they clear-cut the forest? Build a housing development? It's wise to talk with adjoining landowners early on, to make sure they won't try to block your plans. One resentful neighbor can make life miserable.

Check with your local planning department to find out whether there are plans to widen the road or change the zoning. Different jurisdictions have different land-use policies and varying abilities to enforce those policies. Certain areas are zoned for specific purposes, such as residential, forestry, or light industry. If your plans include agriculture, manufacturing, multiple residences, or building with alternative materials, and you pick the wrong location, you may find yourself fighting your neighbors and local government. Try to find a region where people are already involved in the sorts of activities you would like to do, and ask their advice. (See appendix 2 for more on building codes, etc.)

Spend Time on the Land

Next, go onto the parcel and head for the highest point—climb to the top of a hill or a tall tree or get on a roof. From up high, you can take in the longest views and get an overview of the whole area. Where can you see long vistas from? You can usually change the close-up views from your future home, but long views are probably beyond your control. Remember that views are two-way, so you may want to avoid building on a conspicuous ridge or unspoiled hilltop, yet at least one long view is very valuable.

Use all your senses. Not only landscape, but soundscape and smellscape are important. Listen carefully. Road noise is objectionable and deadening to the senses, and it masks the tiny sounds of Nature so important to well-being. Insects buzzing, a leaf falling, distant songbirds—these are the fine-tuned stimuli that keep us aware. Sniff around, too. If you're considering buying the land, ask all the neighbors—you might hear about a summertime stench from a hog farm that's a problem only in atmospheric high pressure when there's a rare north breeze. Or maybe there's a noisy factory that only works at night.

For millennia, the Chinese have used feng shui to place buildings, but you don't need to study feng shui to know how a place feels to *you.* Investigate all the property's potential building sites; hang out there. Where do you naturally gravitate? Which places pull you back to them? Are those because of ephemeral attractions (e.g., daffodils in April) or more permanent features such as the shape of the land, ancient trees, old buildings or homestead sites, sunny spots, rock outcrops, or a pond? Would adding a building with all its attendant clutter such as power lines, parking, fences, a paved road, tool sheds, and so forth destroy the magic? Building on the places already *most* impacted—logging clearcuts, old parking lots, the site of a ruined building—will damage the ecology least. A beautiful new house could improve the worst damage and upgrade the site.

It's always useful to know the history of the land. In recent times, have people used chemicals that might still be present in the soil and water? If there's a history of manufacturing, agriculture, or even previous buildings, you may want to test the soil for poisons. Who were the original human inhabitants of this place? Are there sites of archaeological or religious significance that it would be better not to disturb?

SELECTING THE PRECISE SITE

After you decide on the property, locate your building. Here are some guidelines to help you choose exactly where to build on the piece of land you have chosen.

Create a Master Plan

You need to have a good understanding of overall land use before you site your house. Look as far into the future as possible. What buildings, gardens, orchards, pastures, and ponds might you eventually want, and where does it make the most sense to put them? Where should woodlots and wild areas be? How can you sensibly position these functions relative to one another so that each part of the system meets the needs of the others and of the whole system? For instance, can you dig a pond that will provide earth for your cob house and water for fire control, be a home for ducks and geese, help with erosion and drought control, be part of your graywater system, and irrigate your fruit orchard? Design of this complexity takes a lot of thought and careful planning, but it is immensely satisfying and valuable. For help, take a Permaculture design course or consult Bill Mollison's *Permaculture: A Designer's Manual.*

If you plan to use building materials from the land (such as earth, sand, stones, trees, straw, or water), consider where they are located and how you could transport them. It's much easier to roll boulders downhill than up. Would the extraction of materials benefit overall land management, or would it cause problems? Turning problems into opportunities,

OVERLAY MAPPING

Suppose you want to build on a site that:

1) IS NOT TOO STEEP
2) IS NOT TOO HIGH
3) AVOIDS HOT WEST SLOPES
4) IS NOT ON THE BEST SOILS or a wildlife corridor.... then

- MAP & SHADE each area.
- Add other features you wish to consider

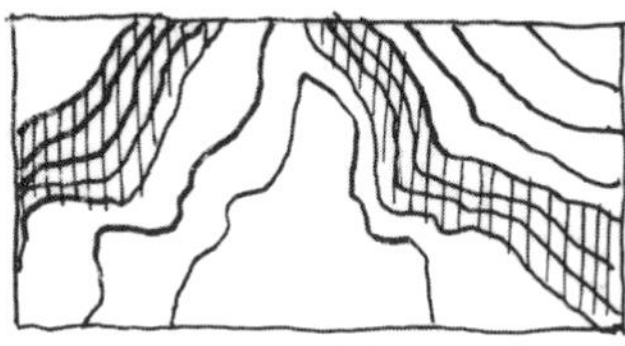

TOO STEEP

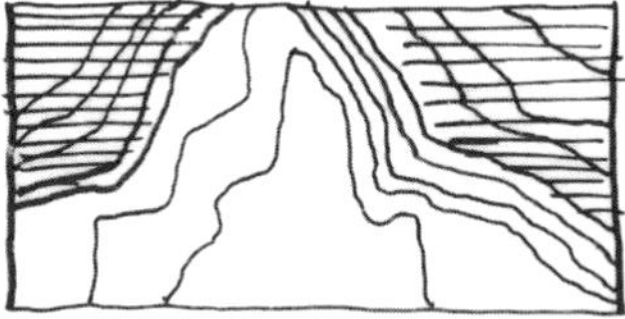

TOO HIGH

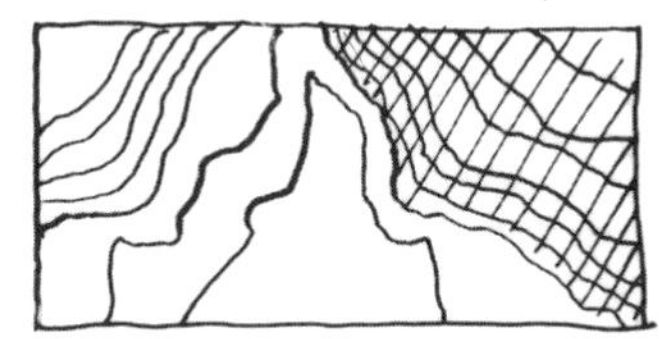

HOT WEST SLOPES

BEST AG. & WILDLIFE LAND

- Then OVERLAY these patterns on a single map. This will indicate both the best (Lightest) and worst (darkest) building sites.

OVERLAY MAPPING

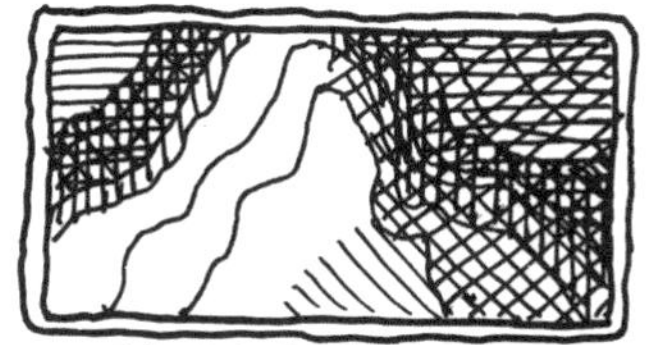

Light = BEST darkest = WORST

could you fell a tree to improve the view *and* provide lumber for your roof?

To help pinpoint potential building sites, you might want to make models, maps, or mockups of site characteristics. You could map all the areas, for instance, where you don't want to disturb the ecology, then make another map of all the areas that are too far from the road to be a practical house site. Overlay the two and together they will present a third story—all the places that are close enough and where you can't do too much damage. Compounded overlaps of several factors can reveal site characteristics not otherwise apparent.

Minimize Ecological Disruption

Any kind of construction (including roads, leach fields, yards, fences, and even gardening) creates havoc for the plants and animals that are already there. Such damage is often obvious and dramatic, but the damage caused by the ongoing existence and use of the building after it is finished may cumulatively be even worse (or it may be healing, if done right). Be sure to consider both levels of ecological effects. Think through the lifetime of the building, how it will affect, destroy, alter, or improve its site ecology over several hundred years.

Practice the Deep Ecology of construction—pay respect to the existing occupants of your building site and take precautions to disturb them as little as possible. Avoid building in places where you need to drain wetlands or even tiny ponds; such wet spots are essential to keystone species such as frogs, dragonflies, snakes, and bats. Look for evidence of plants or animals that may be present but not visible at the times you visit: tracks, burrows, rare spring wildflowers that have left seedpods, or the castings of owls beneath a night perch. You may want to map out and avoid disrupting obvious wildlife corridors, such as thin bands of woodland connecting two big blocks of forest, or a year-round stream that animals move along at night. Keep away from plants known to be scarce in your bioregion. Consider building on thin soils or rocky outcroppings, which have less life associated with them.

Don't disturb special places! A well-situated, beautiful building can improve an unattractive site, yet the same house can destroy a mountain meadow. *Resist the temptation to build in the most pristine places.* Build so there's a glimpse of the creek, but don't destroy the streamside by building there.

Even with natural construction, without mechanical tools, and on walk-in sites, there will be a lot of foot traffic, materials storage, compaction, and changes to the ground levels, for instance, due to excavation. If possible, build where someone has built before. Probably the most damage you'll do to your site comes through providing for motor vehicles—clearing roads, creating turnarounds and parking areas. *If you can avoid creating new roads, so much the better; locate where there is existing access if you can.*

If you plan on electricity service, phone lines, or piped water, remember that installing the supply lines will cause considerable disruption. Access to existing services may affect your choice of site, especially on ecologically sensitive land.

Although it's romantic to build on a remote site with no vehicle access, ask yourself these questions: How will you transport materials to the site during construction? How will the inhabitants get themselves, their babies, or bags of groceries and laundry to the house at night or in rain or snow? What about emergencies, getting sick people out or fire trucks in? Our experiences hauling heavy materials such as sand, cement, and foundation stones uphill via wheelbarrow make us recommend that you seriously consider building close to vehicle access.

Consider Soils and Drainage

How is the soil—where is it soft, hard, deep, shallow? Take a shovel and dig test pits in the most extreme locations, such as hilltops and valley bottoms, damp hollows or disturbed areas. Look at what's beneath the surface. You could find high bedrock, a perfect clay deposit, toxic waste, or rich, well-drained soil for a vegetable garden. Digging holes carefully, by hand, is very revealing at both conscious and subconscious levels. Read the section in chapter 9 on determining a soil's suitability for cob.

Remember that one of the few conditions a cob building can't stand is being submerged or having its walls become saturated. *Don't build on flood plains or in gullies!* Carefully check the local records of floods as far back as you can through the local planning department or the Army Corps of Engineers. A cob building in Buda, Texas, was destroyed in October 1998 when a thousand-year flood poured five feet of water through it. Look at flood prediction charts for hundred- and thousand-year storms, then add some caution, as their predictions may not account for changing weather patterns. If your site has poorly drained soil and a rainy climate, build on a slope so you can create artificial drainage around the building.

Cob is heavy, so keep out of quicksand, swamps, or peat bogs. *Build on the most solid subsoil you can find.* Heavy buildings will subside over the centuries if they are built on any but the firmest ground.

Determine the Microclimate

Are you in a valley that channels cold winds past your site, increasing your future heating costs? Are you on a ridge with a spectacular view of the ocean, but with no protection from whipping gales? Wildfire runs uphill, up gullies particularly, so ridges and hilltops are the most susceptible to burning. Which side of your house would a heavy rain hit? Don't speculate; be there in several windy rainstorms. Then imagine the hundred-year rainstorm, and project surface flow and flooding from a four-day deluge.

Because cob stores heat well, you will want the winter sun to fall on the south face of your building from midmorning to midafternoon. The sky should be 80 percent visible from southeast to southwest, above the line of the midwinter sun. You will need winter sun outside your building too, particularly

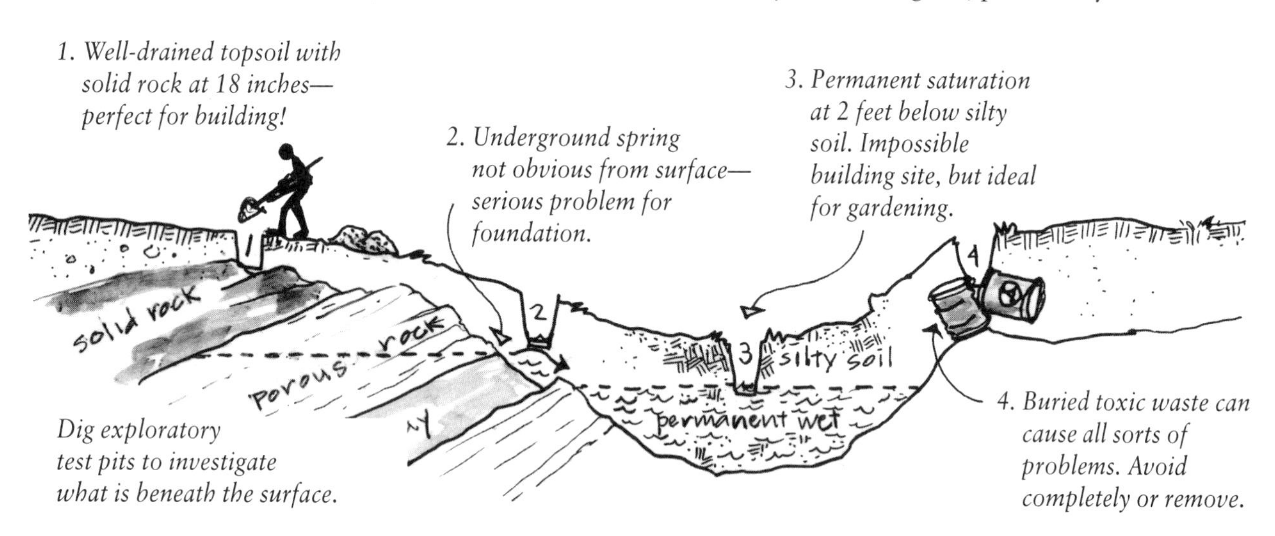

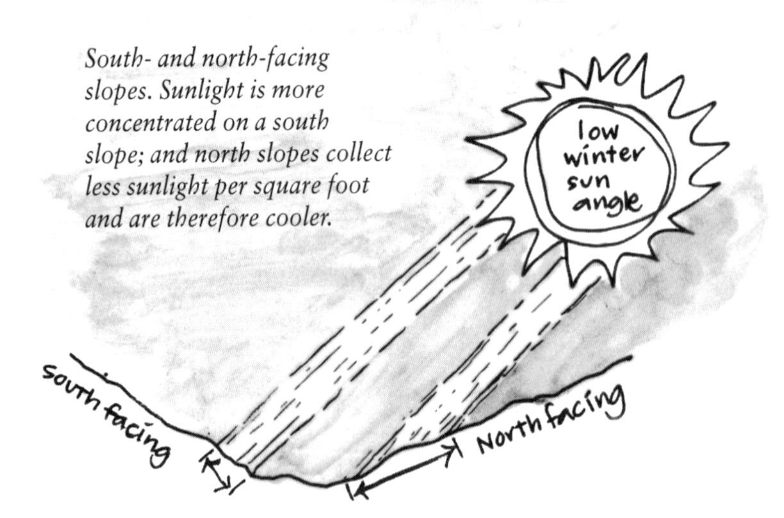

South- and north-facing slopes. Sunlight is more concentrated on a south slope; and north slopes collect less sunlight per square foot and are therefore cooler.

to the south and east of it, to enjoy being outdoors when it's cold and sunny. Also try to predict shade at midsummer noon and at the hottest times, such as August afternoons at 4 P.M., when the sun is already well to the west.

In general, trees and vegetation around a site will keep it cooler and moister. In hot summer climates, afternoon shading can make the difference between a cool, comfortable retreat and an oven. Look for sites with trees close in on the southwest and west. Deciduous trees are especially useful because they block the summer sun but let the winter sun shine through.

If a potential site is shaded by trees to the south, consider respectfully harvesting or substantially pruning them. The number of other trees you will save by decreasing your heating needs over the lifetime of the building may very well compensate for their loss. In all except the very hottest areas, you want clear sky to the south for winter warmth.

Look for a Far Horizon

Eighteenth-century English landscape gardeners suggested that any house should have three kinds of views: foreground, middleground, and background. You can create or manipulate foreground or middleground views, but not so with the long view; you can't create the horizon, so look for sites where you can see a far horizon, with whatever that view contains. Sometimes pruning or felling just one small tree will make a great difference to how your house feels, how much winter sun it gets, how much of the sky you can see. Most people are not happy unless they can see the edge of the sky, regularly, from their home and workplace. Not the whole sky, just a part of it, preferably wherever the weather comes from—storms, gales, changes. Try to site your house so that you see the longest view from your desk, from where you relax most, and perhaps from the bathing place. Being able to see at least part of the horizon can rescue you from "cabin fever."

Look for a Dominant Permanent Feature in the Landscape

There's a big difference between a new building set out on a flat field and one that is connected visually with the history and features of the place. Don't create the sensory isolation of a building unconnected in space and time. The building will assimilate better visually if you build on a change of level, the edge of woodland, an ecotone, or tight up against an ancient tree, exposed rock outcrop, or running water. Relate your design to these forms,

Views of the foreground, middleground, and background. Look for a site with a far horizon.

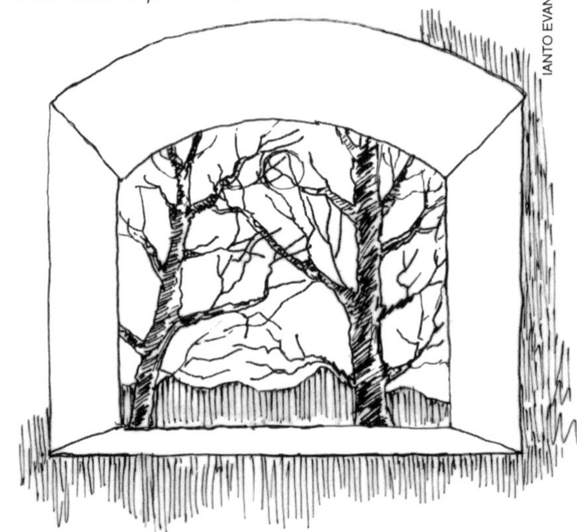

which will help introduce a new building to the assembly of existing features, like being introduced to strangers at a party by a friend who already knows everyone there.

On a flat site, if big trees are absent, what other permanent site features are there? Big rocks, streams or ponds, changes of level, existing roads, buildings, neighboring trees, and so on? Use them to anchor your building psychically.

And if circumstances allow, build onto or near an older building rather than pushing into a new place where the ecology is undisturbed. Your building will connect with history better that way and will be less of an affront to the place and its accumulated equilibrium. In town streets, a new building linking two older ones can sometimes create a physical, historical, and psychic continuity between all three buildings.

Build on Slopes If Possible

Although the building industry seeks out (or creates) flat sites for ease of construction and to maximize profit, you could do well to build on a slope. Houses on slopes have many advantages, such as views, good air drainage, and good water drainage. And, in general, sites on slopes have fewer other values, such as for agriculture. By building on a slope you can leave the level ground and best soils for growing food.

It's harder to make a building sing in a flat place—it will tend to be a flat building. A house is a three-dimensional series of spaces. You can play the symphony of those spaces best where the ground is already helping you by being sloped. Steps, sunken courtyards, drains, raised berms, indoor level changes, and view decks all come more easily.

A sloped site can avoid the need to pump sewage uphill to a septic system or leach field. You can plan to use wastewater downhill from the building site, in an orchard, garden, woodlot, or pond. All this suggests not locating your home at the lowest point of the property.

New houses can be anchored to existing features of the landscape.

If your site is on a slope, make sure that cold air can drain away downhill. On clear nights, wherever it is exposed to the sky, air cools off and becomes denser, flowing downhill like a viscous liquid. Wherever there is a rise in the ground, or a line of trees, or even a building, the cool air slows down, creating "frost pockets" of much colder air, which can be 15°F colder than their surroundings. Valley floors are often the coldest. These are the places that freeze first—not a good location for your tomatoes, or for a cozy home.

Which direction a sloped site faces makes a big difference in ground temperature. South-facing slopes, with their surface more nearly perpendicular to the rays of the sun, collect more heat. In cool climates this can translate into substantial energy savings, and makes

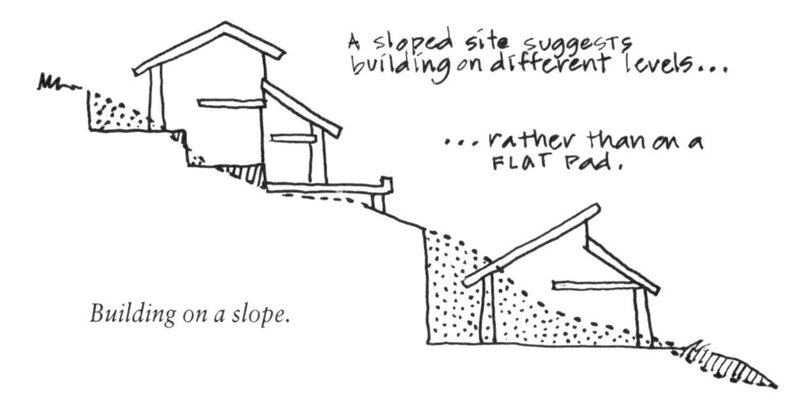

Building on a slope.

Frost pockets occur where cold air flowing downhill collects.

your outdoor environment more useable. In cool climates, southeast and east slopes usually get sufficient sun, and in warm zones they cool off much earlier in the day. Avoid north slopes in all but the hottest places. Also avoid west slopes, which often overheat on summer afternoons and evenings, even in colder zones.

Respect Your Neighbors

Many of us imagine that as private landowners we owe nothing to the existing landscape, ecosystem, or neighborhood. We own the title to the land, so we have a right to change it and thus change how it affects its surroundings, without consulting anyone, even if we just moved in from across the country. If you choose to build a big, conspicuous structure in the normal noisy, disrespectful manner, insulted neighbors may retaliate in a variety of ways, giving you the cold shoulder, or even calling officials about this or that infraction in your construction project.

Respect your neighbors. Perhaps from their viewpoint your new building crowds the neighborhood and gives them no benefits. Unless they know and like you, they may try to make life difficult for you. Design, site, and construct your new creation as inconspicuously and considerately as you can. Talk with surrounding residents, even before you settle on a site. Explain what you're planning. Invite their involvement; put on a special tour for them, with refreshments. Take their concerns seriously and try to address them. They may have good ideas, having lived in the vicinity longer. Show how you're trying to fit in. Point out potential advantages of having you and your building there. With a natural building you're already at an advantage: natural materials recede visually and ecologically into the landscape they came from, and the philosophy of natural building is to build small and quietly, in a visually unobtrusive manner. Yet neighbors may not be acquainted with natural techniques, so take the time to ease their uncertainties or suspicions.

TAKING TIME

ON OLD FARMSTEADS in Europe, when a newly married couple moved into their farmhouse, they would start assessing house sites for their (yet-to-be-conceived) first son. Thirty years later, when he married, they might have finally decided on the site. Every day, walking the land, the farmer would be quietly watching just where the sun hit on what day. He would observe where trees fell in the great gales once in a century. He might likely discuss the matter with his own father, and try to discern patterns over time. Sometimes the farmer himself would die before the son's house was ever built, so the son would move in to his parents' house and postpone the decision for another generation.

Testing Your Choice

If this site is to hold your home, can you imagine living there? Return to the site repeatedly. Each time, your understanding will be expanded; each time you will have new questions to research. You may well change your mind about building on your chosen site, after being there in all conditions. First, live on your site in your mind, considering every possible situation rather than rushing to build, only to discover later that your siting is less than perfect.

The house will be the better if you can watch your land for a whole year's turning, see where the snow melts last, feel how shade helps on hot July afternoons. Maybe you could camp there, year-round, cooking outdoors, sleeping where you think your bed might go. Can you buy a travel trailer and be there for a year or, at the very least, representative seasons? Try to be present in big rainstorms, in gales, and when the temperature is lowest, early on winter mornings.

Without living outdoors year-long on a site, it is hard to project what happens in the time when you're not present. If you can't be there, you'll have to apply astral projection, time travel, and imagination. For example, if you can't be on-site at midwinter noon, estimate where the frosty shadows and winter sunlight may fall. Where might undesirable lights shine at night, and from which direction could bad smells or seasonal noises come? A lake view from a window is wonderful in chilly March sunshine, but try to imagine mosquitoes and powerboats in steamy August when the lake view is hidden by the leaves of deciduous trees. Bugs and noise could make life miserable when you most want to be outdoors.

DEALING WITH PROBLEM SITUATIONS

Some problems can't be fixed, and shouldn't be tolerated. You should altogether avoid building a house in a place with the following problems:

- Steep north slope
- Flood plain
- Obviously incompatible neighbors
- Sensitive ecological conditions
- No solar access

Also be very wary of places that have no possibility of long views and sites in areas of rapidly rising property values where land taxes could impoverish you over time.

Rather than more drastic clearing, try to gain access to sunlight and views with subtle, very selective pruning.

On the other hand, it's rare to find a building site that is perfect in every sense. Following are some of the most common difficulties a site can present, with a few ideas of how to tackle them.

Short of Sun

Does your site only get sun in the mornings? Then orient most of the windows and the longest side of the house to the southeast, even east.

Your own trees are blocking sun? Prune them to let in cold-season midday rays. Prune when the winter sun is shining, if possible, after the building is up, to remove the least possible vegetation. Prune very gently at first, as you can't put back what you remove.

A neighbor's trees are blocking sun? Site your building as far as possible from those trees. Negotiate pruning with your neighbor; offer cash for the trees as lumber.

A cold north slope, sunny only in the warm season? Build exterior walls of strawbales or other highly insulative materials. Limit glass to good views and for natural light (consider skylights, clerestories, and small, high windows).

No Trees on Site

Plant some trees ASAP! Plant species that are natives, fast-growing, deciduous. In cold regions, plant evergreens where they will form a windbreak but not block sun. In towns, avoid tall evergreens if when full-grown they could block a neighbor's light, but plant the north and west sides of your lot with big transplants as soon as you can. In hottest zones, shade trees to the west will be a priority. An absence of site trees is an opportunity to sculpt a treescape that will best suit your needs.

Our eyes read the highest parts of a landscape first. The tops of power poles, giant billboards, treetops, whatever sticks up into the sky will be what our subconscious digests as the essence of the place we're in. If the highest thing against the sky is a roof ridge, chimney, or TV aerial, it dominates our subconscious and affects how we feel about being there. In any town with big trees along the streets, the ugliness and disorder of the buildings is absorbed by the silhouettes of the treetops. Test it for yourself in two streets with similar buildings, one with trees higher than the rooflines, one without big trees.

As the top of a landscape is so visually commanding, on open sites the roofline of the building needs special attention. Uncompromising shapes that are alien to Nature are most difficult here. Power poles and projections up from the roof surface will need to be kept beneath the skyline. Shiny stovepipes, vents, or roof flashing should be avoided.

Noise

Is your site near a freeway, under a flight path, on a busy street? Cob is the ideal medium for sound absorption—thick, heavy, rather soft, with an irregular surface. It is better when left unplastered on the noisy side, and best if you deliberately design irregularities in the surface sculpting. Baffles or sound blocks can be added, such as fins extending out from doorways and windows. Try to face opening windows away from the sound source. Double glazing will halve the sound problem, but opening windows need a very

Where to plant trees.

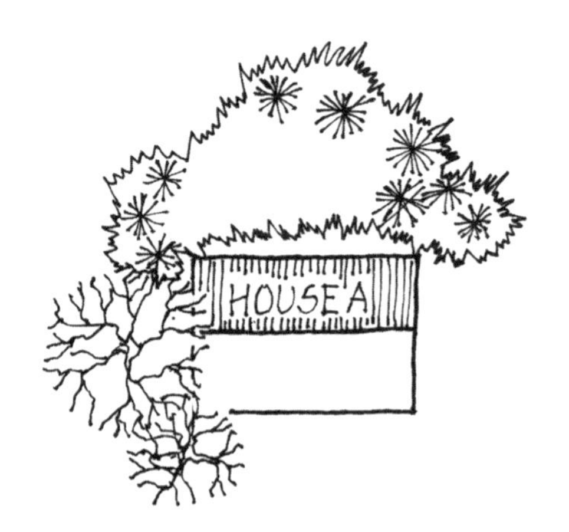

A: Has cool to warm summers with cold winters. Use short, dense evergreens to the north to block winds, and deciduous trees to the west to block hot summer afternoon sun.

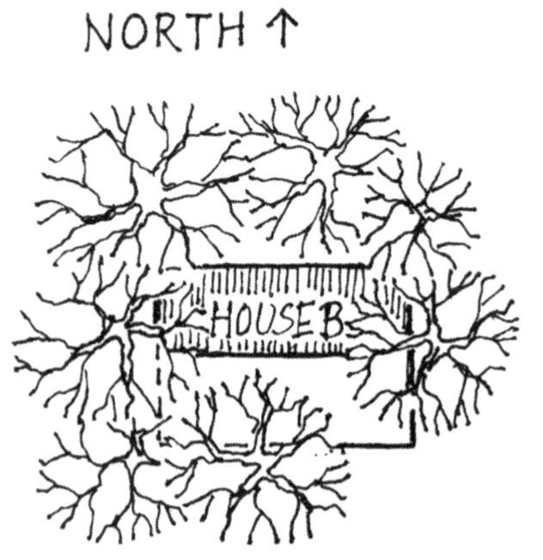

B: Has hot summers, mild winters. Use tall, deciduous trees all around, close, to allow breeze to flow beneath them, shading the south side of the roof.

C: Has hot summers, cold winters. Use short, dense evergreens to the north with tall, spreading, deciduous trees to the south and west.

IANTO EVANS

Two comparable streets. The ugliness and disorder of the left-hand street are absorbed by the silhouettes of the treetops on the right.

tight fit. Think about an earthen roof; airborne sound will travel easily through most kinds of lightweight roofing. Build cob or straw bale sound-absorption walls around your patio or garden.

If noise comes from the sunny side, the source of your free solar heating as well, consider creating a solarium by placing a double-glazed skin over a room on that side. This solarium will help provide indoor privacy too, and it can be a toasty place to hang out on crisp snowy days. Alternatively, berming, or sinking the building down into the ground, will reduce heat loss from the sides, increase the heat storage available in the surrounding earth, and of course muffle the impacts of noise.

Not Enough Privacy

There are special problems associated with building in a dense neighborhood, but you

INVISIBLE ARCHITECTURE

IN ROLLING OR HILLY COUNTRY, a curved roof ridge helps ease the building into the landscape, and a living roof, of sod or other green material, is easy to overlook. I once took out a group of people for a tour of one of our buildings. We stopped on an open hillside, perhaps thirty feet behind the house, in direct view of the sod roof. I took ten minutes explaining the context of building with natural materials, then asked if there were any questions. A woman who had all the while been looking past me down over the building said, "I don't have any questions but I need to leave soon so I was wondering where this building is that you're taking us to see." She hadn't recognized the grassy hump in front of her as being in any way connected with a house.

The Cherry House, Cottage Grove, Oregon, 1995–96, from south and north. Design and construction: The Cob Cottage Company.

can use the principles of passive solar to advantage. Passive solar design recommends that you put most of your glass on the east and especially south sides, with the north walls lower and protected (by earth berms, evergreens, or a well-insulated roof), and the west walls likewise minimal, without large windows and screened by vegetation. If you employ these principles, not only will your building be more comfortable when the weather is extreme, but close neighbors to your north or west will be less likely to have visual problems with your house. You in turn would feel less crowded by neighbors close to your south and east, if you had only views of blank short walls and grass roofs.

Lack of visual privacy inside a building due to big glass windows can be solved by mirror-faced glass; fast-growing vegetation (start with a seasonal bean trellis, sunflowers, Jerusalem artichokes, or hops, all of which can protect you that first summer while more permanent

NOT A SITE FOR A COB BUILDING

THE MANAGERS OF a summer camp and retreat center in southern Oregon phoned. They had been reading up on cob building and were interested in hosting a workshop there. To extend their guest season, they were considering building a new year-round dormitory. Could it be done in cob? "Of course," I answered. "Cob does really well in this climate, no problem." So one day in early March, with crocuses poking through the damp ground, Linda and I went to assess the situation.

Camp Latgawa is at the end of ten miles of gravel road, up a creek off a creek, off the Rogue River. It's an old Forest Service place, on an alluvial fan where two crashing streams come together. The mountains are young and still very steep. The entire site lies at the bottom of a thousand-foot canyon, on the north-facing slope where year-round groundwater lets the firs grow two hundred feet tall. The site was clearly chosen for keeping cool. The trees have been carefully protected since the early 20th century and form an impressively dense canopy.

We searched for clay. Nothing! The creeks have scoured this canyon clean, leaving only boulders, driftwood, and a little sand. Very little sand.

Five strikes against cob: no sun, north slope, absence of clay, flood zone, and a building needing to be warm in winter. "Build it with bales," I said. "You have tons of beautiful rocks for a foundation and enough deadfall trees to make a lovely timber frame, and straw bales would be the perfect insulation."

The lessons? Never give an opinion until you've been to the site, and don't jump for a familiar solution. Even had a perfect mix of sand and clay appeared right on the site, cob is a material for storing surplus solar heat, not for the good insulation needed here. If it were a summer-use building, with hot days and cold nights, cob could work well even without sun, but here at the bottom of a dark and dripping forest, a winter cob bunkhouse would need too much supplementary heating.

A terrible site for cob: no sun, north-facing slope, no clay, flood zone, and difficult to heat in winter.

screens of grapevines, ivy, or dense hedges grow up); upwardly tilted/angled windows, relatively easy to incorporate in thick walls; or windows set toward the outside of thick cob walls, with cob mullions to block low-angle afternoon sun at hot times. Another solution is to build a trombe (heat-storage) wall immediately inside the glass. It can be floor-to-ceiling, waist-height, or with windows in it, providing a view through both panes of glass (see page 84).

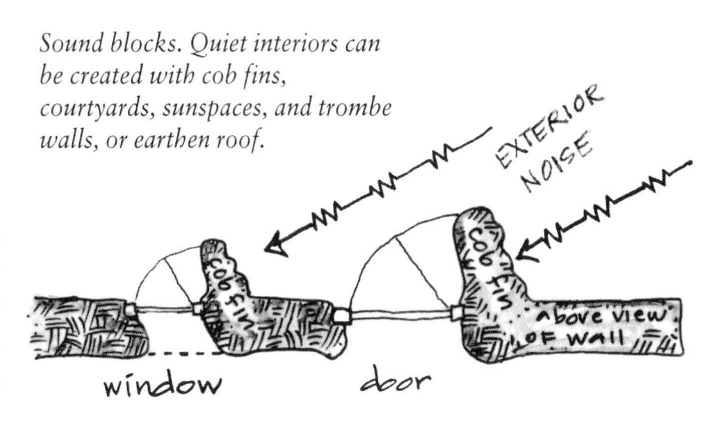

Sound blocks. Quiet interiors can be created with cob fins, courtyards, sunspaces, and trombe walls, or earthen roof.

Often you can avoid being conspicuous merely by careful location of a building on the site. A few feet one way or the other can make a difference—nudging the building up against a group of trees, attaching it to an existing building, dropping the roof ridge so it doesn't break skyline from an adjacent road. Sometimes a cob structure can be built inside a wood or canvas yurt, a "temporary building" that is often permitted under local laws. Casual inspection doesn't reveal that there is a genuinely permanent building inside the more ephemeral one.

Once in Mexico, passing a derelict factory, I peered through a chink in the door to find to my astonishment a complete two-story house under construction inside the metal factory building. Due to legal complications, the builder explained, he couldn't openly build a new house on the site, but later when he demolished the factory, the completed house would miraculously appear!

INTERVIEW: ERIC HOEL'S HOUSE

Eric has been a professional builder most of his adult life. He is building a cob house near Salem, Oregon, on land borrowed rent-free from the owner. Ianto talked with Eric about the process and his tenure agreement.

IANTO: Could you describe the *place* where you decided to build?

ERIC: We're on the very edge of Salem, a town of 120,000, where it used to be all small farms. There are maybe twenty houses on this dead-end road that has a very agricultural feel to it still. It has a very open feeling. This place we're on is mixed woodland: many cherry trees, a few old oaks, an ancient maple, and some fir. There's one huge fir tree at least two hundred feet tall. The land we look down into is very pastoral, sheep and cows grazing between big old trees, a Capability Brown look to it. There's a hill to the east which cuts off long-range views, but on the other side we can see down into the Willamette Valley. This land is in five-acre parcels, and my friend George owns two, together, where his house is.

IANTO: Is all this space what made it easy for George to let you build right next door?

ERIC: Our house is a hundred yards from his house and has its own driveway. There's an oak tree and a whole copse of scrubby cherries, plums, and hazel trees between us. There's no direct view of any of the neighbors.

Eric Hoel's house. Salem, Oregon.

Nobody has to cross paths, even if they're on the same plot. George was counting on us not to intrude on his house.

IANTO: Tell us the story of how you made an agreement to build, rent-free.

ERIC: I have a friend with an interest in cob houses, natural building, land sharing, all that. We have had a trusting relationship all along, maybe for four years before I came here. He had trusted me enough to leave me with checks into his bank account, a $60,000 fund for rebuilding, when he went out of the country, and I had been building for him.

We have an agreement now for five years. We have nothing on paper. It might be scary for a lot of people, but we feel very secure. I will build the house, pay for all the materials, even the gravel for the driveway, in exchange for use of the house for five years. After five years we may extend the agreement, but until then we pay only for utilities. If we want to leave and move elsewhere for a time we can lease out our house. It's ours to do what we want.

George is so excited about the building he's been one of the most reliable workers on the site. He loves this land, it feels good to him to create this kind of a house here.

IANTO: What made you decide to build on a friend's land instead of buying your own?

ERIC: The decision was quite simple—I had no money resources. I wanted to get the satisfaction of building from start to finish for the sheer pleasure of building a house I would live in. Also, I wanted to experience specifically living in a cob structure.

IANTO: What have you learned that you never would have expected?

ERIC: I never expected that planning and sequence would be so critical. When and where to store building materials was very stretching for me. I wasted my own and everyone else's time. Also the foundation is *so* critical, you've got to get it right. If you screw up the foundation you can't go back later to fix it. No amount of workshops had prepared me for that. It's gone faster in some ways, slower in others, than I expected. For a hand-built house it's gone quickly, given it has two-foot-thick walls that were nearly all mixed by foot, applied by hand. It's not tiny, 720 square feet, and all the first floor got built in a summer. I never expected the number of people who have come along to help. We counted a total of seventy-six people who have helped out, a good thirty on a regular basis. I was never prepared for the energy this house has brought. Materials kept on just showing up at the site—long beams, the foundation stones, doors, lumber, and windows. It has almost brought itself into being. I feel more like the conductor of a fine orchestra, rather than a builder.

IANTO: At only seven hundred square feet, some people might feel "Oh, that's so tiny." How is your perception of the space?

ERIC: After we built the model we went out with flour and drew it on the ground. It felt smaller than we anticipated so we moved some of the flour lines out. Then I made a final floor plan. As foundation trenches went in, digging by hand, we adjusted, shifted the lines again. When the foundation went up it felt bigger, then the walls made it bigger again. Now, with ceilings in, everything dark, all the dark earth, it feels smaller. When we plaster the walls and finish the floor it will probably get bigger again.

The wall thickness makes it seem much bigger. When you go through a door it's like going through a passage, which makes you feel you've traveled, though you only took one step. And the rounded shapes. It really feels like more space than if the walls were rectilinear with corners I can't quite get into. It is a very spacious 720 square feet.

IANTO: How is the space feeling now you have the first floor up and ceilings in?

ERIC: It's more beautiful than I expected. The reality is more beauty than I can draw on paper. I went in at midnight on the full moon in November, to see where the sun would be in May. The moonlight coming in, shadows on the walls, it felt like a fairy-tale setting, very protective, very safe. It already feels like a home with no doors, no windows, no floor. Even the foundation felt great, it felt like a finished product. Someday when this house is gone, when it's a ruin, the foundation is going to be a lovely place to be.

6 Designing with Cob

❧ Nature begins with
a genotypic template . . . then creates
endless diversity by responding to every
nuance of the surroundings. ❧

Building with cob doesn't require a professional builder, and designing cob buildings doesn't demand a professional designer. What you need is common sense, inspiration, and understanding of what your materials and techniques can do. You can develop a brilliant house without ever setting foot in architecture school.

The special properties of cob mandate special design approaches. This chapter, together with chapter 7, will help you design a magical, intelligent house. We explain here why most modern buildings fail and how cob's unique properties can set you free to build the house of your dreams. We also tell, for your enjoyment, edification, and inspiration, the story of how we designed and built the Heart House, the cottage where Linda and I live in the Oregon rain forest.

THE BOX POLICE

In the 1960s I spent seven years in design school, and in the 1970s another six as a professor in schools of architecture. I learned architecting, then I taught it. It was not a prerequisite for teaching that I should know anything about building (nor about teaching

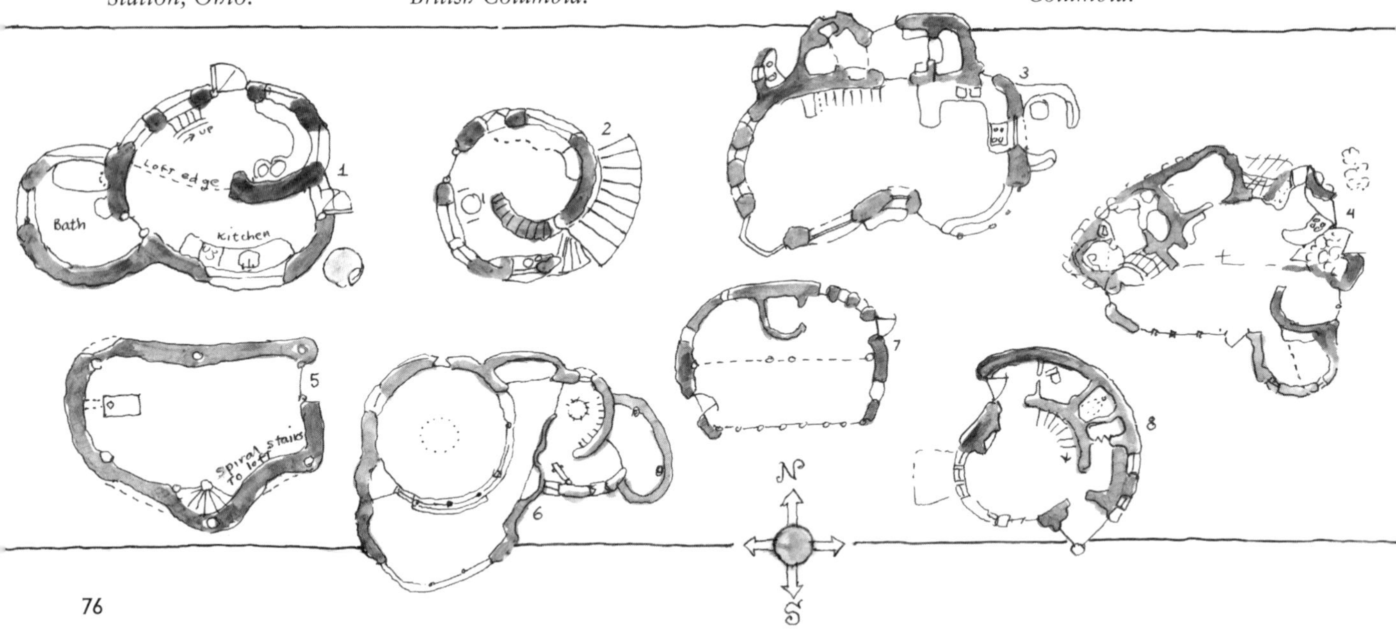

1. House by Jan Stürmann in Hermanus, South Africa (note sun in north!)
2. Garden Studio by Meka, Wolf Creek, Oregon.
3. Family House by Eric Hoel in Salem, Oregon.
4. Family House by Elke and Steve Cole, Courtenay, British Columbia.
5. Cottage by Mark Hoberecht, Columbia Station, Ohio.
6. Sanctuary by David Shipway for Hollyhock Retreat Center, British Columbia.
7. House by Katie Jeane in Willits, California.
8. House by Hilde Dawe and Patrick Hennebery, British Columbia.

IANTO EVANS

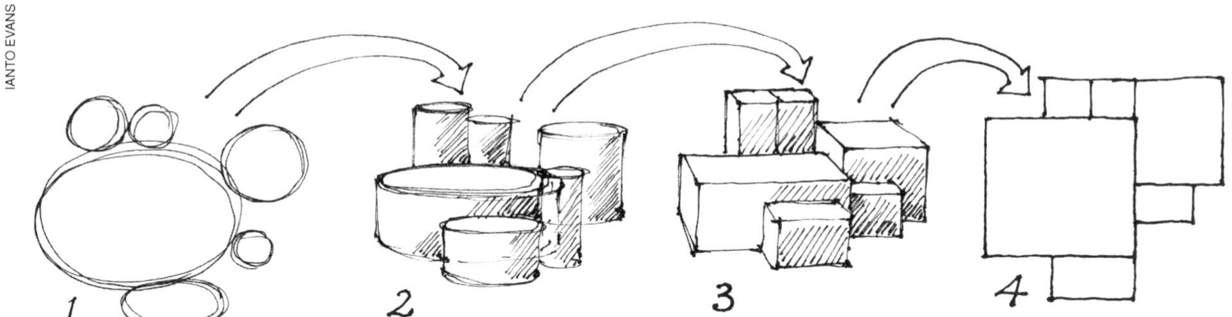

Bubble diagrams are used by architects to explore relationships of building elements in layouts. But why abandon rounded forms in favor of square ones?

for that matter), and the same had applied to my own teachers, and probably to theirs. Nevertheless I was at first confident that I knew a fair amount about building design. Quite soon, that cockiness was replaced by a great humility as I realized that our built environment is almost completely dominated by the designs of men and women who have never themselves built a thing. I myself was one of that category, so I quit architecture and started building. What a revelation!

Architecture-school design revolved around creating two-dimensional gridded fantasies without a clue as to how they might actually get built, who would really pay for them, what industrial processes were used to create components, nor how such strictly rectilinear containers might be conducive to life. Admittedly, a very small proportion of all our buildings are designed by architects, but they and engineers, developers, and bankers are influential in how we all think buildings ought to be. We have in fact, almost no alternative model to their way of arranging buildings. The results surround you.

One aspect of architecture schools that always fascinated me was the Bubble Diagram. The theory goes, you get a rough idea of the spaces you want in a building, then draw a series of bubbles, their differing sizes representing those spaces. The bubbles are free to move, so you can cut them out and slide them around in different configurations until you get the best compromise. Then you tape them together and turn them into a plan. It's a process a bit like that shown in the diagram above.

But do you see the logic jump here? If figures 1 and 2 are roughly the shapes of spaces the people really need and want, why do we end up with figure 4? We're told the bubbles need to be translated into boxes—"rectilinear rooms, for convenience." Whose convenience? Not the user's, clearly. To whom could it be convenient to live in a series of rigidly squared rooms when the bubble diagram looked snug and comfy, just the shapes it was? Could it be for the convenience of the building-products industry?

It seems that most architects, engineers, bankers, developers, code writers, and local building officials are all part of a giant tacit conspiracy to fill one another's pockets at the expense of the householder and of our hard-pressed environment. The system works best for them if we buy their rectilinear, industrially processed building components. The shape and size of our buildings is controlled by the standardized sizes, modular dimensions, and Cartesian geometry of building materials, the use of which is enforced by an army of inspectors, code writers, architects, engineers, and building officials, whose function is to keep the wheels of industry spinning, hastening the progress of precious, non-renewable resources to the dump.

WHEN THE MATERIALS DESIGN THE HOME

In our own cob cottage I can honestly say that I wake every morning thinking, "What a wonderful place to be, how fortunate I am, what an inspiration to make the most of my brief life!" I'm never trying to just get through the day. The day is never long enough, I am so well nourished by the quality of my surroundings.

At first I was arrogant enough to imagine that *I* had somehow created this magic, because *I* designed the building, but soon it became clear that the materials themselves contain magic and that the act of marrying them together creates a hallowed union. To my utter humility, the natural buildings I have helped organize are beautiful in proportion to how *little* I have done, and in almost every case industrially made materials that I had introduced compromised the feel of the building.

Skill in ecological architecture may be measured by a talent for using materials in as natural a state as possible, and by an ability to encourage "uneducated" clients to do their own designing. Natural materials dictate their own shapes and colors and textures no matter who the architect or builder. Now, I know with confidence that *anyone* can build with cob and create a work of art. It is really hard not to. Linda-Marie Luna (a colleague and builder in Colorado) says, "You have to work hard and spend a lot of money to make buildings ugly."

Consider the integrity of riven wood, split along its fibers. By contrast, in sawn wood the end grains are cut and exposed. The tiny vessels that carried fluids up the tree are now open to bacteria, fungus, and insects that can nowinvade and weaken the wood.

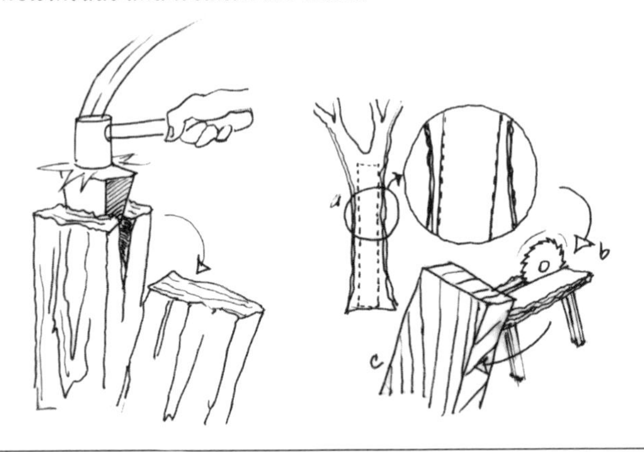

A CURRENCY OF PENNIES

Manufactured building materials impose geometric and dimensional restraints. The very process of manufacturing building components with industrial machines creates predictable standardized building units, based on the "International Building Module," which is 100 millimeters (roughly 4 inches). The results are an 8 foot 2 × 4, a standard concrete block 16" × 8" × 8", sheets of plywood 4 × 8 feet, and so on. Most components can be cut or broken, but such maneuvering creates waste and takes time. The result tends to be, for simplicity's sake, a modular architecture of deadening sameness.

It's as if for the convenience of banks all coins and bills less than $50 were withdrawn, so that everything we bought would have to be in standardized units of $50 or $100. We would in time get used to it; we would adjust the quantities of everything bought or sold to lots of those values. We might then even sneer at the silly old days of quarters, nickels, and dimes. Our economy would be less flexible, and would favor only buying and selling wholesale: it would favor capital and the Big Guys.

Cob is like a currency of pennies. It encourages small detail and slow transactions, but its flexibility of use encourages us to tailor make each building to suit the needs of its occupant, in a way that is difficult or impossible with nearly any other building material. It is, therefore, a more democratic material. In building incrementally by the handful, almost everyone can create a uniquely suitable house.

Cob is like a currency of pennies, encouraging small details and slow transactions.

COB'S NATURAL GEOMETRIES

Cob shows some characteristics that set it well apart from any other construction material or technique. It owes its unique potential largely to its plasticity—meaning an ability to take up many roles easily, because of its heterogeneous makeup and fluidity of form—and the freedom of expression that these traits allow. With cob, we can make a completely different *shape* of building. In architectural and aesthetic terms this in itself would be significant, to be able to create whimsical structures of curvilinear and irregular form. But to talk only of aesthetics cheapens a more fundamental asset—the possibility of making buildings that nicely fit the elegantly curvilinear movements of social humans and that adapt ecologically to the world we live in.

What kinds of forms does cob naturally suggest? Thickness, solidity, taper, bas-relief, sculpture, modeling, integral monolithic connectedness, arches, vaults, rounded corners, and gentle curves connected by tighter ones. This is cob's essential geometry.

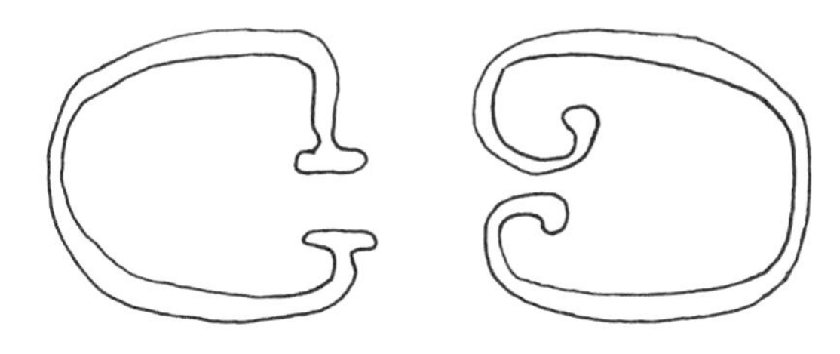

The sculptural freedom that you're given with cob frees you from the tyranny of building boxes. There's no need to fit into the stultifying modular grid dictated by all materials being sized in multiples of 2 feet.

It's easy to create natural buildings exactly the right size and shape for the job. Who needs every door in the house to be precisely 32 × 80 inches, given that they all have differing uses? Does a sleeping space where you're mostly horizontal and passive need an 8-foot ceiling, the same headroom as the place where you cook, standing and active? In fact, do you feel better with a flat, level ceiling, or should it dome or slope up slightly? And do you want every wall straight or vertical? Notice the difference when a room has a bay window added, projecting out. The room feels bigger and has an added dimension. By adding a bay, you add another *place* to your room, at almost no extra cost in materials, maintenance, heating, or taxes. Now imagine a completely curved bay window, then a curved wall, then a whole curved room . . .

Three-part window, faceted, extends out only 10 inches across 9 feet, yet makes the room much bigger.

The curves and irregularities that are universal on Earth are reflected in cob building. It's easier and cheaper to build in tune with the fluidity of a medium rather than forcing it into an unnatural shape. We are sometimes asked, "Is it possible to use cob to build *normal* [!], square, straight buildings with totally flat walls and floors?" Our basic answer is "Yes, of course. You can build boxes with almost anything, but that kind of precision will cost you extra." Why? Because Nature doesn't make the world that way, it costs extra to straighten the mud, just as there's a cost to taking a round tree and making it square. This is a big ecological cost, first in manufacture, then in construction, and finally in trying to keep everything rectilinear, against the tide of

FOR YOUR EDIFICATION

BEFORE BEGINNING a new task, it is sometimes revealing to consult an etymological dictionary.

When we build an "edifice" (*aedes*—a temple or house; *facio*—to make, Latin), even a small one, we edify, "build up in knowledge and goodness" *(Chambers)* ourselves by so doing. In the act of building is self-improvement. Constructing our house then is self-edifying—"to instruct or improve spiritually" *(Webster's)*.

In other words, the process of arranging, constructing, and living in a house can be spiritually uplifting. Remember, *aedes* also means "temple."

entropy, which will eventually bring it all down. Normal to us is what we have only very recently become accustomed to. Square and straight are never normal in natural conditions. Builders manufacture straightness with great effort and expense.

Handformed earth can take on so many forms that distinctions blur between wall and furniture, sculpture and windows, floor and wall and ceiling. Try to design three-dimensional *volumes* rather than assembling a series of named components; cob's chameleon character can help you.

THERMAL MASS

Thermal mass means the total amount of heat a solid body can contain at a given temperature. In the heat of a sunny day, cob provides mass for the heat of sunlight to be stored until the temperature drops at night. Earth can store that free heat for hours or days. This capability is most useful where there are hot days and cool nights, or in mild maritime climates, or where cold nights are interspersed with reliably sunny days, even if the days are cold.

To exploit (or better celebrate) its true potential, like rammed earth and adobe, cob needs to be sun-heated and naturally cooled. Heavy masonry buildings, which incorporate thermal mass in their primary materials, are most effective when the occupants let the sun's rays in, then store the heat. As direct sunlight comes into or strikes the walls of your heavy house, heat is absorbed into the structure. This thermal sponge prevents the building from overheating as the day warms up. Heat trapped is then released slowly, providing gentle background warmth in the cooler evenings, when you most want it. We must be clear: earthen walls don't prevent heat loss very well, but that doesn't matter much because they can store a lot of heat, for days or up to several weeks, depending upon thickness. Heating the mass with ordinary sunlight is a basic principle of passive solar construction, viable in any building built of earth and in almost all nontropical climates. The long-term dividends of utilizing passive solar design are so huge you would be shortchanging yourself and robbing your descendants if you built any other way.

There is a distinction between cold weather, a cold climate, and feeling cold. Houses don't feel cold; only people can do that—houses just don't care. When we say the problem is the cold climate, we usually mean it sometimes feels too cold for *us*. Many of us live in places where it is warm or sunny in daytime, too hot perhaps, then gets uncomfortably cold at night. A heavy building, oriented south, can collect and store that freely available daytime heat, warming us at night at no cost to checkbook or the environment.

Suppose, just for argument, you feel most comfortable at 65–75 degrees. Suppose that today the temperature swings from 45 to 95 degrees. If you were outdoors, there would be a period in the afternoon when you were 20 degrees too hot, then a time later at night when you were 20 degrees too cold. A heavy cob house can protect you from those extremes, so that your building always stays between 65 and 75. If it's really heavy, it will average about 70 degrees without extra heating or cooling, even without windows. If the structure also has south-facing glass, the sun it takes in will raise that average, both day and night. But to do that, the thermal mass at the core of the building needs to be receptive and heavy enough to absorb and store heat.

In North America, most of us live in wood frame houses. When it's chilly outside, we turn on the heat. When it's too hot, the air conditioner automatically kicks in. We buy the energy to heat and cool from a big corporation. The corporation is there to make profits, so it sometimes engages in activities we dislike and protest against, such as building another nuke plant, damming a pristine river, or stealing coal from the Navajos. Yet still the air conditioner clicks on, even if nobody's home. In an attempt to reduce the amount of purchased energy we use, and at the urging of the industry that makes insulants, we install more and more insulation, so the heat leaves more slowly during cold times, or gets into the building more slowly when it's hot outdoors. Lightweight frame houses don't store much heat, so although they can be too hot in late afternoon, by early morning they're too cold unless we continuously adjust the heat supply. Some people feel a need to use the heater and the air conditioner both in the same day!

A passive solar school building built of poured concrete has worked perfectly since 1961 in Wallasey, England. At 53 degrees north (the latitude of the southern Aleutians), in the industrial smog of a Liverpool suburb, Wallasey gets only 1,500 sunhours per year, less than anywhere in the continental United States. The winter midday sun barely creeps over the surrounding buildings, being only 13 degrees above the horizon at noon in December. If passive solar can work there, surely it can heat and cool *your* home. In most North American, British, Australasian, and South African climates, a heavy house should need very little purchased energy for heating or cooling.

ESSENTIALS OF PASSIVE SOLAR DESIGN

In the past thirty years, more books have been written about solar construction than you can shake a stick at, but if one functional solar house got built for every five books it would be surprising. Yet every architect we know has a shelf full of these books; some bookstores have whole sections devoted to solar architecture, all located in buildings heated by nuclear electricity, Persian Gulf oil, and strip mining in Wyoming. You can read endless statistics about BTUs and R-values, ASHRAE and AIA, kilowatt hours and infiltration, pumps and collectors and power towers and

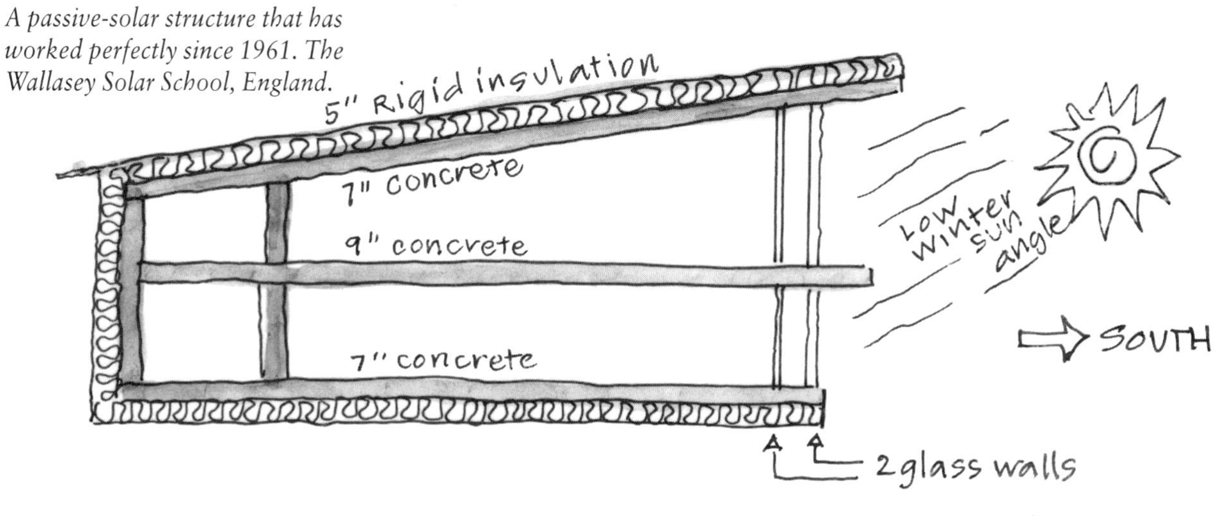

A passive-solar structure that has worked perfectly since 1961. The Wallasey Solar School, England.

South-facing sites get full sun midday during the cooler season.

gravel beds. You can even spend $189 on a fancy series of see-through graphs that tell you at any time at any latitude exactly where the sun is, without even having to look out of the window. With so much detailed information, why are we not applying it? Perhaps its very complexity is daunting, so most people avoid the whole issue. We turn on the furnace instead.

Keep it simple! Toss out that shelf full of solar building books. Most of us don't need them. The principles are so easy it's a shame to confuse matters with unnecessary details. You can learn all you need to know in about ten minutes. So—get ready to discover how to halve your heating and cooling bills with no extra construction cost. Here are the stripped down *Five Essentials of Solar Houses.*

1. South-Facing Site. When choosing a place to build, pick a site that has cool season sun in the morning and the middle of the day. Do not choose a north-facing slope, and avoid northwest-, northeast-, or west-facing slopes, in that order. Look for a site sloping down to the south, southeast, east, or southwest, in that order, although in *very* hot climates a gentle north or northeast slope might work.

2. Unobstructed Sky View. Make sure you have an almost unobstructed sky view. Stand facing solar south; this means face the direction in which the Sun reaches its highest point in the sky each day, around noon in the winter, 1:00 P.M. during Daylight Savings Time. (Note that this not quite the same as magnetic

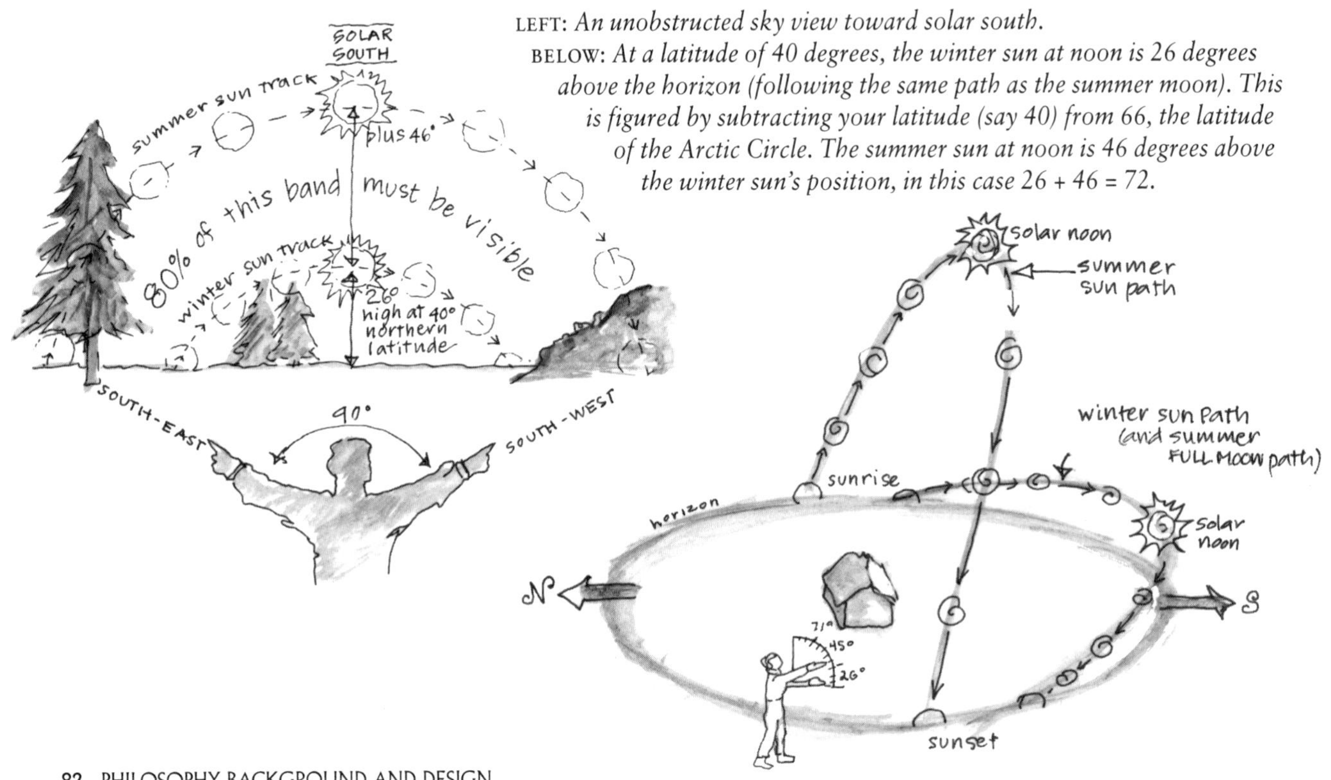

LEFT: *An unobstructed sky view toward solar south.*
BELOW: *At a latitude of 40 degrees, the winter sun at noon is 26 degrees above the horizon (following the same path as the summer moon). This is figured by subtracting your latitude (say 40) from 66, the latitude of the Arctic Circle. The summer sun at noon is 46 degrees above the winter sun's position, in this case 26 + 46 = 72.*

south.) This should be the direction that the south wall of your building will face. Extend your arms straight out, at *right angles* to each other. Your hands will point southeast and southwest. Within your arms, you need 80 percent of the sky to be visible above the arc of the winter sun's path and below the line of the summer sun's path (which is about 45 degrees above the winter arc).

If it's not presently winter, how can you predict the line of the winter sun? At midwinter noon, the Sun will be due south. At the Arctic Circle it will be on the horizon, so it is one degree up into the sky for every degree in latitude your site is south of the Circle. The Circle is at 66 degrees latitude, so if your site is at 40 degrees, midwinter sun at noon is 26 degrees above the horizon. If you're at 30 degrees, the noon sun never drops below 36 degrees. Or observe the arc of the full moon in summer, for the Moon follows the path that the Sun will follow exactly six months later. How can you determine where in you vista the winter sun arc passes? There are 90 degrees in a right angle, between the horizon and directly overhead. Half that, easy to estimate, is 45; half that is 22½. The rest you can guess.

3. Put Windows in the South Wall, Not Too High Up. Put most of your windows in the south wall of the building, some in the east, as few and small as possible in the north and west. Orient the building's long dimension east to west. The sun you want to trap will come mainly from the south, southeast, and sometimes east. "But my best view is to the northwest." Fine, you don't need to lose the view; see the section on design for peep-windows in chapter 14. Why no large west-facing windows? Because unless you have *really* cool summers (high mountain areas or northwestern coastal islands) you'll overheat the house by adding low-angle west sunlight at the times of day and year when you're already too hot.

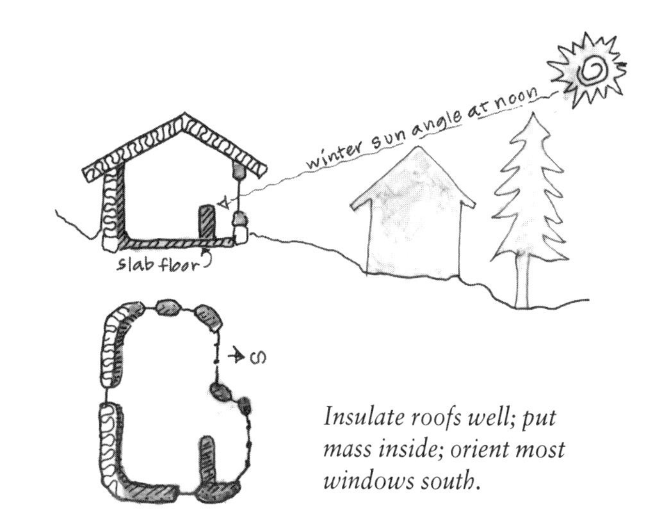

Insulate roofs well; put mass inside; orient most windows south.

In hot areas, limit east windows too, or you'll overheat early in the day. Make south windows low in the building if it is more than one story, to heat the lowest parts. Within the building, warm air will rise without help. An overheated upstairs cools the ground floor by sucking up heated air too fast.

4. Use Weight to Your Advantage. Build weight into the interior of your building, especially where direct sunlight will hit. Floors on the south side, interior walls, thick interior plasters, and trombe walls all need to be *heavy*, of geological materials, not biological. Any floor where the sun will hit should be a solid slab, on grade (with direct ground contact). The slab can be of any heavy geological material—earth, rock, brick, or concrete. Remember that what counts is *mass* (weight).

5. Insulate Well. Insulate a natural passive solar building well. In climates where daytimes stay below freezing, that means straw bale (or another natural insulant) for north and west walls; if midwinter nights drop to 10°F, you'll need straw all around. Or "outsulate" cob walls with straw/clay, straw flakes, thin stud and cellulose, or by other means (see chapter 8). Insulate outside your foundation, and its trench, too, and be sure to insulate the

Creating comfort with different types of trombe walls and snug spaces.

roof particularly well. Use thermal drapes or shutters for all the windows.

Heating and Cooling Your Building

Here are some additional strategies for making your cob house enjoyable and snug.

❧ As you are planning, think thermally. In cold times, the core of your house will retain heat longest. In very hot times, the converse is true—the core will heat up most slowly. Put sources of more expensive, non-solar heat toward the middle of your house (often baseboard heaters do the opposite, delivering heat to the outside walls and windows, where most of it immediately escapes). Your non-solar heat will create secondary warmth en route, as it moves out from the middle of your house toward the building's perimeter. Locate the parts of the house where you want to be warmest (bathing, typing, and handwork like sewing) near your nonsolar heat source. Design in a couple of heavily insulated, snug spaces there. Sitting on a built-in padded bench with insulated back, relaxing or sedentary work can be cozy, even when the room temperature is low.

❧ A big, slow-turning Casablanca fan will prevent all the hot air lodging at the top of your cathedral ceiling, and at higher speed will cool you in summer.

❧ In hot summer zones make a cool place, well away from windows and low in the building, where you can sit to work or relax on hot afternoons. Sometimes the winter snug space can double as a summer cooler, for instance, having a wide cob bench where you can lie down with a lot of body area exposed to the cob's coolness.

❧ In cold weather, a tall building is a thermal chimney, sucking cold air through microcracks in the window frames, around doors, at skirting boards, and at electrical sockets. That valuable warm air then exits through cracks at the top of the building. Be extra punctilious about sealing skylights, upstairs windows, ceiling cracks, and vents; check carefully for drafts along the tops and bottoms of exterior doors and downstairs openable windows—any gap through which cold air can come in.

❧If you will have any form of combustion heating—wood, oil, gas, or coal—provide a direct draft from outside to feed your stove. An open fireplace sucks so much cold air into your house that with an outdoor temperature below 25°F, the fire will be incapable of adding more heat than it robs. Seal off open fireplaces tightly in extra cold conditions and use other sources of heat.

Be Stingy with Glass, Unless Your Climate Is Mild

In thermal terms, glass is a big expense, so economize on its use. Consider that even expensive double-glazed, argon-filled windows lose heat about twice as fast as a two-foot thickness of cob, and probably four to six times as fast as a straw bale wall. With age the windows' performance will drop even further, as seals break, the argon leaks out, and moisture gets between the panes.

Make sure every square foot of glass in your house will serve several functions. A good general review of your glazing plan starts with considering whether a specific window will help to heat the building, that is, if it is low in the building on the south- or southeast-facing side. In cold climates, this is the only place you can afford much glass, and then only if you have good solar access.

Any glass not adding to solar heating is constantly losing heat when you want to keep warm. Conversely, when it's too hot outside, even shaded glass is constantly adding heat to your building by conduction.

Large expanses of glass can be quite uncomfortable to be near. They are cold when the weather is cloudy or cold, dark and chilly at night, or too bright and sometimes too hot when the sun is shining. Try a half-trombe wall or an internal trombe wall with perforations to see through and a glass skin covering it. Many permutations are possible. And don't build without easily closed thermal drapes or shutters. On cold nights, close them, religiously. You don't need sun or view at night. Then be sure to open them all to let sun in.

In harsh climates (too hot or extremely cold), it may be better to thermally zone the parts of your house that have more glass. For instance, in Winnipeg or Minneapolis, an attached greenhouse is a lovely asset when it's chilly outside, but by 10 P.M. in December you probably want to be able to close a door to isolate it. In Tucson or Bakersfield, those picture windows need to be in a place where you don't tend to sit, from 10 A.M. to sunset, from May to October.

Natural Lighting

Thick walls let you direct where sunlight (and therefore heat) enters your building, by giving you control of how far out in the wall thickness the glass will be set (see chapter 14). Aim window openings toward the south to southwest, not west or northwest, and south to southeast, not east or northeast, in order to encourage winter but not summer sunbeams.

Wherever there are multiple windows beside one another, try to mount them to form a convex outward curve. The feeling of extra space is remarkable. In Linda's and my cottage, a 9-foot-wide bay window is only ten inches out of alignment with the surrounding walls, yet from inside the extra space feels expansive and grand. This is a lesson applicable to tight city conditions. Many homes in Amsterdam, for instance, not only have small bay windows for sitting beside, but also mirrors attached outside to see better onto the sidewalk and reflect in more light, expanding the apparent interior space.

Windows are inefficient at lighting indoor spaces. On a winter day in western Oregon,

One square foot of skylight can provide more light than sixteen square feet of vertical window.

most houses have on artificial lights, all through the day, even if the sun is shining. A much better source of illumination is skylights, even more so farther north, where winter days are so short, or in very cloudy zones. The lack of access to natural light in our homes can contribute to Seasonal Affective Disorders (SAD) and depression.

Why are skylights scarce? First, because skylights added to existing roofs are sometimes ineptly installed and therefore leak. But if instead you install narrow skylights at right angles to the ridge when the roof is first built, you should have no trouble; less water accumulates on the ridge of the roof, and a narrow skylight is much less likely to leak. Second, skylights may seem to be impractical when suspended ceilings are far below the roof, necessitating a deep lightwell connecting the roof with the room beneath. Instead, amend the roof truss design to accommodate

CATNAP RESEARCH IN A COB GREENHOUSE

BY LINDA

ONE SUNNY WINTER AFTERNOON, I went prowling around looking for the warmest outdoor spot from which I could hear the sounds of the little waterfall pouring into our pond. Like Goldilocks looking for the "just right" place, I first sat in a wooden chair. "Hmm, looks warmer over there," I thought, so I went and sat in a chair on the south side of the garden courtyard wall. Nice, but what about the cob greenhouse? "Oh yes! This is just right." A cozy, warm spot to research the health implications of cob construction and to see just how comfy, relaxed, and happy you can get while working in a solar-heated cob structure in January.

Settling in, I began to see what needed to be done to complete this greenhouse. I can't wait to put in that last window, make the door, pour the earthen floor, and plaster—and oh, it's time to build the seed-flat table. Oh right, and replace the plastic roof and south-facing wall with glass. So this is what happens when you're in a hurry to cover the greenhouse before the rains come. But how nice that we designed in that window seat on the east wall overlooking the pond, so lovely to hear the natural waterfall into the pond in the wet season. Watching the condensation's water droplets smear the ink on my paper, I shifted my focus to the lush salad bar growing in the ground and felt myself deeply relax into the warmth of the sunlight.

I could put up the hammock or swinging bench (yawn), I thought, while not budging from my chair. Like a cat, I had found the warmest spot. What do cats do when they find the coziest spot? Catnap, of course. When I awoke from mine, I felt like I was in sunny California, so comfy and warm. Research, I thought. I'm working, remember? I went for a thermometer to take the air temperature. Outside in full sun, on the south side of the wall it was 64 degrees, and it was 96 inside the cob greenhouse. Not bad for January, a few cob walls, and a big sheet of plastic thrown over the top.

The results of my catnap research project on the healing implications of cob? Refreshment. Well-being, bliss, relaxation, enjoyment of the qualities of sunlight, water sounds, and green plants, and the joy of being, not doing, in an unfinished cob structure.

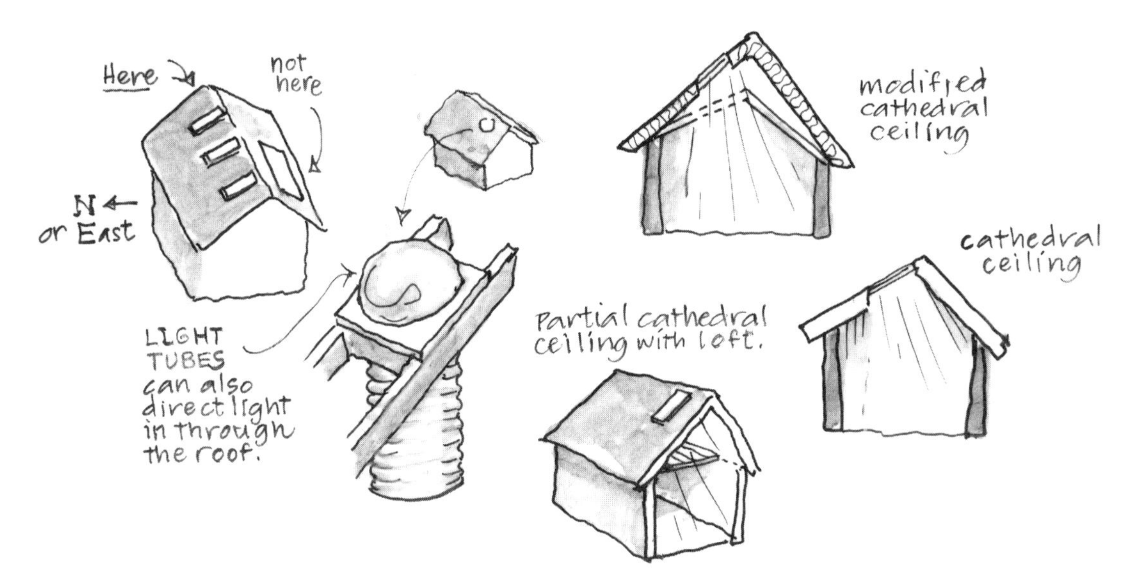

Criteria for successful skylights: installed high on the roof (as near ridge as possible); narrow; facing away from full sun; several smaller ones better than one large.

skylights from the outset, splaying the lightwells in both directions or creating a partially cathedral ceiling.

A common belief is that skylights lose huge amounts of heat, but this doesn't need to be so. You can economically build in extra glass layers to reduce heat loss; the lower layer can be removable for cleaning. It is not uncommon to see a triple-glazed unit with two additional removable panes beneath. Also, because a skylight provides many times more illumination than a similar-sized vertical window, you can make your skylight quite small, so the total heat loss can be proportionately small, or invest in "super-window" highly insulant manufactured skylights, instead of making giant windows, which lose so much heat.

THE HEART HOUSE

When Linda and I had to leave our first cob cottage, wonderful people came to our aid. We had nine offers of places to move, and by March 1993 we were settled in with our good friends Lew Bank and Joan Levine, who offered us a building lot on their land. So came the opportunity to build the Heart House, where we have lived these past nine years.

In May of 1993, Michael Smith journeyed from Costa Rica to join us, and The Cob Cottage Company was born—a true company in the model of a theater company, a fellowship of friends. Immediately we set out to build, using the experience we had gained from our first cottage.

Linda and I needed a place to cook, eat, sleep, entertain a few guests, and store clothes and books. We wanted this "playhouse" to be no larger than the 120 square feet our county would allow without a building permit, and planned for it to have a phone, electricity, and running water, with a composting toilet outside. A shower and washing machine were already available in an existing nearby building. It took the back of an envelope (to be honest, three envelopes) to develop a rough design, and then we could start to build.

Although our house became heart-shaped, the floorplan was not based on the concept of "heart." We never designed a heart as such; the shapes of the human dance do not confine themselves to labelable geometries. Nor did we build the 10-foot × 12-foot box so readily suggested by 120 square feet. The design grew, like lilies on a pond, in response to the activities we needed to have enclosed.

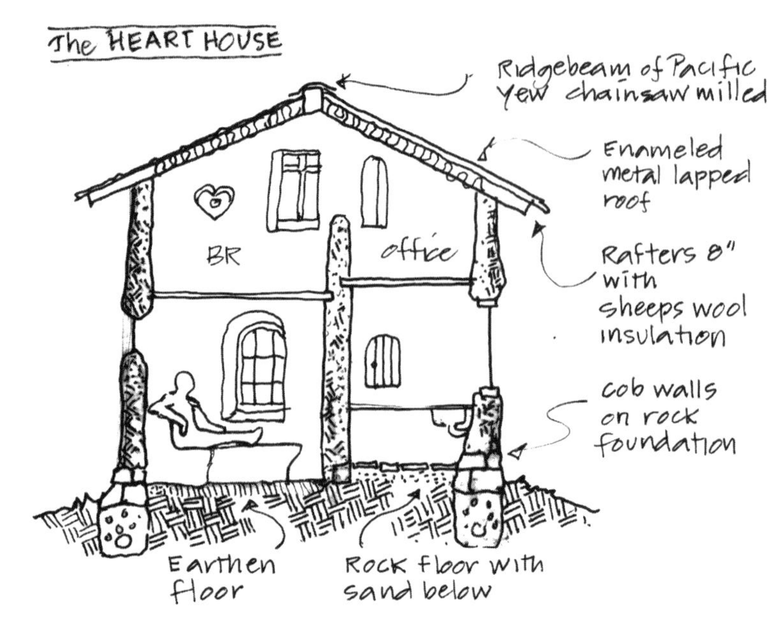

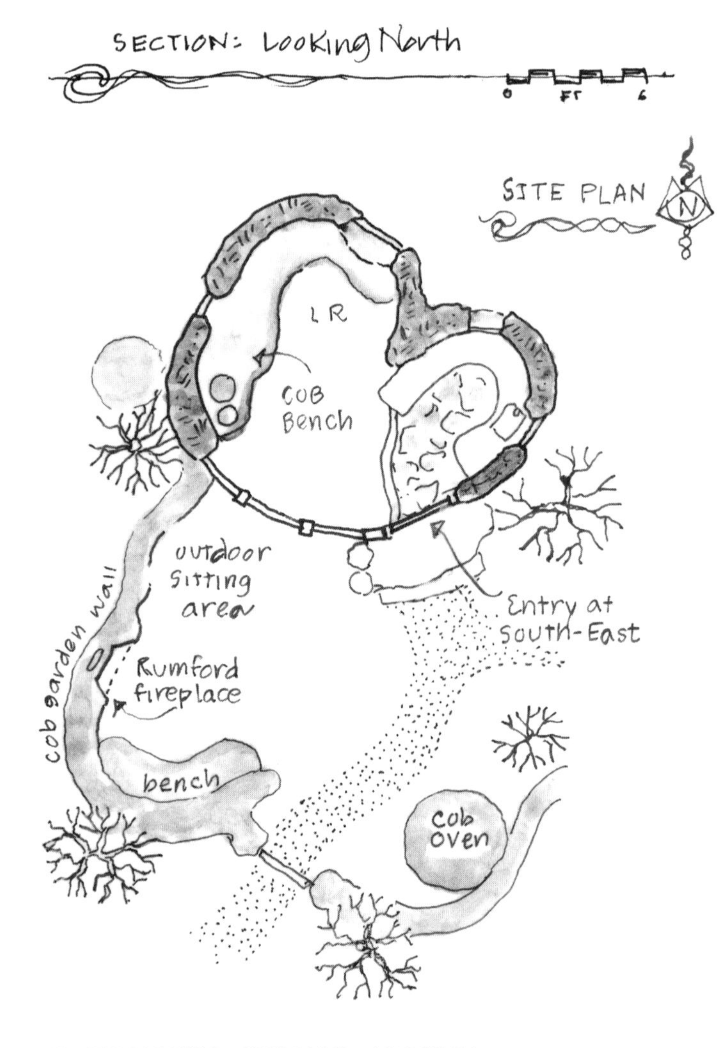

Space to cook in? Well, standing at an imaginary wraparound counter, spread your arms. You really can't stretch easily beyond that span, so the kitchen ends there. I am 5'6", Linda a little smaller. My arm-span is 5'7", so the kitchen is 5'8". Within reach, without walking, are fridge, stove, sink, dishes, pots, food, storage, and counters. The "desking space," immediately above the kitchen, uses the same plan with a lower ceiling, as one doesn't stand up to desk.

We designed by verb and adjective, by the planned activity and the qualities of the place. Nouns have relevance mostly to realtors—you can't snuggle up in square feet. We tried to fit the spaces to activities, as the shell fits the snail. And what emerged, those three envelopes later, was a sort of tiny, quirky heart, separated into several tinier places by a step, a buttress, and the edge of a loft. All other design decisions were made ad hoc, as the building progressed.

We chose to build on a 15 percent south-facing slope at the north end of a one-acre clearing, with rain forest at our back and on both sides, and a quarter-acre garden in front. We were building in spring 1993, and, of course, 1993 had the wettest spring in Oregon's history. The only site with good solar access was awash with surface runoff. When we dug pits to test for drainage, they promptly filled to the top and ran over! So step #1 was digging a 3-foot-deep curtain drain to circle the site. As an experiment, we mixed the drain rock with thousand-year-old Douglas fir bark off the stumps the loggers had left, on the questionable logic that if it had stood a millennium on the tree, it could last another in the ground.

Next we built a temporary roof on poles and covered this with clear plastic tarps. Then we dug into the heavy clay soil—first a level floor pad, then foundation trenches. Since frozen ground is almost unknown here, we dug the trenches only about a foot deep, to where the clay consolidates, and laid a little gravel in. Here's a lesson: Always lay drainpipe in a rubble trench beneath your foundation. We didn't, and had to deal, two years later, with a wet floor following the only six-inch storm in decades.

For the foundation plinth, rock came from anywhere we could find it—roadsides, an abandoned quarry, the excavation for a neighbor's house. For mortar we bought four bags of cement and used it in a ratio of 1 to 8 with commercial river sand.

We made mistakes. Somewhere we'd gotten the notion that earth needs to be screened for building, so we wasted hours shoveling heavy clay soil through a ¾-inch screen. We piled our excavation material too close to the building, cramping our work area, and we forgot to separate topsoil from subsoil. We even left the planned-for rigid foam insulation out of the foundation trench. But overall the work went smoothly.

Every day we would mix clay soil, sand, water, and straw, turning with a shovel and treading with bare feet. Then we would hand-form it into loaves, toss them up to each other, and work them into the wall.

As we formed the walls, we built in furniture. In a tiny space, it's good to place sedentary objects toward the edges. Against one wall, we built in an S-shaped bench, wide enough to sleep on, heated by the stove flue. We built in a cob bookshelf, integral with the wall. We made an alcove for the phone, with wooden shelves. We made niches for candles.

When the walls reached waist height, we smoothed flat the windowsills. Cob walls can be thick enough to contain places within them: The north windowsill is so broad it's a window seat. We laid brick, mortared with lime and sand, for the window seat. In the kitchen, there's a window reveal angled through the 16-inch wall at 45 degrees, to give

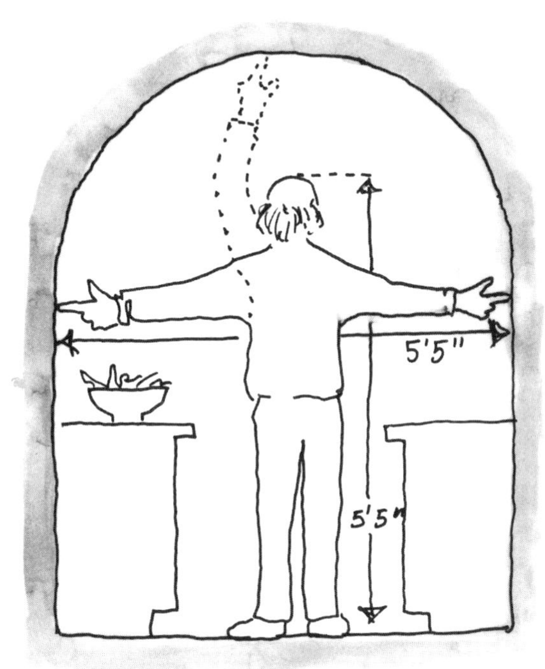

The dimensions of a kitchen determined by the cook's armspan.

interior privacy. Then, most fun of all, we mounted free-form windows, oval, Norman-arched—all were just plate glass, offcuts and shards, edges taped for safety, built directly into the cob. Serendipity tweaked our well-laid plans, and an oval window became a yin-yang, but only visible as such from one spot

The Heart House from the east side. The south face is mostly glass, for passive solar heating.

The Heart House sitting nook. Heated cob bench below, corbelled cob bookshelf above.

on the approach to the cottage. Design happened daily, almost constantly, by whoever was working.

Walls went up steadily at three to six feet per week, depending on how dry the weather was. Progress on the Heart House was slow, partly because there were just two of us most of the time, partly because we were still amateurs with nobody to teach us, and partly because it rained all through June and some of July. We build much faster these days, at twice or three times the speed, with less effort.

Ledger board to support loft floor.

When the walls reached six feet, we set in ledger boards, just 2 × 6 offcuts built into the wall; loft floorboards would then sit on the exposed two inches of wood.

Out in the woods there were treasures, unwanted logging refuse—the smooth poles of madrone, shittum, western dogwood, and a curved dead fir for the loft edge beam. In Oregon, they generally burn these precious trees; we built with them.

We heeded advice from Alfred Howard, the old English master cobber from Devon: "Don't fix them beams in the cob. They got to slide loose-like. Just set 'em in, or sure as fate they'll bust your cob up when the house moves." We notched the edge-beam—only a five-inch pole, but with the strength trees attain if they grow curved—to take loft boards. It projects only 2 inches below ceiling height. We laid in no joists, simply using recycled 2 × 6 tongue-and-groove fir flooring, good quality and a bit springy. That's how you keep overall height down. Snuggle the building into the landscape, build less wall, burn less firewood, get more headroom.

As we worked on the walls above loft height, there was a loft floor to stand on. The loft's bed alcove demanded special treatment, so what better in a heart house than a heart window, symbol of a love bed and Linda's Valentine's Day birthday? A wood-framed heart window is a challenge if not a nightmare, and you have to be very good with a glass cutter. With cob it was easy, a shard of quarter-inch plate molded into wet mud. The putty was that same mud, smeared with a thumb.

When the wall was solid we would trim it down with a machete or a flat-ended shovel, ax, or handsaw. We buried pipes and wires as we went. The interior we plastered; the outside was left exposed.

Roof design was ad hoc. At the last minute, when the walls were almost finished, we switched the direction of the ridge completely from east–west to roughly north–south. The sod roof we had talked about all along suddenly became enameled steel, and I was out in the woods searching for a ridge beam until it became obvious we had totally abandoned the idea of a sod roof. At first I was defensive about the last-minute changes. After all, I'm an architect, and we are supposed to have a complete picture of every little detail, right?

Mercifully, we did remember to deadman the walls, burying small bolts of firewood into the cob every three or four feet, about two feet down from the rafters. To blow the roof off, that hurricane will also need to blow away the top several tons of wall. And blocking was easily added, as we packed cob into the spaces between rafters where they sit on the wall.

An important goal of the Heart House was to push the limits of our resourcefulness in using as few new products as possible, a demonstration that one doesn't need to rape the world to make a dwelling. As dollars spent roughly equal resources exploited, we kept to a budget of $500 for the whole building. That money went chiefly to new steel roofing, a few straw bales, two nine-yard loads of sand, some cement, gypsum plaster, and some hardware and electrical fittings. We estimated a budget allowance for running the truck and chain saw, but we didn't include our labor. Extreme frugality is time-consuming at first, but the practice was invaluable as it forced out hidden creativity and made us inventive. Our desire to avoid purchased manufactured products also kept the building almost completely toxin-free.

We used no treated wood and no paint at all, instead exposing wooden surfaces. The cob interior we plastered with bright gypsum, later adding a coat of *alis,* a mixture of white clay, flour paste, and mica. Later still we experimented with a one-coat kaolin plaster over the *alis*. We avoided plywood. The solid rock and cob floors have no fixed carpets; we use pure wool and cotton throw rugs. For roof insulation we used sheep's wool, a pioneering venture in Oregon, home of three million sheep. Local farmers, disgusted at wholesale wool prices, were happy to donate fleeces, which we then washed and carded, but that's another story (see chapter 15). The wooden parts at risk from weather (sills, threshold, lintels) are of untreated incense cedar, hand-milled from the only tree we cut for sunlight. Caulk and window putty were made from horse manure and red clay.

Our finished building is tiny. Overall height is only about thirteen feet, but because of careful design you can stand upright everywhere you need to. By making use of the site's changes of level and a very low ceiling in the hangout space—it's 6'2"—we managed to provide a sleeping loft and a tiny wraparound desk loft for my books. The 3 × 5-foot skylight, facing southeast, over my desk was another mistake. I freeze in winter and fry in summer, and the damn thing leaks. Next time, I'll face the skylight east or north, and keep it tiny.

"Wasn't it an awful lot of work?" people ask. Yes, of course it was. Building a house is a lot of work. You have some choices though. One is to work for someone else to make money to pay a builder and a bank. That can mean hard work for thirty or forty years, possibly doing something not altogether joyful, fighting the traffic to get there, compromising so as not to lose that job.

As a culture, our industrial urban lives now lack peasant satisfactions, the steady rhythm of ritual handwork, or listening to the wind in the fir trees, the frogs croaking by the pond. The beauty of cob is that you work silently at home, on your own terms, with your bare hands and feet. It's a daily rhythmic meditation, connecting with the primordial. One of the great disappointments of my life was the day Michael and I finished the cob

walls. For days I stalked around the site like an abandoned lover, wondering what on earth I would do with the rest of my life, feeling the emptiness of not cobbing.

The Heart House was a lesson in abandoning classical notions of Design, in which every last detail is figured out in advance and carefully drawn. For me, architecture-school trained with British precision, this experience opened the gate to a completely new way of designing: we created a minimally skeletal plan, then immediately began building. The Heart House was the prototype for what is now becoming a new school of design thinking—design more as Nature does it. Nature begins with a genotypic template, which is a fairly rigid idea of what's possible, then creates endless diversity by responding to every nuance of the surroundings. In a similar, parallel way, once our house's foundation was built, nearly every design decision was made by intuition—how it felt at the time and how we imagined it would feel finished. The only fixed points were the foundation stemwall and the break in it for the door opening.

In November we moved in, though the walls were still a little damp. The lime stucco intended for the exterior has never been applied, as we like the exposed cob so much. Like all good houses, the Heart House is still unfinished, years later, though the earth floor finally got a top coat. Our home has been seen by hundreds, probably thousands, of visitors who have responses ranging from curiosity and predictable questions to complete awe and love at first sight. The Heart House, all 120 square feet of it, has brought Linda and me more joy than any home either of us of us had ever known.

Redefining "House" 7

❧ A HOUSE . . . SHOULD TELL YOU, "STAY HERE, BE COMFORTED, RELAX, BE YOURSELF, YOU'RE *HOME NOW*." ❧

IN THIS CHAPTER WE INTRODUCE YOU TO a step-by-step way of making decisions about creating a magical house. You won't need drafting instruments or very much paper, and the computer will be altogether irrelevant. Although the design method is laid out in a set sequence, there are parts that are optional, and some of the stages can be rearranged for particular situations. There's a lot of detail here, so we would suggest skimming the topic headlines, then first reading the parts that attract you most. For this chapter to be most useful, you will probably need to switch back and forth between close observation of your intended site and analysis of what kinds of physical spaces you want to create and live within.

In some ways designing for cob is more like designing a space capsule than a house as we know it. Cob buildings, like spaceships, fit snugly around the activities they protect. Both require more planning than construction; in both, the quality and shape of the spaces provided is more important than how the assembly looks to observers. Designing your own home and building with your own hands, you can make a complete response to your need for enclosure rather than a stereotyped reiteration of what society expects a house to be.

INTUITIVE DESIGN

Do you remember the places you designed as a child? Remember the forts in the living room when you used anything you could find to create a sanctuary totally your own? Remember the spaces you made in closets, basements, attics, on rooftops, in garages, in backyard tree houses? Maybe you built rafts, dams, nests, driftwood beach huts, igloos, or tipis—or stole away to tunnels, rock caves,

quarries, or ice caves. Maybe you turned old outbuildings or chicken coops into your own secret abodes: GROWN-UPS KEEP OUT.

Each of us has woven our significant places into a unique "blanket of memories." Every thread of this blanket, rich in remembered shapes and textures, symbolizes places that have had a powerful role in shaping our lives. As intuitive designers we can open this memory blanket anytime to rediscover and integrate its sacred qualities into our designs.

In the Onword of this book, you will find Linda's description of a process of creating your magic spots. As you go through the exercises that follow below, keep in mind that there is another dimension available to you in designing your house—your intuition.

Dispense with Convention

Try to disassociate your mind from the bad models that surround you. No, your house doesn't need to face the street or be built "as an investment." It doesn't need to be square on the lot, drearily colored, as big as possible. It may not need a concrete driveway or a double garage. It may not even need a front door—perhaps you'll have no door at all. Include only the elements you want, not what is conventional.

"But what about resale value?" You probably don't choose your *clothes* for their value secondhand, so make your house how *you* want it now. We need to live well, not die wealthy. Don't worry about making your house conservative so that someone else will want it. Take the courage to satisfy your own peculiarities. Unless you're very unusual, much of what suits you will suit other people, too, and if not, so what? Suit yourself; it's *your* home. If you build your own house with love, and live in it with joy, there is its value. You may find, too, that people often pay premium prices for the unconventional, the wacky, the whimsical, the artistic.

Any building, but especially a house, is a continuum of planning, building, inhabiting, demolishing, recycling, and reuse. At all stages other stages are beginning or phasing out. A satisfying house can never be a finished product where you "live happily ever after." It needs to change with your needs, not just superficially (e.g., repainting the bedroom, putting photos on the walls) but functionally, as you create new *places*. There's an old Chinese proverb that says, "Man finishes house. Man dies."

The process of design is that of matching needs with resources in the best possible way. First, *list* and analyze the needs—what the building needs to do—in detail. Then *inventory* your resources: building materials, skills, money, time, advantages of the site's climate and social milieu, views, slopes, and so on. Then go back and *refine* your definition of what the building can do. The resources inventory will have opened up new possibilities. In turn, revisit the resources list, as the new needs will suggest resources not yet listed.

Designing a house is not the linear process suggested by blueprints. It involvse making clay models, changing them, drawing details, cataloguing useful resources, changing ideas and methods, trying unlikely solutions, all in whatever order that particular building requires. It is a continuous story, from how the site was before you ever considered building to long after you have left, as other people inhabit your house, constantly changing it.

Brainstorm a Lot

Creativity sometimes needs to be freed from the limitations of everyday logical thought processes. Brainstorming can help wipe your mind clean of assumptions you have accumulated as to what a house should be, how big, what shape, what materials. A house can be an inspiration, or silly, frivolous, humorous, surprising. It should feel good to the child inside you; it can be the playhouse you always wanted.

Brainstorming works best when there are no constraints and everybody is relaxed. The crazier an idea the better; its function is to stretch our horizon of possibilities to where new thoughts come more easily. Feasibility of any given idea is not initially an issue, until many ideas have been aired and recorded somewhere so everybody can see them. Like a smoldering fire, these ideas ignite one another as you breathe into them, creating new combinations you never could have generated otherwise.

Brainstorm especially with family and those who may share your house. If you're building solo, brainstorm with a friend, then switch—try another friend.

"What if?" games can stimulate new ideas and unasked questions. "What if we needed to have a self-sufficient water supply?" "What if the door is a tunnel like an igloo has, so you really appreciate getting inside?" "What if my next husband has thirteen dogs?" "What if we lose our jobs and have no source of income?"

The Site Was There First!

Your site was there first and will be there long after you and your building are gone. Whatever changes happen as a result of your building need to be reversible when the building has vanished. In the long view, even a thousand-year building is merely borrowing the site for a tiny fraction of the life of that place.

Pay respect to the site's natural systems, inhabitants, and processes (see chapter 9). Consider the creatures and plants that lived there before you came along. It's hard to justify moving or exterminating them, so be respectful of the site and of all of the beings who need it. Even if your site is in the heart of the city, there will be life there.

Meditation on your site should be the first step in design of your home (see chapter 5). Hang out—alone, in silence—where you are planning to build, for as long as you can, under every condition of season, time, and weather. If you can, live on your site for a full turn of Earth around Sun before you make any irreversible changes to the land. Carefully observe what the site has to offer you; make an inventory. Observe the diurnal cycles of human activity and how they affect you. The location of your house will almost certainly change as you study its site over time.

Share Your Home with Nature

For real satisfaction and connections, plan to bring real Nature into your house; it will

HENRY KUNOWSKI

Land clearance for mansions in L.A.

Wildflowers, not weeds.

come anyway if conditions are right. Try to call the plants that come to you *wildflowers* not weeds, the insects *spiders* or *beetles* or *wasps,* not just bugs. Learn their names if you can, for you will dignify both them and yourself in so doing. Leave space for them all: cracks in the paving, gaps under eaves, loose rocks in your walls. For a more thorough discussion of this notion, see Wildlife in the Home, appendix 6.

Keep It Small!

Don't build any bigger than you absolutely need. Almost any option is possible if you build small enough—you can afford extravagant materials and finely crafted work. Think of quality over quantity. The majority of first-time cob builders admit they started too big. The results are often spectacular ruins, vast foundations, unfinished masterpieces, wherever people began building with surplus ambition.

Build no more than you can guarantee to finish in a single building season; you can add on later if you like. If you're not used to day after day of hard physical work, think tiny. *How tiny?* A hundred round feet (see the box on page 97) would accommodate sleeping, dressing, some storage, and a kitchenette with eating alcove. Sculpt volumes to just fit your activities; sleeping spaces, for instance, can be snug, low closets.

Plan to phase the construction so you complete one enclosed space at a time. You can include a knock-down temporary wall (straw bales are ideal) or a window that can later become a door to another room. Think out well in advance a roof system that is expandable, so you can extend it as the building grows. Or you can later connect small, discrete cob buildings with covered walkways, winding paths, or garden courtyards. Be sure you locate phase 1 so that phase 2 won't block solar access, cut light from your windows, cross the lot line, or necessitate cutting down your best trees.

Identify what you most need to shelter first. A rural homesteader on a remote site might need to build a sleeping place first, just a bed alcove with space to store clothes. Sometimes it makes most sense to build a kitchen first, with enough floor space to roll out your futon. Add an outdoor composting toilet, and you can stop paying rent and move in to your home immediately.

PLACES, SPACES, AND ROOMS

Consider the dynamics of how we use most rooms. There are basically two ways we use a room: we *go through* it or we *be in* it. A corridor, a garden path, or a doorway are "goingthru" places; an armchair, a bed, or a bathtub are "beingin" places. Ideally, goingthru places should lead to beingin places.

The essence of a HOME is being in it—*human being* not *human going.* Being is the essence of humanity; going through is just a means to an end. But our culture thrives on going—movement, transitory activities, almost to the exclusion of being. We live in a public world designed almost exclusively for us to pass through rapidly, a world of freeways, airports, corridors. Our very terminology lauds

CURVED SPACES FEEL BIGGER: THE CASE FOR ROUND FEET

PERCEPTUALLY, curved spaces feel bigger than boxes, so by making cob buildings curvilinear you can build them smaller. I call this phenomenon *round feet*. Our conclusion from questioning dozens of experienced professionals is that one round foot perceptually equals about two square feet. This tentative conclusion seems to hold true not only for cob but for any enclosure with curvilinear geometries—domes, yurts, even the barrel vaults of gothic cathedrals, though the effect is most pronounced in *irregularly* curved buildings. Another interesting feature of the geometry of all rounded structures is that they look smaller from outside than when you get inside. This is especially true if the inside is mostly undivided.

Learn to think in round feet. Deciding how big you want spaces to be has nothing to do with square feet. Buying a home like cloth yardage encourages quantity before quality. What if *clothes* were sold by the square yard, instead of tailored to fit individual humans? This concept of *square feet* serves the real estate industry, not people. We call them "realtors' feet."

Fairly predictably, visitors to Linda's and my cob cottage (especially conventional builders) have a common first question: "How many square feet is this place?" Questioners of this type usually have the germ of the answer already in their mind, so I tend to look surprised and answer, "Gee, I'm not sure. I've never measured it. How big do you estimate?"

The answers usually come with some assurance. "Two hundred seventy." Then they'll debate a little between themselves: "Nah, more like two fifty-five." The guesses for my house usually range from 180 to 400, with an average around 250.

This is all very interesting in a building that has been measured quite precisely at 126 square feet.

If round feet were accepted generally as a way to increase space without additional construction, they could have a big effect on the building industry, which loves to find ways to increase its profits. If builders and developers could charge the same price for a smaller house, that might have a significant effect on building design, materials extraction, and construction technique.

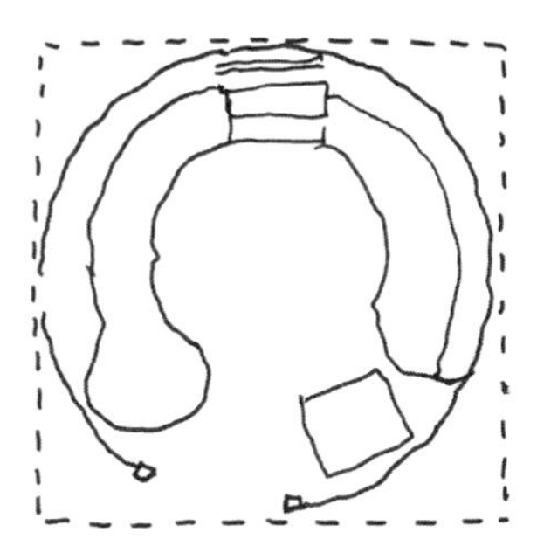

Squaring a circle adds 28% in area and 21% more in perimeter with no additional usable diameter, only more or less useless corners. With three-dimensional volumes this effect is even more dramatic: a cube has 50% more volume than a sphere, again because of those inhospitable corners. All this is to say that round spaces do not usually feel smaller than square spaces of comparable diameter, though their footage is less.

You can encourage goingthru or beingin with ceiling height.

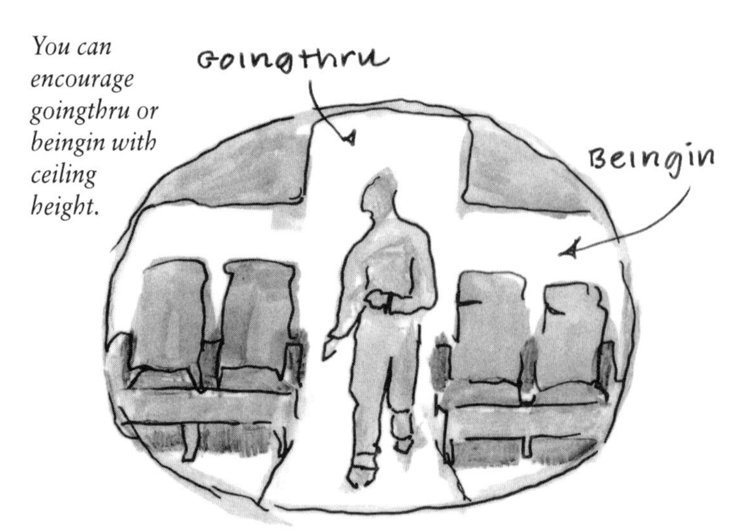

speed of movement: *express*ways, *rapid* transit, *Kwik*-Stop Market. Even parks are not for relaxation anymore; they're set up for people to walk, jog, bicycle, or drive through, and sometimes there isn't anywhere to sit. Even houses are designed around a "circulation diagram" of "desire lines" to facilitate movement in every direction as easily as possible, so they end up being a series of invisible corridors, with almost nowhere to be in, nowhere to feel secure or relax.

The price of a goingthru house, a house laid out for ease of access, is that there is no longer space to just be; there isn't really relaxed space. TV viewing is constantly interrupted by people walking between viewer and set. Kitchen frictions are common—crabby cooks collide with children carrying cookies. Evaluate your own present living space. How much of it is not walked through at all, and how much is completely protected from goingthru access? How might you change the balance? Would you want to make these changes?

In a goingthru house, being in places becomes secondary to movement. We assume all sorts of rules: for instance, that furniture must all be moveable too, so that we can shift it around for access; that all doors should be wide enough to accommodate a grand piano; that every ceiling must be high enough for a six-foot-ten person carrying a potted palm, all over the house. Wall to wall, every inch of the floor is treated as potential corridor. This makes us crave the security of the nook, the cubby, the snug enclosure.

A house should be an antidote to the freeway culture. When you're in it, it should tell you, "Stay here, be comforted, relax, be yourself, you're *home now*." How can you make this happen?

❧ ***Control goingthru using ceiling height.*** You can control where people walk by ceiling height. Goingthru places need to be high, usually taller than a person; beingin places are sometimes quite low, sitting- or lying-height. As in airplane or bus design, control the goingthru by making the ceiling lower closer to the walls.

Bringing a ceiling down to exactly head height (say, six feet) over a relaxation space or a private desk will discourage casual invasion. It is psychologically uninviting to duck into a space lower than your height, yet it won't exclude you if you've chosen to go there. Use this space for built-in seats. It will reduce the need for space-greedy chairs.

❧ ***Keep down the number of doors.*** Dead ends make great places to be in. You can work without interruption, snuggle up with your lover, leave that half-done jigsaw puzzle

With too many doors, a room becomes just a corridor.

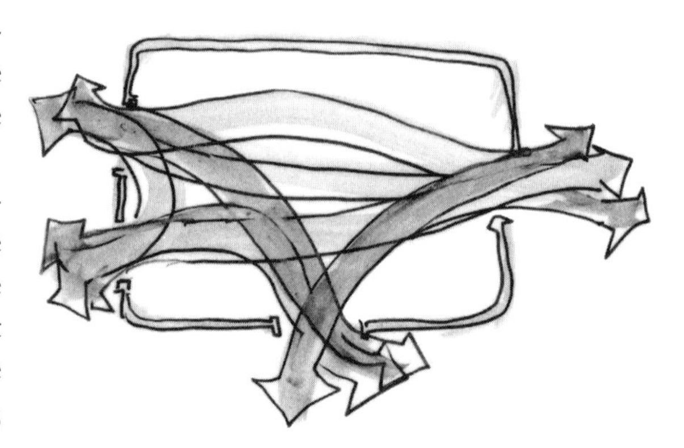

all over the floor. Any room with more than one door becomes a corridor, sometimes so much so that the beingin spaces are too small to be effective, or are transitory—"Can you move your chair please, I have to get through?"—and therefore insecure. Instead of yet another door consider a window from knee height to eye level, which can be a sometimes exit. Normally nobody will go through—it is not inviting—but in emergencies or for sheer fun, this opening makes another access.

A window can be an occasional door.

❧***Build-in alcoves.*** Alcoves are small spaces recognizably separated from the main body of a space. Alcoves work best if their floor is higher or lower than the goingthru space that serves them. They can be indoors or out. Sometimes an entire house can be one big gathering/hangout space with a bed loft above, an alcove for cooking, perhaps a snug and heated alcove for being comfy, and a window seat alcove for dreaming.

❧***Build-in level changes.*** Level changes help define changes of mood. A step down into a hangout space tells people to slow down, take off outdoor shoes, encouraging them to respect entering another domain. In Linda's and my cottage there is a 7-inch step down from the cooking area to the hangout.

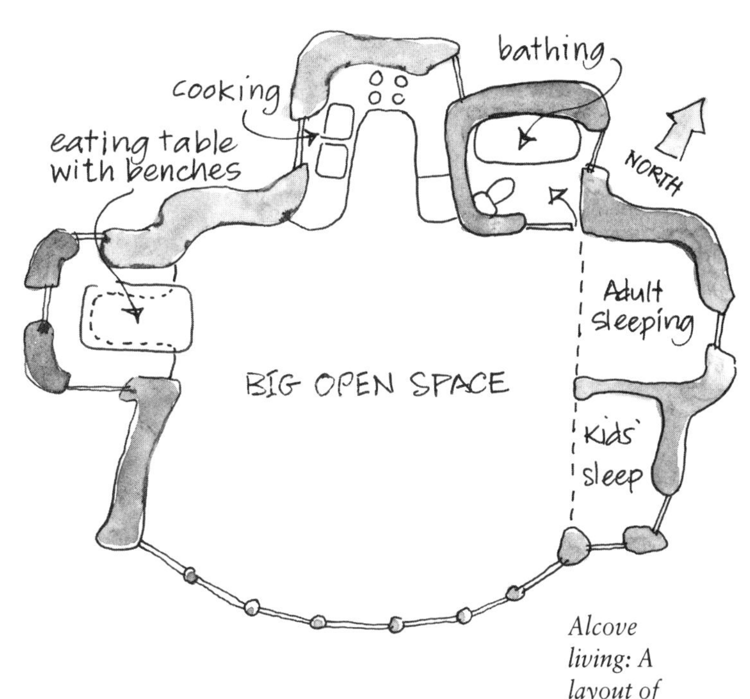

Alcove living: A layout of larger living space surrounded by alcoves.

Almost nobody steps down into that sanctuary with their shoes on.

❧***Make a place for spectators.*** Where food is prepared, make a place for spectators to sit, an out-of-the-way seat (in an alcove perhaps with a little table for coffee, or a food-prep counter). Otherwise it is inevitable that onlookers will cluster in the kitchen to chat, snack, or anticipate meals, obstructing the rhythmic ritual of cooking.

Make a place for onlookers to sit and watch the cook, or inevitably they will congregate in the kitchen.

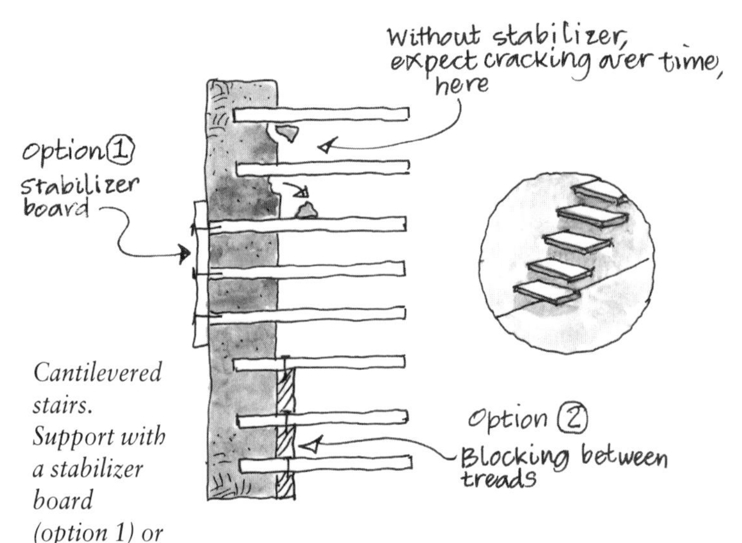

Cantilevered stairs. Support with a stabilizer board (option 1) or blocking between treads (option 2).

❧ ***Consider using ladders.*** Instead of stairs, a ladder can be pushed up into a loft when not in use, freeing floor space. Stairs can be space-intensive, complicated to build, and expensive, but if you build to code or with a permit or mortgage, stairs may be required.

Lightweight wooden or bamboo ladders are easy to make and satisfying to climb. A ladder can be fixed in place, moveable, temporary, or retractable. A fixed ladder, attached to a wall for instance, will be much easier to climb if it slopes, even slightly. The rungs will be easier on the feet if they are fairly deep front to back, flat instead of rounded, and set out a little way from a wall. Make sure there's a strong, secure handhold one to three feet above the top floor, to pull yourself up with and to provide a solid grip when heading down.

If you do use stairs, you can make them out of boards, roundwood logs, cob, flagstones, or brick; be creative, but keep the risers a consistent height for safety. It's sometimes possible to cantilever stairs out of a massive cob wall, merely by setting stout boards into the wall as it is built.

Banish the Old Noun-Rooms

In early 21st-century America, we have come to accept that our homes should be rectilinear boxes, divided internally into a series of conventionally named smaller boxes. The industry that sells these to us insists upon giving each little box a room label (such as bedroom, sitting room, family room), as if to convince people of the need for more *rooms*.

What we do in these rooms is often quite different from their labels. Bathrooms often don't even contain a bath. As for the living room, don't we live in all parts of a house? Most traditional cultures don't even have bedrooms—people sleep where they're warm or cool, where it's comfortable, wherever they fall asleep. Small children don't sleep apart from their mothers in separate rooms; often the whole family sleeps together, even in one bed, as of course did our own ancestors. Most garages are no longer a "shelter or repair shop for automotive vehicles" *(Webster's)* but a storage shed for things we don't use very often.

Once we have abolished the box idea, the concept of House becomes more interesting—a series of fairly flexible enclosures and part-enclosures leading into one another, all of which are useable for different activities depending on season, time of day, weather, and personal whim.

When you design your home, you're arranging a stage for the play of your own home life. As you describe how you want your house to be, use verbs instead of nouns, so that you are describing as accurately as possible what activities you hope to accommodate. Each space needs to respond to its special uses, and to be the right size, shape, mood, smell, and sound. Don't think of bedroom, bathroom, kitchen. Say to yourself: sleep, bathe, cook, eat. Verbs will help you remember that you don't need a box, but a place to nourish whatever you do. If you tell yourself, "I need a place for bathing," rather than "I need a bathroom," you break your lifetime stereotype of that ugly little cubicle with mold in the corners where you can't shower without looking down the toilet.

Imagine a series of detailed verbs that describe what you do in each place. When you *bathe* you also *take off* your clothes. You might also *store* dirty clothes and *choose* clean ones, *select* soap, *contemplate* the view, *listen* to music, *light* a candle, *soak, drink* a cup of tea or glass of wine, *adjust* water temperature, *dry* yourself, and *dress*. List the sequence of what you want to do, then give all the parts adverbs, describing *how* you want to do it—quickly, contemplatively, casually, carefully.

Describe your experiences in this place. Use adjectives (words for *quality*) to describe the smell, the views, the feeling of fresh air flowing through, the coolness or warmth. Explain what kind of place keeps your clothes dry and orderly. Describe to yourself all the qualities that the realtor's term *bathroom* fails to convey. Your bathing place might be calm, warm, sunny, private, quiet, tall, fragrant, meditative.

As you describe to yourself one by one the spaces you need contained within (or without, attached, or near) your house, consider what should be close to what. It's often good to put bathing close to sleeping, and clotheswashing near to the clothesline. Eating is better situated close to food prep; we may snack as we prepare, live on finger food, eat right out of the fridge. Imagine your eating place within reach of the food-prep area, so that the cook can slide dishes right across to diners.

Act Out the Dance of Your Daily Life

Once you've imagined the scenes in the play that is your home life, it's time to act them out. Use your whole body, realistically, so you can observe how much space you need, how high or low certain features need to be, the relative positions of things, in order to imagine the spaces flowing one into another. Deal first with the activities that most determine your mood. You shouldn't have to just "get through the day." Design your house to help you be happy all the time.

Use your whole body to translate your imagined layout into actual steps in space.

The way the spaces are arranged in a house can make life inspiring and effortless, or abrasive, a slow-burning fuse of irritation due to explode someday. We have enough annoyances in our lives; home needs to be a sanctuary where as much as possible runs smoothly.

Let's take showering as an example. It involves many actions other than merely standing under a showerhead. Act out each part of that sequence, ideally with a critical audience to remind you of parts you forget and to suggest elements they themselves would add. Where will you keep clean clothes and towels? Take off your shoes. We've all been in shower rooms where you can't stoop to get your shoes off without hitting your head on the sink. Plan to avoid those frustrations. Will your bathing space have a door? Do you need to contain steam or keep the kids out? Do you need visual security so nobody walks in unannounced? Where is the door handle, are the hinges left or right, does the door open inward or out? Feel your way through that imaginary room, with an armful of dry clothing and a book and three bottles of toiletries. Act out all the parts of your showering, from start to finish. Be as relaxed as possible; don't rush. Enjoy every part, exactly as you would enjoy the luxury of a hot shower.

As you give movement and voice to every action, record what you really want. Try not to be constrained by habits and expectations such as: "I have to hang a translucent plastic shower curtain because (a) that's what everyone has, (b) I've been used to it, (c) when Granny comes to stay she'll think we're weird if we don't have one."

Talk aloud about the *qualities* of what you want to see, hear, smell, and feel at every stage. "I'm taking off my sweaty shirt, looking out of an eye-level open window into the wisteria vine. I can smell beeswax from the candle sparkling in the golden candle niche above the tub. The ribbed texture of the rock floor is massaging my bare soles, but the big slabs of black schist are heated. I'm sitting on the smoothest of cedar benches taking off my socks, listening to the rain splattering on the skylight. A little green frog just hopped out of the rock waterfall in the corner."

Describe with your body its needs in space. When showering, our eyes are sometimes closed; we need elbow room so as not to knock the cold faucet full on or rub shoulders with a frigid tile wall. And knowing we won't slip in a soapy bathtub is the difference between relaxation and tension.

Standing, we're widest at the shoulders, but perceptually we need more space in front of our eyes. Be aware of the exact places where you need to accommodate only the volume of your body (behind a seat for instance) versus where a psychological need demands more space in front of you. Be creative about incorporating visual and auditory space, beyond the physical envelope your body fills, using openable windows, strategically placed mirrors, skylights, plastic glazing that lets in birdsong (since glass deadens sound), the filtered sounds and smells and light of another place drifting into where you are. You can cunningly constrain the scale of your practical commitments while enjoying the world beyond.

For each household activity, repeat the role play. Explore in detail how your body acts out eating, sleeping, preparing food, making love, meditating, hanging out with friends—all the daily actions that we need to do with care in the highest quality surroundings. To this list I like to add "desking:" "all of the things I do with paper, telephones, computers, books, and pencils." The word

Attend to visual as well as physical needs. Greater volumes of space at eye level are more comfortable.

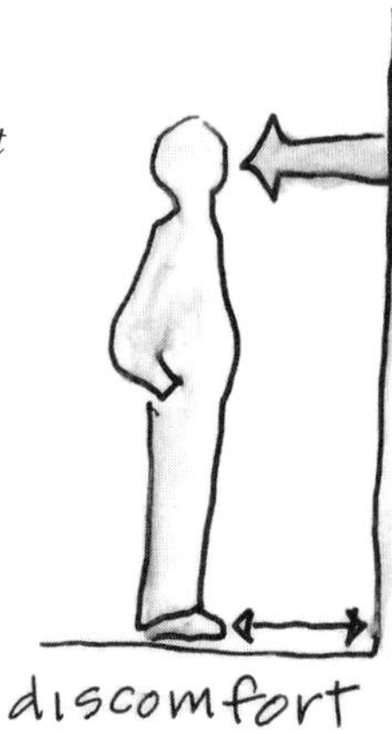

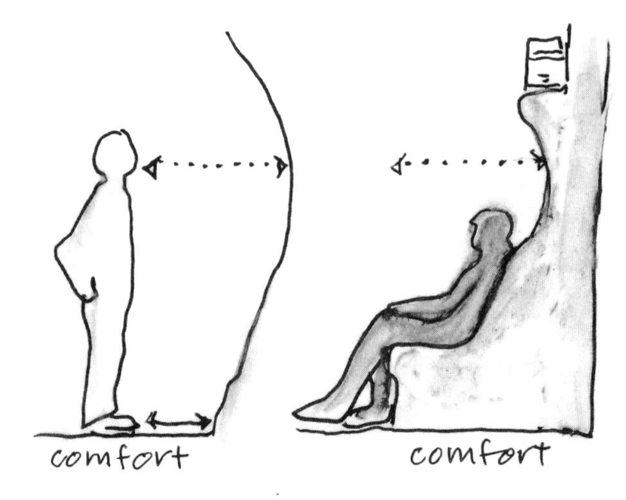

office seems grim and businesslike and embodies too many bad memories. I need my desking time to be reflective, serene, exciting, and connected directly to the wildness that surrounds my house.

Think Gloves Not Boxes

Think of the house as a second skin. You don't wear boxes; why live in them? The walls, floors, and ceilings of your home can wrap around the shapes of the dance of your life; let them surround your activities in ways that will contain specifically those activities. Otherwise your new house may end up feeling like a block of storage containers—most houses are exactly that.

Be receptive to surprises revealing themselves. Your sleeping dance occupies a space that is low, wide, and long, higher at the head end, higher still where you climb out of bed, highest where you approach from or undress, stretching up to pull off a tight shirt. Dressing and sleeping resolve themselves into two different activities with completely different spatial needs. Suddenly the eight-foot ceiling is unnecessarily extravagant, and hard to keep cozy and warm. Who ever needs to sleep under six feet of empty space?

A physical expression of a place for one or two people to sleep might be 4 to 6 feet wide, about 7 feet long, and from a couple of feet high at the foot end to about 4½ feet over the shoulders. The approach place, where you climb in, needs about 7 feet in height, with a place to put clothes that can be lower. Some people like to sit up in bed to read with the wall as a headboard. Others want more space to make up the bed. Act out your preferences, and discuss them with other people. The variety of ways to make a house work is infinite; with careful planning, you can build exactly what works for you.

Sculptured cob presents unusual opportunities to shape spaces to accommodate the life lived there.

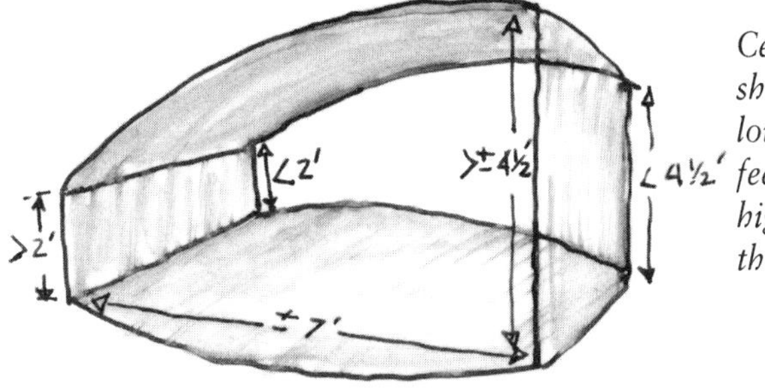

Ceiling should be lower at the feet and higher above the head.

You don't wear boxes. Why live in them?

Cob buildings, with sculptural spaces so easy to make, present unusual opportunities to eliminate unnecessary volume. Walls are especially stable when tapered, wider at the bottom. That taper can be exaggerated to accommodate body shapes and to change a viewer's perception of space. Because we are widest at the shoulders, and narrowest at the feet, a room wider at shoulder level feels bigger, and actually has more useable space. A built-in bench or chair with a sloping back can be very comfortable to lounge in. A rounded connection between wall and ceiling makes the ceiling feel less oppressive and avoids the visual obstruction of an empty right angle.

Your Workplace

Settlers came to Oregon in the 1840s, trudging on foot across the deserts, plains, and mountain ranges. On arriving at the beginning of winter, having walked from Missouri, they threw together a temporary shelter from the rain and started to construct their first permanent building—the barn. They understood that you can live in a barn but you can't keep all your grain, hay, and animals in a house. Without a barn or a workshop, there would be no livelihood, and they wouldn't survive.

Like those early farmers, more and more of us work at home. The benefits of working at home are obvious—no commuting, with its corresponding eco-destruction; being able to live more in tune with natural cycles; being available to safeguard and maintain our homes, to water our vegetable starts and feed the chickens; having visible means of support which our kids can understand and involve themselves in.

Work has the potential to go beyond just making a living. Work largely defines who we are, and so should bring us as much joy as a hot shower or a good meal. Likewise, designing a workspace should involve as much care and sensitivity as we would devote to creating our most magical places, for it needs to be inspiring, serene, and accessible. The view from the place where you work needs to be at least as good as the view from the place where you relax, the visual connection between indoors and outdoors should be strong.

If you plan to work from home like the settlers, build your barn first. In other words, make your workspace part of phase 1 of your building plan. Consider carefully:

❧Will this space always be your office, your shop, or your warehouse, and, if not, what might it evolve into later? If this is temporary, where will the permanent workspace be?

❧Should you build a separate work building (to accommodate noises, fumes, dust, public access) or incorporate your workspace into other parts of the house? With cob, remember, sound transfer is minimal, so a noisy work activity might not disturb late sleepers, nor would music elsewhere in the house necessarily bother you in your office or studio.

❧What views do you want from where you work? What natural lighting should fall on your workdesk, and from what direction?

❧As with other parts of the house, how can you *build around your needs,* to minimize construction costs and maintenance? Remember, "Think gloves, not boxes." Let the shapes of the workspace surround your activities in ways that will specifically contain and enhance those activities.

❧Are other household uses compatible with the workspace? A nine-foot ceiling will accommodate a bunk above a desk, as a now-and-then guest bed. The woodshop could perhaps have a dartboard and a freezer, doubling as the kids' rowdy space. Can you design a computer/phone desk to close down at night so you can hang out there to watch movies without being reminded of work?

Outdoor Rooms

A house doesn't stop at the front door. The woods outside, the street, the space next to your neighbor's house, all are part of your territory. A secret to building small is to make use of outdoor space, if not physically, at least visually connecting it to your home. Outdoor places can be as valuable as indoor, cheaper to build, and enchanting to be in. Design them and the walls, levels, steps, ramps, retaining walls, and earthworks while you design the indoors, to create a counterbalanced group of outdoor/indoor rooms and gardens.

In designing your home, you'll find that some of your activities are best practiced outdoors; some can be done outdoors but need wind protection or a roof, but no heating; some will have to be warm, dry, and enclosed, or cool, shaded, and breezy. In even the harshest climates, minor shelter can extend the time frame for using outdoor rooms—a fence, a hedge, a cob wall, a lean-to roof extending out from an existing building.

Make sure each outdoor space is optimized for use as much of the time as possible. *Lighting* extends an outdoor place's use on summer evenings. Running water, a table, and a stove mean you can cook in the garden when the weather's good. In hot climates, a screened outdoor kitchen can become the main food-preparation place, to keep extra heat out of the house and so you can work where odors will disperse naturally in a breeze.

In many North American climates, a popular indoor/outdoor space is an *attached greenhouse,* solarium, or sunroom. Attached to the east or south wall of a house, it is an inexpensive way to add quality space, plus a means of heating the building, a place to grow light-demanding plants (winter salad greens, for instance), and an insulant layer between your house and the weather. A solarium enables you to concentrate an expanse of glass on the sunny side of your building without tremendous winter heat loss. In the Maritime West, where climates are generally mild, an unheated attached greenhouse can be comfortable the majority of daylight hours, year-round.

On the other hand, a south-facing greenhouse can overheat horribly unless it has summer shade and massive heat storage. A cob west wall for the greenhouse and a cob *trombe wall* between house and greenhouse can satisfy these requirements. You will probably want to be able to look into the greenhouse from inside the house; the trombe wall can have windows, or can be only waist-high (see diagram, page 84). Such a wall can be very thick, offering an opportunity for dramatic lighting through the greenhouse into south-facing rooms. While

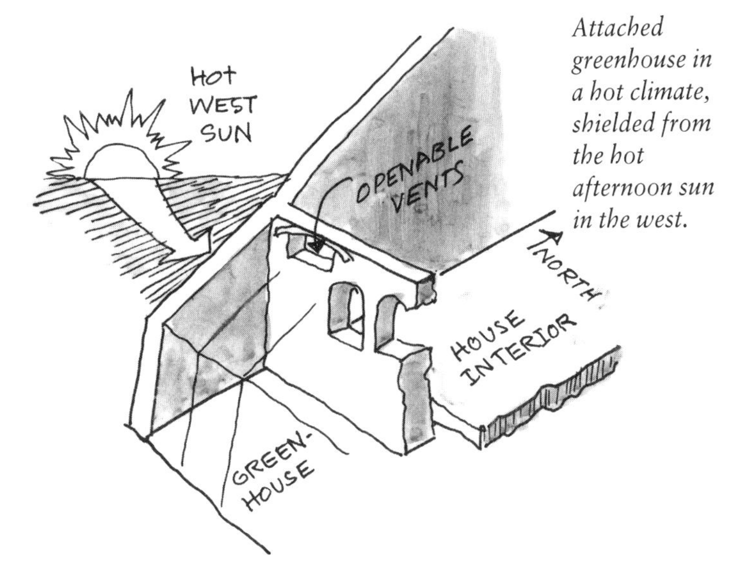

Attached greenhouse in a hot climate, shielded from the hot afternoon sun in the west.

Cob wall to left, grapes overhead, glass wall of cottage ahead.

Simple structures create "outdoor rooms," extending the season for being outside.

perhaps not ideal for a study, the greenhouse space can be wonderful for a bath, a cooking area, or a hangout.

Create some *outdoor suntraps* (secluded places warmed by the sun in cool weather) by building south-facing arcs of cob wall to absorb and reflect solar heat. Plan for a morning coffee suntrap outside on the east side which will also be the first place to cool in the evening, so in hot regions put your hangout place there too. Remember that warm air rises, so you will want to be in a low and shaded place when the weather is hot, a high and sunny one in the cold of winter. Watch cats—they find the first dry place to warm up in the morning; they're snoozing on sunny days in winter right where you should put that comfy wooden bench.

Also remember those cold-sensitive plants you'd love to grow; plan some suntrap space for them, warm corners facing southwest for fig trees in cold climates, close against the thermal mass of a thick cob wall. In very hot climates save the coolest, dampest spots to grow lettuce and those flowers that just fry out in the heat.

Study summer breeze patterns. In hot zones, don't block the *breezeways;* these are great spots to put seats or a clothesline.

Morning suntrap.

Orient to Earth's Tilt and Spin

Here is an exercise to help you locate your daily home life in relation to the movements of the Cosmos, to place the different activities in your new home such that your daily life is synchronous with the tilt and spin of the planet. This will help you describe graphically your daily relationship with sunlight, to provide light exactly where you need it and reduce your utility bill. This technique helps you develop a solid rationale for designing and locating spaces in a house. Your life will resume that contact with season and time of day that we all had, once.

On a big sheet of white paper, draw a circle with the points of the compass. Get each member of your household to draw one. Superimpose the 24 hours of the day. Now you have a map of where the sun shines from when, a quotidian cosmic clock. Other than at due east and west, and for a while around the summer solstice, this diagram is quite precise. Now chart the times you are occupied in daily activities in your house. The result might look like the following digram.

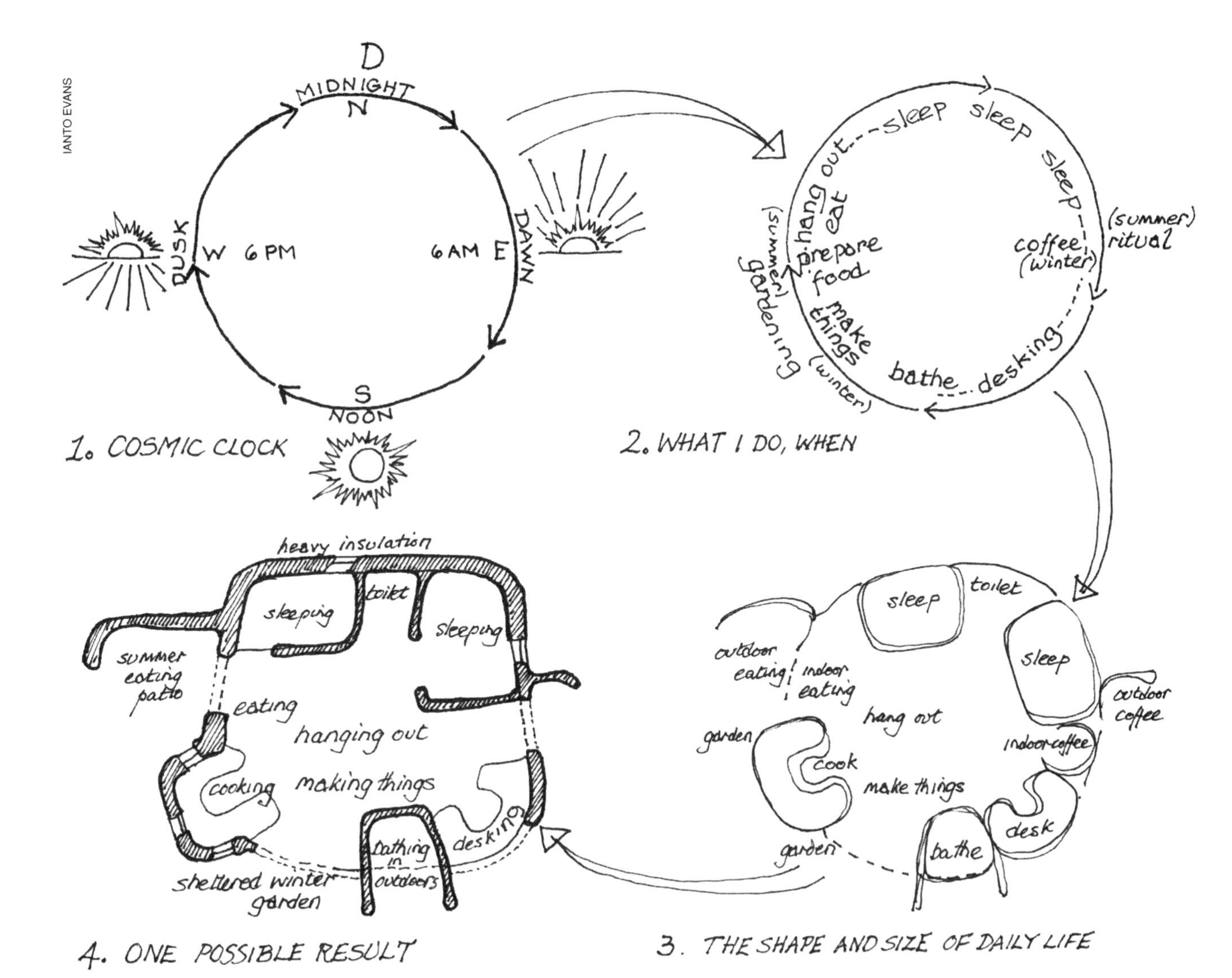

When you have the shapes and volumes of each dance of your home life plotted out, turn them into paper cutouts or 3-D clay models and play at arranging them.

Next move the models or paper cutouts to correspond with your activity clock. In the example you see an average day in the life of someone who sleeps from 10:00 P.M. to 6 or 7:00 A.M., makes a ritual cup of coffee, then does desk work all morning. The spin of Earth dictates that the desking place be in the southeast of the building, the bed in the northeast (it's important to have eastern light at your bed to wake by) and a hangout space in the west–northwest. Food prep seems to be mostly early evening, in the southwest, and it would be perfect if the garden were attached to the southwest side, so you could step out to pluck and snip things for supper, but also so you could sit outside in the afternoon sun of winter, late fall, or spring.

MAKE A MODEL

Two dimensionality is an illusion we can entertain only because we're sighted. If we were blind, we would have no such illusion; we would know that real space is all 3-D. Most of us have poor ability to translate two-dimensional drawings into imaginary spaces in their full volume and substance, so keep away from flat plans until you have decided volumes you want to create. Otherwise you many end up with dead-feeling spaces you then feel obligated to decorate in order to give them life.

So, don't start with paper; begin by making a model to contain the shapes and the volumes that your body occupied in your acting-out dance. Your model can be conceptual, exploring the shapes and relationships of spaces, without precise dimensions; or can utilize a fixed scale beginning with a model of yourself, for example, using an inch to represent a foot. It can even be a full-size walk-in of sticks, string, and cardboard to try out the volumes and surfaces before you commit to building a permanent structure. Best to do all three.

My favorite way to make a model is using sticks and cob at about one inch or two inches to a foot. I'll walk you through creating such a model.

In the Onword, Linda explains how to make an *intuitive* model, using the starting point of meditation on your childhood "magic spots." You might want to start with her exercise, to loosen up your hands and inner space.

You will need: materials for cob (fine sand, clay, fiber); small pieces of glass or transparent stiff plastic; sticks the right size for beams and rafters, lintels, and so on; pebbles to represent foundation rocks or paving; and cardboard or stiff paper for floors and roofs. As a base for the model, use a really rigid piece of board, two to four feet across, depending on the size and scale of the building. The board should be big enough to leave space all around your model—you will want to examine how the building interfaces with the surrounding land—yet narrow enough you can get it through doors. Set the board down on a level surface where you won't have to move it often.

Choose a place to build where sun will reach your board, ideally in an open-fronted garage or on a porch facing south. Keep south to the south; mark clearly on your board where either north or south is. Check it at noon by the Sun or at night by Polaris. As the model advances, you can tilt the board, raising up the north edge to simulate summer, raising the south for winter. By calculating the angles you can simulate how the noon sun will fall on your building in both winter and summer and can adjust window placement, the length of roof overhangs, and wall heights.

Using clay or modeling clay, make several scale models of yourself and anyone else who will use the building often. Give yourself a fixed scale to work to: if you like feet, go an inch to a foot or half an inch to a foot; in metrics 1:10 or 1:25 works well. Measure your little people carefully or the building will fit only giants or dwarves. At one inch to a foot, I am 5½" tall, Linda is 5³⁄₁₆", Michael a full 6". Make one standing, one sitting, and one sleeping model of each person. You will sculpt spaces *around* these little people, so don't start shaping the building first or you will find you're designing arbitrary spaces that don't fit their needs. Try to inject your personality into your model.

On the rigid board, pile up earth or sand to the shape and slope of your building site.

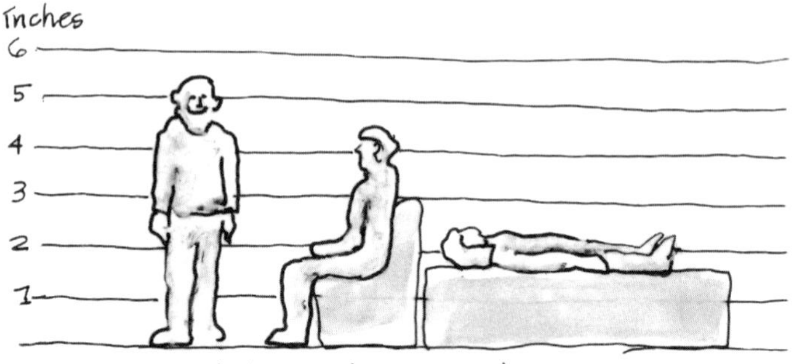

First... sculpt people to scale. Shape them all standing, then bend one to a seated position.

Make sure to model permanent features more or less to scale. Measure to the nearest couple of feet any trees, other buildings, roads, water, abrupt level changes, exposed bedrock, and so forth. Represent features with materials that feel real. Certain kinds of bushes make great ½-inch or 1-inch scale trees, rocks make the best rocks, a mirror makes a wonderful pond. Measure the distance between these features quite carefully, because that measurement determines the limits of where you can build.

Now, using the verb list of those activities you want to enclose, begin to create enclosures. The material you model with will influence those spaces and determine their shapes somewhat, so we like to use very fine cob, with fine sand and fine-chopped straw or old horse dung or cow dung for fiber. Build your model walls as you would build your house walls, little by little, letting them dry. You can move them around, changing shapes and sizes. Avoid building too high, too wet.

It's easy to get so involved in making your model really cute that you miss out on using it as a tool to explore different options. Try building three or four quite different models, all to the same scale, then you can compare. Half an hour each is more than enough time. Force yourself to stop, move on to the next board, and build another one.

The closer the modeling material is to the reality of what you will use for an actual building, the fewer mental gymnastics you have to go through in imagining the final result. Make the roof to scale and fairly realistic, but removable. Then push a small mirror on a stick through a window or door so you can see back out, to get an idea of what the windows will look like and how interior spaces will be. For a smaller model, a dental mirror works quite well.

Sometimes it's worthwhile to make a fast model, at a quarter- or half-inch scale, then a much more detailed one at an inch, after some overall decisions are made. You may even want to model detailed parts at bigger scale—the kitchen at 2 inches to 1 foot, for instance.

Once you think you've got it right, relax. In chapter 9 we'll explain how to turn your model into a full-scale, on-site mock-up.

Make an Outline Plan

Using the finished model, make a two-dimensional scale plan the same size. Your plan should be drawn on graph paper at an easy scale to upsize, such as ½ inch = 1 foot or 1 inch = 1 foot (in metrics, 10:1 or 20 or 25:1) the same scale as you made the model. Measure the dimensions of the model carefully and mark them on the plan, first in pencil, then pen them in when you're sure they're right. Doorways, drains, exterior walls, and level changes are all critical. It is possible to lay out a building from a model, but a paper plan is much more convenient to transport. Make several copies of the plan, cover one with waterproof transparent plastic, and you're ready to lay out the building.

ADJUSTING THE TIME-DEPTH

In this culture of the ephemeral, we crave stability and revere anything that has time-depth, the resonance of age. People feel comfort not just in timeless materials, but in an overall feeling of rootedness in buildings, even new ones. Most new buildings lack connection to time and place. They display no visible history.

Our senses react upon seeing, smelling, hearing, feeling the characteristics of a new building, probably through a lack of genetically encoded signs of familiarity and comfort. The genetic response to new conditions is alarm. ALARM! What our subconscious tells us is that here is an untested proposition. We are on instant alert, ready for flight or fight.

No wonder people act bizarrely in airports and fast food joints; they're utterly disoriented by the newness and obvious unnaturalness of the place, which isn't rooted in time, so is confusing and psychically uncomfortable.

In creating a home, alarm responses are incompatible with the snug, nurturing, secure atmosphere we need. The alarm stimuli in most conventional buildings come from several sources, chiefly unnatural geometries, chemical smells and mechanical noises, raw new materials and a refusal to let them age, plus the building not being obviously connected to the ground beneath it.

We need to bring a new structure quickly into alignment with its temporal, spatial, and ecological surroundings, so that we subconsciously recognize continuity of history and natural cycles of time. Here are some tricks to help your new house feel solid and permanent, even early on.

❧ ***Welcome the olfactory landscape.*** The fragrances and aromas of the region and immediate environment are important to our feeling connected to the place we're in. They need to be allowed to enter your building. Three ways to achieve this are: (a) keep new, toxic, manufactured materials out, so they don't mask local, timeless, comforting odors; (b) invite the natural smells of the place into the house by having many openable windows or by locating close to an archetypal local smell that you like (e.g., the front door beside a bay laurel tree, the smell of sun on rock, sassafras growing just outside the window, wild mint in the path to the kitchen); (c) use local aromatic materials in construction (unvarnished red cedar in your shower, exposed underthatch showing in the ceiling, an orange tree growing inside the building, flowering sporadically all year).

❧ ***Embrace landscape rhythms.*** Look closely at the shapes and rhythms of the surrounding landscape; analyze them, then try to follow them in the overall form of the house, particularly the rooflines. This works equally in town. Try to keep the roofline of your building, as viewed from a distance, below the skyline. It gives a subtle feeling that Nature still dominates.

❧ ***Build in slow phases.*** The smaller you build, the easier the building will be assimilated. Start very small then adjust the plan as the site speaks to you, adding small sections only as you need them over several years. That way, too, only a small part of the ecosystem is affected at any one time, giving a chance for other parts to repair themselves.

❧ ***Shape the site as you build.*** As you build your house, make the necessary changes to the surrounding landforms so that your house continuously fits in. Don't wait until the house is done. By changing the site synchronistically as you build, you will be less

Blending into the landscape: try to build a house so its roofline conforms to local land forms.

likely to make drastic errors, you'll take more care of the surrounding land, and the building will grow with its environs. You can use the earth from leveling floors, patios, or paths to make your cob walls. By the time the building is complete, the changes will have settled in and mellowed, trees will have matured, vines will be growing.

❧ ***Plant perennials.*** Plant perennial vegetation right up against the walls as soon as you can—even before the building is complete. Espaliered fruit trees, vines on trellises, ivies climbing the face of the building, all physically connect the structure to the ground. Seed the cracks in outdoor paving, right to the door. If you're not ready to plant frail young trees in a construction zone, or you haven't completely worked out what will go where, haul in trees in tubs, for the time being. For dramatic summer growth, hops planted close against the building will grow twenty feet the first year, and pole beans shoot up seasonally. For rapid fruit trees, peaches get the prize; in western Oregon we can get seven feet of growth the first season just by directing kitchen graywater onto their roots. Bamboo provides tall instant evergreen cover.

Limit how close you grow plants that will splash rainwater onto walls, in particular where the wall can't easily dry out. In New Zealand we saw a 19th-century cob sheep shearer's hostel at the point of collapse because of close shrubbery on the shady side.

❧ ***Use visual anchors.*** Deliberately create visual anchors that hold the building to its landscape. Snuggle the structure up tight against something old—an older building, a giant tree, a huge rock outcrop, perhaps a creek bank or a change in ground level.

In town, a wooden fence attached to the building, matching local fence styles and complementing the building's materials, connects your new place to another fence or building. In the country a connected cob garden wall works wonders—a physical extension of the building out into the surrounding ecology, like the buttressed bases of ancient cedars in the Oregon rain forest. If your site has exposed natural rock, a foundation of the local rock holds the building *down* visually; brick has a similar effect in a neighborhood of brick homes, streets, or walkways.

❧ ***Choose materials that are aged or that encourage aging.*** With time, building materials naturally age and decay, providing a foothold for new life. You can accelerate that process by choosing aged materials, either pre-used or with the signature of life written upon them. Repairing gaps in drystone stock walls in Wales, masons were taught to face

"I DESIGN OLD BUILDINGS"

ONCE ON A FLIGHT from Los Angeles to Reno, I sat by a nice lady with dyed hair, butterfly sunglasses, and two-inch fingernails. She was a realtor in Las Vegas and spent most of the flight chattering on about the New Property market. My head was unable to process the clutter of numbers, sales jargon, and immense dollar figures. Finally, as we came in to land she asked, "And what do *you* do?" Without thinking I said, "I design old buildings." Her eyelashes didn't even move.

The purpose of *faux pas* I suppose is to help us reflect, and I did. In a way it's true. The buildings I help design often look ancient in a year or two. Within a year, visitors to my own house were asking whether it was from before World War II.

Attached fences visually connect separate buildings and the landscape.

Cedar shake roof by Sun Ray Kelley, in the Washington rain forest. Old-growth moss on old-growth shakes.

the wall with 10 to 20 percent of mossy rocks; that way the repair job disappears. To blend a new section of tile roof to the old, using 20 percent of your tiles from an older building gives the whole roof time-depth. Seeing random lichen patches, though they may be on just a few tiles, prevents the subconscious from registering alarm at a completely new situation. To encourage a rapid growth of the crustose lichens that colonize rock outcrops, new slate roofs were historically painted with a mixture of sour milk and cow dung, the milk to form a glue and the dung to provide nutrients and trap moisture. Lichens then get a jump start on colonizing that hard impervious surface.

Where mosses are part of the local flora, providing suitable substrate (cedar shakes, for instance) allows moss to rapidly colonize a roof—first the coolest dampest parts, then gradually spreading out across drier zones. Sun Ray Kelley, creating natural buildings in the rain forest of Washington state, splits shakes for siding from decaying cedar stumps, then attaches them frayed ends outward, exposing where fungus has eaten away the wood. His buildings quickly attain a feeling of complicity in a natural process instead of having pretensions of immortality. "But won't it rot my roof out?" Yes, but your roof will rot anyway. Why not help decomposition to proceed gracefully and enjoy forty years of a magically green softness instead of pouring on chemicals, weakening the roof by walking on it, and ending up with a biological desert? Most recently, Cob Cottage has experimentally built living roofs of EPDM with local moss spread across it. See page 237.

War plane hangar in England. Fifty years later, still covered with grass and evergreen bushes.

Living roofs can green up quickly, ground a new building within days. To ensure greening, seed them with quick-sprouting plants and water often until they're established. When threatened by German bombers, the British in World War II rolled sod out over their aircraft hangars, making them almost invisible from the air. Some are still there, still invisible.

In conspicuous parts of the building, display built-in features that are obviously old—surf-battered driftwood for the front door lintel, worm-eaten with rusting hardware sticking out; a threshold of 19th-century cast iron; or hand-hewn beams over your dining table, recovered from an ancient warehouse. Reuse carved signs, old brick with the imprint of the maker's trademark, idiosyncratic relics from preindustrial times. Put them at eye level, on corners, at the welcome threshold, anywhere visitors will wait alone. Even the nail-stains on secondhand siding or residual paint on an old door all contribute to connecting with the past.

BUILDING THE COB COTTAGE

II

BY MICHAEL G. SMITH

ROB POLLOCECK

Michael's Story

COURTNEY ROGMANS

I GREW UP IN A CONSTRUCTION SITE AND learned some basic carpentry skills at a young age. My father, a renegade architecture professor, spent the first ten years of my life building a large and complicated addition onto our conventional stick-framed house, using almost entirely found and recycled materials. Our weekly ritual was taking the family station wagon to the local dump and bringing it home loaded with discarded windows, doors, furniture, and machine parts, many of which found their way into the Frankenstein's monster of an addition. I grew up believing that it is possible to build for oneself with very little money.

Naturally, at college I stayed as far away as I could from Architecture. I somehow ended up with a degree in Environmental Engineering, although what really interested me were ecology, life sciences, and writing. After graduation, I wanted to flee as far as possible from urban academia. Through a professor, I got an introduction to a group doing agroforestry research in the Costa Rican jungle. I stayed in Costa Rica for two years, working mostly with a grassroots association of small farmers dedicated to rain forest conservation and sustainable economic development. One of our many projects was the construction of an eco-lodge in the jungle. The site was surrounded by swamp and only accessible over kilometers of narrow boardwalks, so all our building materials had to be either found on-site or schlepped in on our backs. We built the lodge using lumber sawn from fallen trees with portable chain saw mills. I was being introduced to the concept of building with local, natural materials.

During my stay in Costa Rica, a large earthquake hit the region where I lived, and 2,000 houses were destroyed. At the same time, my father was designing a second home for a wealthy couple in Massachusetts. I wrote to him that the price tag of that single house (for two people who already had a house) would have covered all the materials necessary to rehouse the 2,000 families left homeless by the earthquake. Although I had gone to the jungle with the naive ideal of helping to stop deforestation, I was starting to see that the greatest threat to that precious ecosystem was not local. Later visits to Mexico, Guatemala, and Nicaragua strengthened my conviction that among the strongest forces eroding the cultures and ecosystems of the "Developing World" are the consumption habits and economic policies of overdeveloped countries such as the United States. I decided that the best way for me to help the rain forests and the conserver peoples of the world was to return to my own country and try to educate people here about the global and environmental impacts of our consumer lifestyles.

Also while in Central America, I first heard about "Permaculture," a set of principles and tools for designing sustainable human habitats. First developed in Australia, but rapidly sweeping the globe, Permaculture offers an integrated, whole-systems approach to providing for human needs such as food, clothing, and shelter with the least possible resources and energy, leaving more space and a healthier environment for our nonhuman neighbors. I have been studying Permaculture

BUILDING A COB COTTAGE STEP-BY-STEP

EVERY COB BUILDING IS UNIQUE, but most require the same basic steps. The order of some of these steps is critical; others can be rearranged as desired. For a first-time builder, it can be very useful to map out the sequence of all construction tasks. Following is a sample sequence to help you think through your project to completion before you begin. The next ten chapters walk through each of these steps in great detail.

- Select your building site very carefully, for best exposure to winter sun and for good drainage.
- Decide in advance roughly what spaces, shapes, and features you will need, both indoors and out, based on the scale of your own activities. Make many sketches and models. Keep the building as small as possible. If you need more than 400 square feet, consider phasing the project to complete the most essential sections first.
- Work on a scale model of the building, including site features such as trees and slope.
- Decide the wall plan in detail, especially where doors connect the indoors with outdoors. Design the roof at this point, including the roof and how it is supported. Make a full-sized mock-up of the building on the site. Spend time in it, and imagine what it would be like to live there.
- Assemble on-site all building materials you will need for enclosure, including glass, lumber, rocks, hardware, pipes, wires, and so forth. Gather your tools, scaffolding, and water storage.
- Check your soil, and assess by making test mixes and test blocks what needs to be added for the best cob mix. Measure your proportions and calculate the amounts of sand, clay soil, and straw you will need. Plan out and prepare mixing spaces close to the building.
- Arrange delivery of sand, clay soil, and straw as needed. Store deliveries close to and uphill from the building, in locations that won't impede work. Make sure straw is stored under cover, off the ground.
- Stake out your design and finished levels precisely on the site using many strong, firmly driven stakes. Adjust for the last time.
- Clear and level as little as possible. Grade away from the building so that water runs away on all sides. Set aside excavated soil for gardening or making cob.
- Dig drainage and foundation trenches, lay in drainpipe, insulation, plumbing, and wire connections; backfill trenches immediately with drain rock.

ever since, seeking out practical experience with organic gardening, sustainable forestry, and appropriate technologies. Natural building is an obvious and important subset of Permaculture. Of course, none of these techniques ultimately can succeed alone in creating a saner, healthier future. They must also be accompanied by advances in global social and economic justice.

I met Ianto and Linda in the spring of 1993, as they were preparing to build their second cob cottage. What interested me most about their approach was that they weren't promoting cob as a universal solution, but as one tool in a larger, permacultural context. That summer, while waiting for each successive layer of cob to dry on "The Heart House," I spent many joyful hours with Ianto planting and harvesting vegetables in the garden, learning botany in the forest, and constructing stoves and solar ovens out of cans and cardboard boxes. In everything we did was a conscious effort to simplify, to use less, to spend less. Ianto's years of work in Latin America and Africa had given him a clear understanding of the costs of the consumer

- Build the foundation stemwall, setting in door frames or frame anchors and joists if floor will be suspended. Leave openings for utilities and water lines in and out.
- Tamp the subfloor, lay several inches of drain rock, and, if you will be using an earthen or other mass floor, lay the floor base coat.
- Frame the roof, either on permanent posts or on temporary posts that will be removed when the walls are complete. You may also wait until the cob is nearly finished before framing the roof.
- Experiment with cob mixing techniques—for speed, ease, enjoyment, and different numbers of workers. Concentrate on gradual improvements.
- Apply your mix, probably at first by fork, trodden, then using "Gaab cob" or cob loaves. Sew it all together solidly with your fingers or a "cobber's thumb." Try to raise all the cob walls at the same rate.
- Make sure you build vertical or tapered walls; don't leave big bulges or hollows.
- Pare each new part with a machete or handsaw before it gets too hard, leaving it ready to plaster as you build.
- Set pipes, wires, outlets, junction boxes, and so on into the walls as you build.
- Build in windows as you go. Bury "deadmen" in the walls, wooden anchors to which you can later attach door frames, shelving, countertops, and so on.
- Sculpt built-in furniture, bookshelves, niches, and alcoves.
- Let the walls dry and settle somewhat, then build loft or second-story beams, joists, and ledgers directly into the cob if the wall is load-bearing.
- Locate roof deadmen one and a half to two feet down from the rafters. Complete the walls.
- If the roof is on, you can now enclose and heat the building. If not, now is the time to build the roof. If the roof will be heavy, allow the cob to dry throughout first.
- Do final ceiling work.
- Finish interior built-in woodwork, counters, cabinets, interior door frames, and plumbing fixtures.
- Apply interior plaster.
- Lay a finish coat on floor. Seal an earthen floor with linseed oil and beeswax.
- When interior plaster is dry, apply natural paint, lime wash, or *alis*.
- Plaster the exterior, if desired.
- Have a big party! Invite everyone who helped you build.

lifestyle, as well as many ideas about how we can create a healthier alternative.

The next year I helped design and build another cob cottage on a Biodynamic farm just down the road. The owners of the farm understood the connection between food, shelter, and healthy forest management and wanted to develop a community based on all three. The two years I spent in that cottage taught me many lessons, including how surrounding myself with natural shapes and materials gives me a sense of calm groundedness and spiritual well-being. But the farm turned out not to be my permanent home. I spent the next several years on the loose, gathering skills, information, and resources with which to root myself permanently.

One of my main endeavors during that time has been learning, writing, and teaching about natural building as a whole system. In addition to cob, I have been experimenting with stonework, straw bales, timber framing, rammed earth bags, light straw-clay, cordwood, wattle and daub, bamboo, natural floors, plasters, roofs and insulation, and whatever else I can get my hands and feet into. I like to walk onto a site without preconceived ideas and select the most appropriate building materials and techniques based on availability, climate, site ecology and history, and the needs of the human users. My next goal is to start integrating these natural buildings into Permaculture designs including sustainable food production, water and energy use, and environmental restoration.

As of this writing, I believe I have found a place to realize that vision. Along with a group of other people with various skills, I am now living on a large piece of rural land in northern California. The redwood forest, meadows, springs, pond, and creek provide nearly all the resources we need, and I have already moved into a new cottage built of trees, straw, and mud. We have begun teaching on-site workshops in not only natural building and Permaculture but also organic gardening, herbal medicine, and primitive technologies. Our intention is to build a living, working demonstration of integrated sustainable living, with a strong experimental and educational component. Maybe I'll see you there!

Materials and Tools 8

BEFORE YOU BEGIN BUILDING, MAKE SURE you have read the sections on design and siting in part 1. Not only will they help you create a beautiful, functional house, they could save you months of work and thousands of dollars.

This chapter describes the qualities you want in your basic building materials—earth, sand, and straw. It explains where to find them and other useful materials cheaply or for free. It also lists the tools you will need, and concludes with our colleague Jan Stürmann's poetic essay on working without machinery.

❧ THE EARTH OF YOUR WALLS TELLS THE HISTORY OF THE LAND—THEY GREW OUT OF THE COLOR, TEXTURE, AND CONTENT OF YOUR SOIL, CREATED AND LONG HIDDEN UNDERGROUND. BUILDING YOUR COB HOUSE CAN BE LIKE TAKING A TRIP THROUGH TIME. ❧

RAW MATERIALS AND WHERE TO GET THEM

Wherever you decide to build, whatever kind of building you need, the ground beneath your feet is a supermarket of free natural building materials. Sand and clay underlie the surface almost everywhere; it's hard to find a place in North America where one or the other isn't abundant within half a mile. They will be lacking only on extensive good agricultural silt or peat soils, but in general we shouldn't build on those soils anyway, as they're too valuable for feeding us.

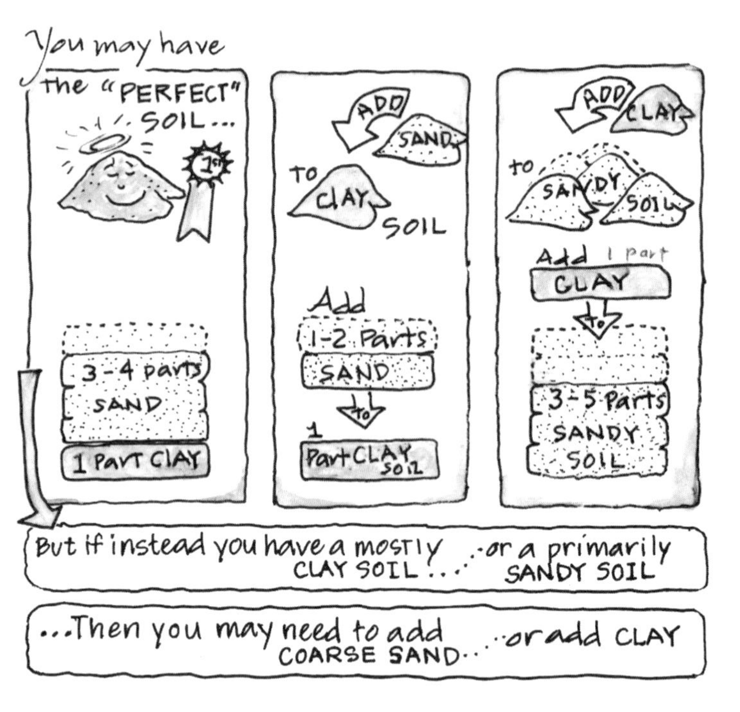

You will seldom encounter a soil that needs *very* much amendment, and with great luck you might have the perfect soil, right there in the ground, ready to mix with straw. These perfect building soils, which we call "ready-mix," naturally contain between 10 percent and 25 percent clay and a high proportion of coarse sand. If you find yourself in this enviable position, mixing cob couldn't be easier. You will be saved the expense and the work of hauling in sand or clay. Ready-mix makes building even on remote, walk-in sites easy. Normally, however, you should expect to bring in some additional clay or sand. You will always need plenty of long, strong straw.

For general building, the ideal mix is heavy on coarse sand and light on clay. It contains just enough clay to stick the sand and straw together—something like 3 or 4 to 1, sand to clay. Most clay soils are a complex blend of sand, clay, and other impurities. You need to understand the composition of your soil in order to decide how to build with it.

Composition of Soils

Most soils have a top layer (usually a few inches to a foot or so) containing life—small creatures and plants, roots, and dead organic

THE GEOLOGICAL HISTORY OF YOUR SITE

BY IANTO

STONE BUILDINGS allow us to see a section of the geology beneath our feet set upright as a monumental reminder of the basic forces that created the planet. Unplastered cob is similar, with its fundamental secrets revealed. In the same way that growth rings in exposed end grain tell the tree's story, the earth of your walls tells the history of the land they grew out of—the color, texture, and content of your soil, created and long hidden underground. Is the sand rounded or crushed, granite or pure quartz? Do you have red or white or yellow clay? What minerals make the mixture look, feel, and taste the way it does? Sometimes distinctly different clay colors can be used to great effect, as built-in red, white, yellow, or blue stripes in the cob. You can incorporate these colors in the walls, leaving irregular boundaries between them. Building your cob cottage can be like taking a trip through geological time.

Chocolate and cream. Different clays from the same site in Texas tell the geological story.

material known as humus. This layer of *topsoil* is usually dark-colored, in abrupt contrast to the primarily mineral *subsoil* beneath it. You may read elsewhere that you can't build with topsoil, but we have not found that to be the case. Topsoil *is* a valuable and scarce resource, though, so if you have a choice, save it for the garden and build with the poorer subsoil beneath it. In addition to living creatures, plants, and humus, soil contains varying proportions of stones, sand, silt, and clay. Most soils contain a mixture, even if they look like pure clay or pure sand. These are the basic particles involved.

Stones and Gravel: We will define these as any pieces of rock bigger than a pea. Gravel and small stones are a major ingredient of traditional cob. Because they can't be easily crushed, they make a good mix with excellent compressive strength, but it's not so enjoyable to tread a rocky cob mix with your bare feet or build with it bare-handed.

Sand: We'll call "sand" all particles of rock from ¼-inch down as far as you can see the individual grains. The best cob contains a proportion of really coarse sand (⅛–¼ inch). Sand is hard, inert, and very stable. Individual grains neither absorb water nor shrink on drying, nor do they expand much with heat.

Silt: Silt is tiny sand particles, too small to distinguish by eye. Avoid soil with more than a small proportion of silt, as silt dilutes the clay's stickiness yet unlike coarse sand is not strong in compression. Often the best agricultural soils are silty; they make the worst cob.

Clay: Although clay is also a very fine-grained product of rock decomposition, it is chemically quite distinct from silt. Clay is a series of tiny microscopic plates, little flat wafers of hydrous aluminum silicates, held together by chemically bonded water molecules. Clay's propensity to bond with water makes it sticky and causes it to shrink as it dries.

Clays shrink on drying between about 5 and 15 percent linear, thus they tend to crack if not mixed with enough aggregate. As clay dries in the spaces between coarse, jagged sand grains, it shrinks tight and locks the sand grains together. Clay may also bond chemically to minerals in the sand and to

Soil cross section.

straw. The result, when dry, is a material of surprising strength.

Determining a Soil's Suitability

Go to your building site. Take a shovel and several clear glass jars with tight lids and straight sides. Quart jam or canning jars work well. You'll need about a gallon of clear water and a couple of ounces of salt or liquid soap to make what is called a "shake test." The shake test can tell you easily how much useful sand you have and the proportion of silt. It will also detect the presence of clay, even in small proportions.

Dig holes where you think you might extract building material. Take samples of any soil you may want to use, crush these samples very fine, and place each in a separate jar, between a third and a half full. Fill the jars nearly to the top with water and add about a heaped teaspoon of salt or liquid soap. The salt or soap will accelerate the clay settling out. Shake long and hard. If there are hard lumps, let them soak for an hour or two, then break them up really well. Otherwise, even with energetic shaking, lumps of dry clay may remain in little balls, which can look deceptively like coarse sand. A couple of marbles or little round stones will help crush them as you shake.

Using soil samples and a jar, make some "shake tests" to see the composition of your local earth.

At the end of shaking, suddenly stop moving the jar. The soil will have been broken down to its separate particles. The biggest particles will fall first, then the smaller ones. Useful sand will fall in 3–5 seconds; make a mark on the jar at that level. Fine sand and silt will fall next, for up to 10–20 minutes. If you leave the jar undisturbed, the clay will gradually settle, leaving clear water above it. Complete settlement may take hours, days, or even weeks, depending on the type of clay. Anything left floating on top of the water will be organic matter.

You will be able to see the various layers through the jar, although fine sand and silt both have grains too small to easily see. As neither is very useful to making the best cob, we lump them together as "undesirables." Often the clay and silt in a given soil are different colors, which makes reading the shake easy. In the event that both are the same color, make a mark on the jar where the sediment has settled 10 minutes after shaking. Anything below this mark is silt; above it, clay. If you have perfect cob soil, you will see a thick layer of coarse sand with a thinner layer of clay—little or no silt or fine sand.

The shake test won't give you an accurate clay proportion unless you dry out the sample completely; as long as the clay is wet, it will stay in its expanded, hydrated state. To dry it out, you can carefully siphon off the water and then leave the jar open to evaporate. We usually don't bother with this lengthy process. The fully settled shake test, even while still wet, gives enough information to evaluate the soil for building. We can easily see how much coarse sand we have to work with, whether there is more silt than sand (which could cause problems), and whether clay is present. More precise information will come when we begin to experiment with a mix.

You can learn a lot from shake tests, but it takes quite a bit of experience to be able to read them accurately. Try practicing on different soils wherever you go, observing the differences in settlement patterns and correlating them with what you know and can observe about each soil. The variation in soils is sometimes astounding, even within a few yards. Compare, for instance, a vegetable garden's soil with subsoil from deep in a roadcut, with a quarry soil, or with soil from a river bar.

If your soil seems completely unsuitable for cob building (containing very little sand or clay), dig more test pits. Go twenty paces, dig a pit, go fifty paces, dig another, and so on. Dig where it will be easy to excavate, where a big hole could be an asset, in a place from where transport is easy. Dig where the shape of the ground changes, at a bench, a tump, a declivity. To find clay in sandy soils, dig deep in the lowest places. We have found good sand on glacially stripped islands and clay in the Sahara. In Indiana we found heaps of clay (overburden) near old marl pits; in Senegal, even at the edge of the sand desert, the spoil tips at wellheads were capped with clay the well-diggers had brought up in buckets from seventy meters down.

Any soil can be amended to make a building mix, though the economies of cob building rapidly worsen as more and more material needs to be hauled from greater and greater distances. In general, be cautious if shake tests show no clay at all or no usable sand. Sand is heavy and can be expensive to haul in, and it is the main constituent of cob by both weight and volume. If the shake test shows little coarse sand, yet the soil has more clay than silt/fine sand, it can make good cob provided coarse sand is locally available. However, if the soil lacks both clay and coarse sand, consider another building technique or site, or find a close-by source of better material.

In those few places where cob soils are *not* easily found, assess carefully what materials are local in that place and adjust your building plans to make use of whatever is locally abundant.

Sand

Natural sand is a product of bigger rocks (and sometimes shells) being broken smaller by the action of waves, glaciers, wind, or rivers. Also, sand is manufactured by mechanically crushing rock for constructing roads and making concrete. The less the sand is tumbled about and thus rounded, the more useful for cob. The best quality would be, in order: finely crushed rock, glacial sand, river sand. Sand produced by wind (in deserts or on dunes) or waves (on beaches) is generally too rounded.

The perfect cob sand is coarse (a size between big sugar crystals and split peas), clean (no impurities, such as silt), hard (crystalline grains, not shell sand), and "sharp" or angular (not beach or dune sand, which is rounded). The ideal is decomposed granite, eighth-inch crystals of quartz and mica, or sand the glaciers left behind. Most commonly, we use river

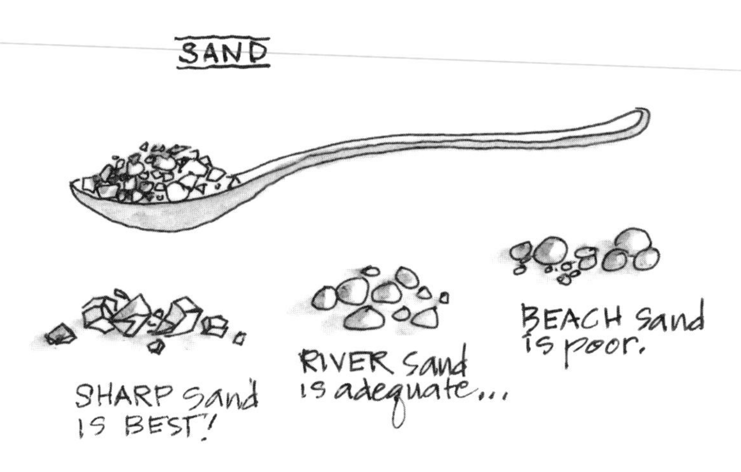

sand, which is slightly rounded but readily available in most places.

Where to Get Sand. Sand is generally inexpensive, particularly in large quantity, so you may choose to buy it. Consider the option of picking up the phone and ordering a whole truckload. Request "fill sand," "rough sand," or "concrete sand," the cheapest grade. If you order by phone, make sure you see a sample before delivery. Calculate in advance how much your project may take, then order a little more. A 9-cubic yard truckload in 2002 generally cost $80–300 depending largely on how far it had to be hauled. If you plan to dig your own sand instead, read on.

When prospecting for sand, try river or streambanks (especially below rapids), glacial ridges, moraines, or roadcuts. Ask locally. Sometimes there are old abandoned sand quarries with easy access. Sometimes sand is revealed by construction projects and you can arrange to get some. And in glaciated country (most of Canada, as well as northern and mountainous U.S.), good sand is sometimes mixed with clay—an ideal combination. Hunt around. If you have no alternative but to prospect on the beach, seek out the coarsest sand and wash it with fresh water before use, to clean out salt or mud.

Sand finer than coarse sugar makes less sturdy cob, so it should be saved for nonstructural detail work, window cob, or plasters. Most sand will be hard enough to make an adequate mix, but beware of sand from very soft limestone, shells, or decomposed mudstone (which can look like sand grains but actually is tiny balls of clay).

To economize on sand, traditional English cob incorporated shattered shale called shillet, but this was trodden by hoof not foot. Sometimes small gravel can be substituted for some or all of the sand. In the Colorado Rockies sand was hard to find but clay and gravel were found together, so we used a ready-mix of grape-sized gravel with clay packed between. The gravel was hell on bare feet; we mixed by driving a truck back and forth over the mixture, then trod it onto the wall in boots, but this made a great building material.

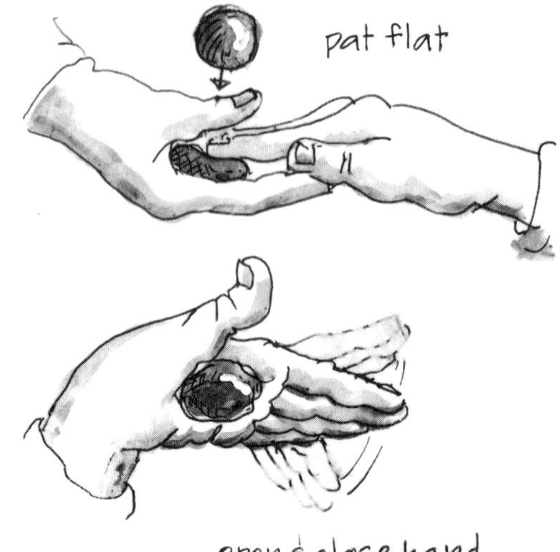

Clay

While identifying sand is easy—it is hard and crunchy, even when wet—recognizing clay can be more difficult. Clay is both the most critical and the most complicated element in the cob mix. Earth builders should learn to differentiate clay from silt, peat, and organic slime, which can all have similar slippery consistencies when wet. Several characteristics can help you recognize clay:

❧ ***First, its stickiness.*** To test whether a soil has a high clay content, make a paste of it by adding water, put a dab between your thumb and index finger, and squeeze. If your fingers stick together at all when you try to open them, the soil has clay in it. Then make a really wet ball the size of a golf ball and flatten it against your palm. Turn your hand

HOW DURABLE? HOW TALL?

Medieval cob houses in Devon (left) and Yemeni cob skyscrapers.

RIGHT: *Permaculture Institute of Northern California's office, by Penny Livingston, James Stark, and Cob Cottage Company.*
BELOW: *Hilde Dawe's house in British Columbia, built by Patrick Hennebery.*

ABOVE: *Cottage in Buda, Texas, by Gayle Borst with Ianto Evans.* LEFT: *Linda and Ianto's Heart House.* BELOW: *Lucy Lerner's cabin at 9,000 feet in the Colorado Rockies, stuccoed with local clay.*

INTENSELY SCULPTURAL

Below: *Sun Ray Kelley's studio, Washington.*
Right: *Jan Stürmann's house in Massachusetts.*
Below, left: *Massage studio by Miguel Salmoiraghi, Eugene, Oregon.*

DEIRDRE DOAN

The quality of light through narrow windows in thick walls. ABOVE: *House in Garberville, California.* LEFT: *The Heart House, Oregon.*

FAR LEFT: GREG LEHMAN; CENTER: ELKE COLE; LEFT: IANTO EVANS

FAR LEFT: SUSA ONATE; LEFT: ALISON FURNEE

WINDOWS, WINDOWS!

With and without frames, opening and fixed.

FAR LEFT AND LEFT: IANTO EVANS

RIGHT: IANTO EVANS; FAR RIGHT: JILL SMALLCOMBE

MALLEABLE, SCULPTABLE.

LEFT: *Interior, Buda, Texas.* ABOVE: *"The Spiral," a freestanding cob sculpture with limewash interior built in March 2001 by Jill Smallcombe and Jackie Abey. Devon, England.* BELOW: *Grand central staircase in new house by Clark Sanders in New York.*

CLARK SANDERS

You could begin with a courtyard wall or a bread oven. Walls in Cottage Grove, Oregon, by Cob Cottage Company; oven by Kiko Denzer. No permits required.

IANTO EVANS

FAR LEFT: LINDA SMILEY; LEFT: IANTO EVANS

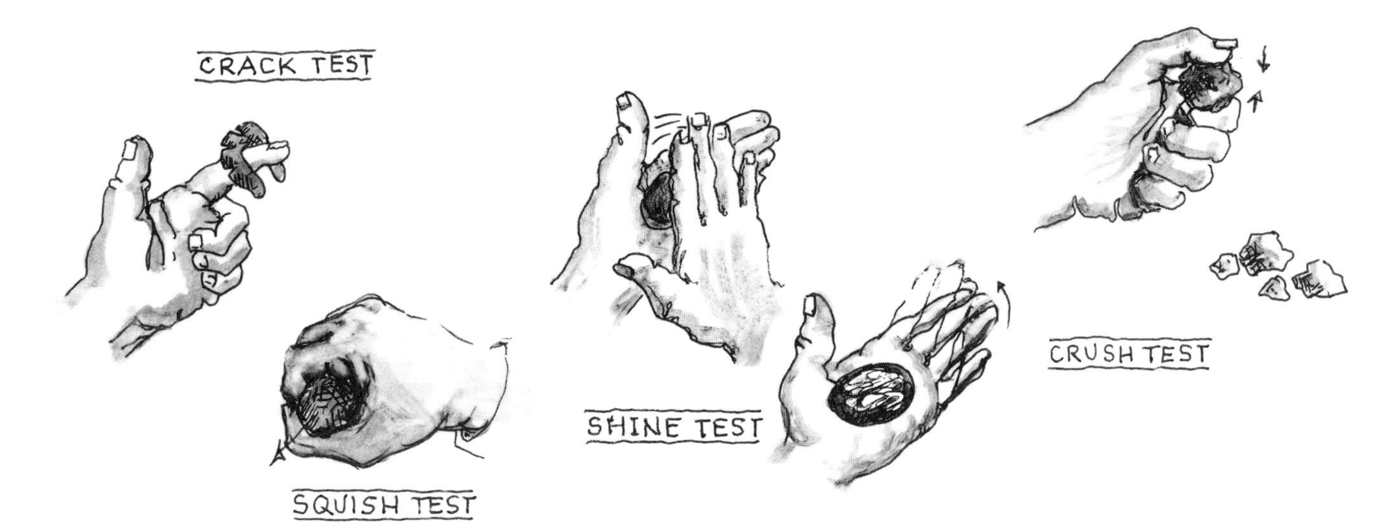

palm down and open and close it a few times. If you have a high proportion of clay, you should be able to open your palm at least five times before the pancake falls, and some clay will remain adhering to your hand; you can't see your palm.

❧ ***Second, its plasticity.*** Work a sample of the material to the consistency of modeling clay. Try rolling some out between the palms of your hands into a rod the diameter of a pencil, then bend the rod around your finger. The more clay the sample contains, the further it will bend without breaking. Potters use this test to evaluate clay soils for ceramics.

Then make another golf ball–sized lump, this time of really wet dough consistency. Hold it tightly in your fist and constrict the fist, leaving a small gap only between thumb and forefinger. Soils high in clay will suddenly extrude a ribbon of wet clay through this gap. If there is a very high silt content, you will be able to squeeze water out, which will drip from your fist.

❧ ***Third, its shininess.*** Take an egg-sized ball of very wet material and flatten it. This time pat it repeatedly in your palm until surface water shines. Watch carefully as you open and close your palm, only about a quarter closed. Clay retains shine as you stretch open; silt goes dull as the water sinks in.

❧ ***Fourth, clay clumps and hardens when it dries.*** If a soil crumbles easily between your fingers when dry, it probably doesn't contain much clay. Dry soils with high clay content are difficult to dig through; they break up into lumps that are impossible to crush in your hands and which sometimes feel nearly as hard as rock. But clay always softens when soaked.

Where to Get Clay: If clay is missing within wheelbarrow distance of your building site, look farther away, at roadcuts, recently dug ditches, old quarries, abandoned reservoirs, or anywhere construction is about to commence. Excavation contractors are usually delighted to get rid of clay soil, particularly if you show appreciation in the customary way. Keep on hand a good supply of cheap beer, and be prepared to make a donation to help things along. For a case of Blitz many a load of unwanted clay found its way to where it would be welcome, rather than ending up in a landfill.

When you go hunting for clay soil, it's useful to be able to recognize it from afar. From a distance, clay often has striking colors such as red, ocher, or blue-gray (though bright color doesn't guarantee that it will be clay). We've also found clay deposits that yielded green, mauve, rose, white, yellow, black, and many

Examine the exposed soil under a blowdown tree to see if there is clay beneath the topsoil.

other colors. We carry a bucket and a shovel when driving through the country in case we come upon spectacularly colored clays, which we use to make colored clay slips and finish plasters (see chapter 17).

Most clays are impervious to water. Best hunting is anywhere water comes up to the surface, particularly springs on gentle slopes. Look for natural springs or vegetation indicative of groundwater, and dig there. Watercress, skunk cabbage, rushes, sedges, most kinds of mint, and some species of willow are all good indicators. Remember the old English ditty: "Sedges have edges and rushes are round and grasses are hollow where willows are found." They are all plants of impeded drainage, often indicative of clay deposits.

Slippage and slumps may be indicators of clay. Check for firm or hard clumps of earth.

Sometimes water-borne clay has settled out long ago, in old lake beds, for instance. These places may be swamps now, with good blue clay buried under peat or silty soil. Excavating in a swamp can be a laborious mudbath, and you should be sensitive to the potential environmental impact of digging in wetlands. Swampy woodland can also be a clay source. Anywhere a large tree has blown over, the subsoil will be revealed, both in the pit the root ball leaves behind, and attached to the roots themselves. Transport from within dense woodland may be tricky, but inspection of blowdowns provides good clues for what to expect in the surrounding area where access may be easier.

Check that section of your dirt road that always gets slippery after a rainstorm, and the places where puddles appear first and disappear last. During dry weather, cracks and fissures in the ground are an indicator of clay soils. In flat country, if there are stock ponds there must be a reason the water stays in them. Check around the banks.

Cruising hilly backroads, be alert for such road signs as "Bump" or "Sunken Grade," or for places where the hillside above has slipped down onto the road. Slippage is a good indicator of a clay deposit. In dry weather, carry water; often the clay in a roadcut dries up into little crumbs that from a moving vehicle look like gravel. Take a handful, pour a little water into it, smush it up in your hand. If it's sticky it will probably work well in a cob mix.

THE ISLAND WITHOUT CLAY

BY IANTO

THE PROJECT was a sanctuary, to be built of all natural materials, on an island off the Canadian coast. To get to the remote site required three ferries, with several hours of driving in between. On small, heavily glaciated islands, the random distribution of glacial deposits sometimes means clay is hard to find, but it's almost always there somewhere.

Our first visit was to teach a week-long cob workshop with about twenty participants. We had suggested in advance the types of locales where clay might be found, but the crew had had difficulties. They had looked all over the island, asked old-timers, gone to the Highway Department, all without result. A whole island covered with glacial sand and gravel! Before we arrived, they had in desperation bought commercial ball clay for the project, dried and powdered, packed in 50-pound bags, hauled at great expense by truck and ferry all the way from the mainland.

When we got there, one of their first comments was that they'd had setbacks in building the foundation because of high groundwater.

"Groundwater?" I asked.

"Yes, flowing down the hill out of the forest."

"Even in this dry weather?"

"Yes, all the time, it's a real drag."

I took a shovel and we walked uphill through the woods. Some fir and maple, but mostly cedar and alder, trees of boggy ground. Skunk cabbage and surface water, a good sign. Within a hundred yards the skunk cabbage ran out and the ground got drier. Just below that point we dug down through the duff, cutting cedar roots, tossing out soft black organic compost. Water, dark and peaty, rushed in to fill the hole. At a foot and a half, a sudden change. The ground got harder, the shovel hit something resistant. Clean, pure, yellow clay!

Red cedar and skunk cabbage need year-round water for their rather shallow roots. On a slope, groundwater drains easily through sand or gravel, so finding a high water table is a sign of impeded drainage. What could impede drainage? Solid rock or clay, not much else. A spring will show where the impervious layer comes closest to the surface, which is why we climbed up to the *top* skunk cabbage before digging.

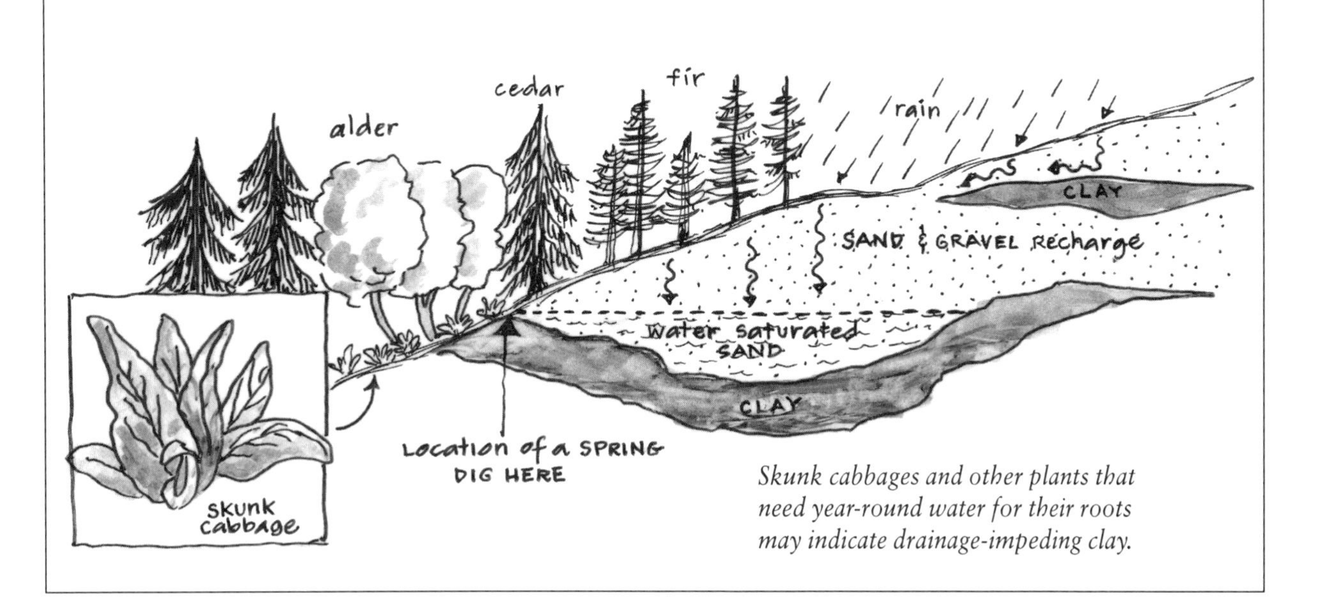

Skunk cabbages and other plants that need year-round water for their roots may indicate drainage-impeding clay.

Long, dry straw is the strongest.

Straw

In addition to cob's geological components, there is a biological one—straw. Straw lightens the cob and adds insulation by trapping air in its hollow stems. It gives tensile and shear strength by making a three-dimensional fabric within the wall. Reinforced concrete depends on iron bars—"rebar"—for tensile and shear strength; each bar is extremely strong, but they are relatively far apart. Cob relies on straw to do the same job; the large quantity and close spacing of fibers makes up for the fact that they're weaker than steel. Cob is a mineral-fiber composite in which integrally mixed straw gives cob walls some ability to move slightly to withstand ground settlement and other shear forces. It follows that to be at its best straw needs to have strong fiber and be in long pieces.

You might assume that the value of straw is only temporary, since the straw in an earthen wall will rot away over time. Our research shows otherwise. There is little available oxygen or moisture inside a cob wall, unfavorable conditions for the microorganisms that cause rot. Essentially, the straw is mummified inside the wall. After its initial drying period (which may take weeks to months), a cob wall maintains a fairly constant low humidity level. Research shows humidities of less than 5 percent are usual, even in the cold, rainy British Isles. In New Zealand we have broken into cob walls 150 years old and found the straw still yellow and apparently as strong as when the wall was built.

Don't confuse straw with *hay,* which is vegetation cut green to be dried for animal food. Hay usually contains flowers and seed heads. Straw is the dry leftover stalks of cereal grains. In general straw is biologically inert—very little nitrogen remains in it after the grain seeds have been removed, so it is less palatable than hay to insects, fungus, or bacteria. If possible, avoid making cob with hay.

After being fed through a combine that removes the seed heads, grain straw is commonly baled into two- or three-string bales or rolled up into 600-pound bedrolls. There are two main types of harvesting machine: a rotary cutter and a sickle bar. Rotary machines cut the stubble high and sometimes also chop the straw into short lengths, while sickle bars mow down the plants close to the ground, resulting in longer straw. Longer straw makes better cob, so get bales cut with a sickle bar if you have a choice. If you plan to use a lot of straw, try to find those giant 600-pound roll-bales—they are always long straw, and each one is equivalent to about a dozen 2-string or half a dozen 3-string bales.

When choosing straw for cob, check the strength of individual stems. Take one between your fingers and try bending and pulling it, testing for strength and brittleness. Then select five good stalks and, gripping them tightly between thumb and finger, play tug-o-war with a colleague; with the best straw it will be impossible to break them. The strength of straw varies a lot depending on how old it is and what variety of grain it is. The most important factor is to make sure that the straw is either very fresh or has been stored in completely dry conditions.

Where to Get Straw: If you live in a grain-growing region, straw will be plentiful and inexpensive. The cheapest way to get some is to drive your truck out into the field after the baling machine has done its work, and pick up the bales yourself.

Farmers sometimes put signs out advertising cheap straw bales, but be careful; you don't want straw that's been rained on. If you can't find a grower to buy from directly, try contacting a farm feed store that supplies straw for animal bedding. Tell them that you can use broken bales or otherwise undesirable straw, as long as it isn't rotten. As a last resort, go to an urban or suburban animal or livestock supply store, though prices may be much higher.

The cost of straw fluctuates dramatically throughout the year, according to supply and demand. For the best prices and quality, purchase your straw soon after harvest, which usually means between midsummer and early fall. If you wait until spring or early summer, you could end up paying several times more for the same straw, and there's a possibility it will have been exposed to water. If at all possible, *store straw indoors, off the ground.* If you store it outdoors, it must be up off the ground, well ventilated, and very carefully covered. Plastic tarps almost inevitably leak and cause condensation, so use steel or plywood roofing. Check regularly for dampness.

If straw is scarce or expensive in your region, be creative. Go to a stable and ask if you can haul away their used straw bedding (if it isn't decomposed). The animal dung will improve your mix. You might have to settle for hay (it's better to use dry hay then wet straw), or there may be other fibers that would work in a pinch, wild grasses you could mow and dry, chopped baling twine, or other industrial products and farm supplies. Traditional cobbers used whatever was locally available—in

MICHAEL G. SMITH

Mud dancers adding more straw toa cob mix. The pit was made by draping a tarp over four straw bales.

Wales, rushes, and in New Zealand, tussock, a wild grass. There seems to be room for improvisation, but cutting enough wild grass for a house is a daunting task if the alternative, baled straw, is available.

Oat, Wheat, Barley, and Rye Straws: Oat, rye, and winter wheat straw make excellent cob. Barley straw is okay, but not the very best, and can sometimes be brittle. Straw grown over winter ("winter" wheat is best) is stronger than spring plantings. Avoid "spring" wheat.

Rice Straw. Rice straw is long and very flexible. In northern California, where I live, rice straw is by far the best-quality straw available. It has good tensile strength but tends to clump somewhat in the mixing process, getting wrapped around itself like a hair ball. Some rice straw is hard and sharp, cutting your bare hands and feet. We have made excellent cob by using half and half, long rice straw and short wheat straw.

Rye Grass and Canary Grass: These are slicker and harder to work into the mix, because the clay doesn't stick to them as well. They are strong and flexible, but make poor substitutes for grain straw.

Other kinds of straw *to avoid,* based on our own experience and the reports of other cobbers, include *alfalfa, hemp, millet, milo, buckwheat, soybean, sorghum, flax, and corn.*

How Much Raw Material Will I Need?

It's useful to have a rough idea of how much sand, soil, clay, and straw you will need for your project. If you're buying any of these ingredients, you must know how much to order, and you need to make sufficient storage space for material piles. If you plan to dig soil from your site, it helps to know in general how big a hole you may have to dig and therefore what size pond or sunken patio to expect.

To calculate the volume of cob in your building, begin by measuring, on your plan or along the stakeline for the foundation, how long your cob walls will be. Multiply that length by the wall height and the approximate average wall thickness. Take into account both the wall thickness at the base and the amount of taper it will have (see page 191). Deduct a little for door and window openings, add a little for floors and plasters.

Meanwhile, you will have done a series of tests (the shake test above, and additional tests described in chapter 11) to determine the proportions of your building mix. To make 600 cubic feet of cob at 3:1 clay soil to sand, for example, you'll need 150 cubic feet of sand (about a 5-yard dump truck). Most soil expands on excavation, as digging will fluff it up with air, but then it will compress a lot, to less than the original volume, when stamped into cob. If you dig the 450 cubic feet of compressed soil that you need to make your 600 cubic feet of cob, you'll leave a hole of about 500 to 600 cubic feet. That 450 cubic feet of soil may measure 550 to 700 cubic feet when freshly dug. Remember, too, to plan on getting some soil out of your foundation trench—a two-foot deep trench contributes to at least two feet of the wall above it.

To calculate the amount of straw you will need, a good starting estimate is that straw will take up 10 to 15 percent of your mix (assuming that the straw has kept the same density it has in a bale). A two-string straw bale is usually 14 × 18 × 36 inches (around 5 cubic feet), sometimes longer. A three-string bale is 15 × 22–24 × 36–48 inches (roughly 9 cubic feet). Therefore, in the example above, you will need 60 to 90 cubic feet of straw for your 600 cubic feet of cob. This calculates out to 12 to 18 two-string bales or 7 to 10 three-string bales. Get extra! Extra bales are always useful to have on hand for temporary seating and scaffolding, not to mention their other uses as bale walls, as insulation, and as a major component of mud plasters and earthen floors.

Here are some figures that can help you out with ordering and transporting material. A cubic yard (known as a "yard" in construction jargon) contains 27 cubic feet (a cubic meter holds 1,000 liters). A "yard" of sand weighs about a ton and a half (a cubic meter weighs two tons). Dry clay is about a ton per yard. North American dumptrucks carry 5 or 10 cubic yards (of sand, for instance). A small pickup generally carries at least half a ton (1,000 pounds) comfortably. A two-string straw bale weighs 30 to 50 pounds, a three-string up to 80 pounds (this varies substantially according to the amount of tension on the baling machine). Any kind of truck can probably carry the weight of as many bales as you can load onto it.

NUMBERS ARE NOT SIZES

BY IANTO

A WORD ON NUMBERS and measurements, which can be intimidating. Numbers are one of the ways that professions maintain mystique and persuade us that we are impotent without their expensive services ("Professions are a conspiracy against the laity," said George Bernard Shaw). The building/permits/architecture/development industry is no exception. In creating natural buildings and especially building with cob, your best defense is to ignore numbers as much as possible. Here are some hints on how to do that.

Build to conform to your own body, anticipating the needs of those who will share the dwelling or visit you there. It is after all *your* body that the building will need to accommodate, and unless you are unusually short or tall, even closely fitting buildings will, like socks, fit most sizes of people. So use common sense about how high you build ceilings, how wide doorways are, the height of a windowsill or the thickness of a wall. If for some reason you need to standardize (let's say, to keep a consistent wall thickness among several cobbers), use a length of string, a stick, or, even better, an actual part of your own body to continuously check that standard measurement.

Remember you are measuring a *size,* not a series of numbers. Size needs no numbers to define it. Never forget that your purpose is to build *quality* not quantity and that, as my family motto says, "Better roughly right than precisely wrong."

To be able to coexist and communicate (roughly) with the numberholics, get to know the measurements of some of the key body parts you will use regularly in construction measurement. Here are some suggestions, from my own practice.

To coexist with numberholics, get to know the approximate measure of your stride.

I like to know the length of my stride. If I step out a little, it's about 3 feet (90 centimeters), loafing it's 2½ feet (75 centimeters), and a very long stride is 40 inches (a meter). It's useful, too, that a meter comes exactly to the top of my hipbone. If I stretch my arms out as far as possible, my fingertip to fingertip span is 5'6", same as my height (165 centimeters).

Measuring in relation to your body.

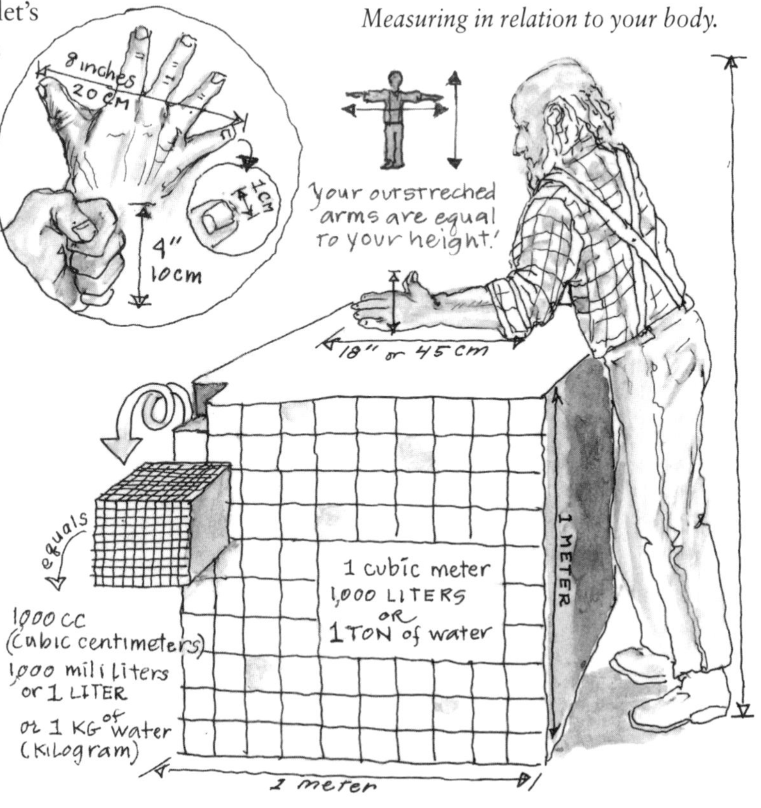

A starter cottage, one story, the size of Thoreau's Walden Pond cabin, with a 10 × 15-foot interior and 18-inch walls 6 feet high, would need about 400 cubic feet of cob. If Walden had been located on good stiff clay soil that needed a 50/50 soil to sand mix, Thoreau would have used 200 cubic feet of sand—about 7½ cubic yards, weighing about 11 tons. His neighbor's ¾-ton pickup truck could have carried it in about ten loads, or he could have phoned to have a dumptruck load of sand delivered. Sometimes you pay for sand by the load, sometimes by the yard, meter, or ton. I'm not sure how it worked in Walden, but he was a thrifty guy so he would have called around for the best price. For his cabin, Thoreau would have needed about a dozen two-string straw bales or half a dozen three-stringers. The borrowed pickup could carry the load easily, though perhaps he'd want about twice that amount, allowing for scaffolding use, and so forth. Better make it two loads, Henry!

FINDING OTHER BUILDING MATERIALS

Suppose you have a site ready but no money at all. How will you acquire what you need to build your house? Start collecting supplies now; it's never too early. As you prospect for clay and sand, look out for other materials you will need, particularly those whose absence might delay the job. You don't need a big truck or an army of helpers. When you run across a good rock, a little pile of lumber, a beautifully curved piece of driftwood, just load it in the car or carry it home. If you need to, borrow a pickup for bigger items, or, at an extreme, buy an old half-ton. Slowly you'll accumulate a stockpile of all the things you need for a wonderful building.

You will need places to store your collections, in some cases covered and dry, in other cases just open space. It's worth creating a rainproof tool shed right at the outset. Plan your supply piles carefully to minimize moving anything twice, yet keep everything close to the building.

To build a house with cob you'll need all of the following (roughly in sequence):

- *For foundations:* rocks, bricks, "urbanite" (broken-up recycled concrete from slabs and sidewalks), drain rock (gravel).
- *For cob walls:* coarse sandy soil or sand, clay soil, straw, water (already discussed).
- *For roof structure, upstairs floors, door and window frames:* lumber, poles, and boards.
- *In addition:* Doors, windows or glass, roof coverings, some sort of insulation, nails, twine, pipes and wires, screws, other hardware such as hinges and knobs, and a few key tools.

If you are building a rock foundation (see chapter 10), you will almost certainly need more rocks than you can imagine. Many kinds of natural stone can be used, though the closer they resemble bricks in shape, the easier they stack. Look for suitable stone in recent roadcuts, old quarries, or abandoned homesteads with rock field walls or stone buildings. Look especially for stones that are as big as you can handle, with flattish sides and right angles. If there are two of you, you can take some larger stones that require both of you to lift or roll. Make sure they are not soft or crumbly. Also collect a good selection of thin, triangular, wedge-shaped rocks called "chinkers," ranging from the size of a book of matches to a thick paperback. If rock is not available, look into sources for urbanite.

You will need lumber, almost any dimension and of most qualities, including round poles, but specifically stout boards for making door and window frames, 2 inches thick, between 6 and 12 inches wide, various lengths. Bits of thin plywood are very useful for making curved forms for cob or concrete.

Glass for built-in windows needs to be plate, 3⁄16- or 1⁄4-inch thick, or recycled car windows. Don't use thin rolled glass unless it has a frame. Glass can be almost any size, intact or broken, as long as the pieces are more than about a square foot. Framed windows are useful; old ones with wooden frames and multiple panes can be very attractive, though frame condition isn't important, as it can be at least partially buried in the wall. Also excellent are tempered, double-glazed thermal units, available in all sizes. Big ones can be obtained cheaply by taking apart used aluminum-frame sliding doors.

Caution: Store glass very carefully, where it can't be knocked or blown over, almost vertically (at about 75 degrees) on wooden runners, with not too many panes in any one pile, being careful to keep every piece parallel with the others. Broken glass is a sad proposition. The other item that needs most careful storage is straw. It will be useless for building once it gets wet, so straw is one supply it may make sense to buy at the last minute.

Doors can be custom-made, or found at demolition sites, in yard sales and junkyards, at the dump, and in reused materials stores. Look for tools at the same time.

Pipes for plumbing, electrical wire, used brick, even concrete block are all useful, if not essential. Also stovepipe, roofing materials (especially used steel), gutters, ceramic tiles, nails and screws, including rusty ones. If you need to buy these things new, they're expensive, but most people's garden sheds are full of old building materials, unwanted and waiting to be hauled to the dump. Offer to do the hauling and they are often yours for the asking.

Scavenging in the City

The wastes of the city are available for enterprising and creative use. Even cities have clay, sand, rock, and earth beneath them. These materials are often available as the unwanted by-products of town construction sites, road building, drainage trenches, or ground leveling. Immense quantities of lumber, windows, doors, roofing, broken concrete, hardware, and plumbing and electrical fittings are all hauled to landfills daily. You have only to intercept them. Raiding dumpsters and reclaiming materials from the landfill is holy work.

For hauling free building materials, borrow a pickup. Cornstalks on a matatu in Kenya.

Prospecting for free materials is a skill at which we all get better with practice. It requires a change of mind-set for most of us, who are used to buying new everything we want. At some point you realize that most of what you need for a house is already there, free for hauling or already on-site. Any junk you can incorporate will reduce landfill wastes and simultaneously provide you with free building material. Examples include bottles, tires, tree stumps, brick, lumber, concrete, and plate glass. Broken store windows can provide high-quality 3⁄16- or 1⁄4-inch plate glass for building windows directly into your walls. Sliding glass doors are real booty, as they are usually tempered and therefore difficult to break. With the frame removed they make dandy skylights, solar walls, and attached greenhouses.

For your foundation, go looking for rock, brick, or broken sidewalks. Urbanite from sidewalks and concrete slabs is smashed up and hauled to landfill sites all over the country, every day. Call haulage contractors, public works departments, or places you have seen concrete being broken up. Chances are they'll be pleased to give it to you, even deliver it to your site free, to avoid dump fees. The small pieces can go in the drainage trench. Bigger chunks, reshaped, will be the foundation for your cob walls. In Salem, Oregon, Eric Hoel got 50 cubic yards of urbanite delivered free, so he selected the best 35 yards to build with. The "dressed sidewalk" foundation wall is handsome and looks like a natural part of his cob house. In Eugene, Oregon, Rob Bolman got an official city permit for a broken sidewalk foundation for his straw bale and cob building.

If you're in an area with brick buildings, check demolition or remodeling sites. Look in the dumpsters; ask if they're dumping old bricks. They can be chipped, broken, bad on two sides, and they'll still make a fine foundation. If there is a brickworks in your community, even an abandoned one, they often have a huge pile of rejects. You might try a masonry products supplier, where there is sometimes a junk pile or dumpster you can scavenge.

Scavenging in the Country

The range of building components available to a rural scavenger is quite different from those a city cousin might find. Materials are likely to be less processed, more natural. Try to find material that's a nuisance where it is, and help somebody out to your own advantage. For instance, sometimes you can find fieldstone sitting unwanted in big piles on agricultural land.

Roundwood (unmilled branches, poles, and even entire trees) is sometimes free and plentiful—commercial forestry operations are very wasteful—and in a sawmill you may find a range of free wood products from sawdust for insulation to "slab," the first piece cut off the side of a log. Sometimes it is possible to thin whole young trees from crowded plantations, and in more remote areas on public lands small trees that have died naturally are there for the taking.

In places about to be dug up for development, sod, sand, and clay are often unwanted. Sometimes you'll find bamboo, carrizo, or small thin poles that could be used for reinforcement, ceilings, and wattle and daub. Thatch can be made of palm leaves or the wild reedgrass that is so abundant in marshes east of the Rockies (although be warned, thatching takes an immense amont of material).

Once you have a good idea of what kind of rock you want for your foundation (see above, and chapter 10), go to fill sites, roadcuts, any kind of excavation. Carry a rockbar or a 4-foot piece of ½-inch steel pipe that you can probe the ground with. If there's no surface rock, look for areas where the vegetation is poor or which dry out quickly in droughts. There may be good rock just below the surface.

For a foundation to stay in one piece if the ground moves, it needs tensile material incorporated. Look for old wire fencing, polypropylene fishnet or baler twine, and random lengths of wire and steel. You can bury this reinforcement in a cast concrete ringbeam beneath your stemwall, or set it into the mortar between courses of brick, stone, or urbanite (recycled concrete).

Free insulation includes sawdust, wool, newspaper, and raw cotton. Be creative; try cattail fluff, feathers, moss. To insulate a concrete foundation, we have used polystyrene packing peanuts as aggregate. Waste sawdust, bark and wood shavings, straw bales, wool, and old cotton clothes have all been used in ceilings.

TOOLS

At The Cob Cottage Company we're sometimes accused of being neo-Luddites, truculently resisting industrial progress. We don't see the situation quite that way. After years of working with tools, mechanically powered, hand powered, new fangled and from ancient traditions, we have tried almost everything. In general, we have found that the simplest tools give the most satisfaction. There's a rule in the Maintenance Department that says, "Breakdowns are proportional to the number of separate parts."

At a certain level of complexity, a tool insulates you from the task at hand. The worst frustration of being on the Moon must be that your impedimentia prevent your feeling its surface, listening to its silence, experiencing how hot you can be in front while staying so cold behind. Cob invites involvement in a very direct way—dirty hands, dirty feet. Choose tools that don't steal that involvement. Many of us do not enjoy mixing by tractor, digging with a diesel ditcher, or performing any operation that requires boots, goggles, or earplugs. Handwork is sensual; that is, revealing to the senses.

There are about half a dozen tools that will ease your work and give great satisfaction: a shovel, a machete, mixing tarps, buckets, a level, a file, and a cobber's thumb. Another twenty or so help a lot but are not essential for the basic work; these include a wheelbarrow, excavation tools such as a mattock or pick, flat-tined garden forks, wire screens, drums for water storage, squirt bottles, duct tape, crayons, a glass cutter, and basic carpenter's and mason's tools.

Your essential tool kit should cost less than $100 if bought new, though almost all of these tools are easy to obtain used. If you buy a wheelbarrow, get a good one, a contractor's model at 6 cubic feet, with an inflatable tire and hardwood handles. Carefully chosen and regularly maintained, it will be a lifetime tool, so plan on spending up to another $100 for the more durable variety. The machete needs to be fairly stiff. The best models are made in Brazil (Tramontina brand) and Central America.

Every time you use a tool, try to put it away in *better* condition than when you took it out. Clean, polish, sharpen, fix any damage. A tool used in damp clay will rust rapidly if not kept superclean. During the workday, a half-full water barrel is a good place to keep any tool with clay stuck to it—shovels, forks, mattock, or digging bar. The water slows down oxidation, and immersion softens the clay. Make sure all tools, including their wooden handles, are dust-dry before you store them; keep special drying cloths (old towels are ideal) for that purpose. The file and glass cutter go home in the toolbox; neither should ever get damp. Keep the glass cutter in a bottle of oil. If there will be more than two workers on a site, a tool-hanging Peg Board helps avoid day-end losses. If you spray-paint all the tools in place, they'll be easier to find and identify, and when you take them down their ghost silhouette will alert you to which tool is gone.

Create a dry shrine for your tools before any other construction begins. Nine bales of straw and a plywood roof make a good tool shed, set up on pallets off the damp ground. Or you can improvise a temporary tool shelter from a big packing case, scrap metal roofing, or the canopy from a pickup truck. Make sure tools are protected from drifting fog and blowing rain.

A Full Cobber's Tool Kit

All the tools in the first group below are practically indispensable; make sure you have them before starting work. If necessary buy them new.

- Excavation tools (mattock, pilaski, grubhoe) for loosening soil and digging trenches.
- Shovels, for shoveling sand and earth. It's good to have at least one small one, too, for small helpers.

- Tarps, for mixing cob (these should be between six and eight feet square, or possibly a bit bigger (see Tarps section in chapter 11).
- Used 5-gallon buckets, for storing, carrying, and measuring cob materials and water and for mixing lime washes and finishes. Get as many as you can! You could use 20.
- Carpenter's levels (at least one should be 4 feet long); also make a "taper level" (see page 191).
- Machetes, for trimming cob and making wooden tools. A short stiff one for hacking, a long flexible one for scraping or sawing.
- File, for sharpening spades, machetes, axes: flat and sharp.
- Wooden cobber's thumbs, for sewing and perforating fresh cob (see page 184).
- Pocket knife, for cutting twine, tarps, and countless other tasks.
- Wheelbarrow: heavy contractor's models are best.
- 55-gallon drums with removable tops, for storing water, soaking clay and lime, cleaning tools.
- Squirt bottles for wetting down cob and plaster. Liquid detergent bottles work well.

These next dozen tools are extremely handy for mixing and building cob; start collecting them now, used or new.

- Spades, for digging, with long handles.
- Garden forks, flat-tined or square-tined digging forks (not pitchforks!), long-handled if possible, for applying trodden cob.
- Screens, for sifting soil and sand (particularly for plasters and floors): ½-inch, ¼-inch, ⅛-inch mesh, and window screen, on rigid wooden frames, about 2 × 3 feet.
- Hoe: flat, swan-necked, for mixing plasters and floor mixes.
- Hatchets, old handsaws, and short-handled adze, for trimming fresh cob. Saws need coarse teeth (up to 7 per inch), or re-cut them with a file or grindstone to 3 to 4 per inch.
- Spud (a sharpened, sawn-off shovel), for trimming walls and peeling logs.
- Jars, for soil tests (quart-sized glass canning jars with tight lids work best).
- Garden hose (if you have piped water), watering can.
- Scaffolding: trestles, step ladders, straw bales, barrels to stand on, and long sturdy boards (see page 195).
- Duct tape, gloves, builders' twine, and rags for cleaning and drying tools.
- Measuring tapes.

These other useful tools are not necessary for the cob process itself but will be useful in completing almost any natural building.

- Plasterer's trowels, either rectangular or rounded. A round-ended steel pool float 16-inches or smaller is an almost indispensable tool for finishing earthen floors and for plastering curved walls. Plastic yogurt container lids also work great.
- Mixing boat, for plasters and floor mixes.
- Wood tools: an ax for roughing out lumber, splitting wood, etc. (keep it very sharp), hatchets, an adze for hewing and shaping wood, froe for riving.
- Chain saw, with a long bar for milling, and with a short bar for rough-cutting wood and chopping straw; plus safety gear. A lawn mower, leaf mulcher, or weed-eater in a plastic barrel also work for chopping straw, as do machetes and screens.
- Carpentry tools: hammers, squares, saws, chisels, drill, chalk line, Japanese pull saw, and block plane.
- Rock tools (for foundation work): masonry trowel (diamond-shaped), trowel, 2–3-pound hammer, cold chisels, rock or crowbar, safety glasses.

- Paintbrushes, for applying lime washes and *alis*.
- Glass cutter.
- Colored crayon, grease pencil, permanent marker, and/or chalk, for marking levels and measurements.
- Colored flagging tape, for marking stakes, directing traffic, and protecting fragile ecology.
- Rope, any length, any thickness. We use ⅜- to ⅝-inch, mostly.
- Cordless screwdriver/drill.

HAND TOOL REFLECTIONS

BY JAN STÜRMANN

The unquestioned creed of modern carpentry proclaims that without the extensive use of power tools we cannot build efficiently, profitably, or well. I question that creed.

I build both conventional homes where the first thing on the site is power and natural homes where power is often added as an afterthought. Working on these natural homes has confronted me with my own prejudices and the pleasures of using body-powered tools.

I get a sense of profound satisfaction from hand tools that I never find with power tools. Entering the tool shed, my hands automatically reach for my favorite chisel, the ax, the hatchet. A need just to touch, caress. I heft the three-inch-wide slick. Found rusting in an old barn, the blade chipped, the handle socket all mushroomed where some idiot whacked it with a hammer. I carried it home like a sick animal, ground the burrs, buffed off the rust, carved a handle from a piece of maple, and sharpened the edge. The slick came alive for me, sings in my hands, shaving off thick wood curls. I love to look at it, hold it. I never feel that with a power tool. My hand never reaches out just to touch those dead weights of plastic and metal on the shelf.

Why when I use the circular saw or the drill or the chain saw for any length of time do I feel like a bionic man, hard, rigid, at war, forced to wear goggles, ear muffs, respirators to protect my fragile body? While after a day of using the bit and brace, chisels, a plane, my body is soft, my mind still, like that listen-to-the-Universe silence after making love? One depletes, the other nourishes. A mystery.

Yet power seduces. The brute weight of a chain saw in your hands. The engine screams, bittersweet two-stroke smoke swirls through the nasal cavity up into the brain. Trees drop like pins. What a surge, a high we lack the responsibility to handle.

Power tools give us power that is not ours. It is lent indiscriminately. But somewhere down the line it demands payback—with interest. Inevitably we end up paying more for nonhuman power than we gain. Gradually I am realizing that this is true, not just abstractly or instinctively, but practically.

While writing this I got a job installing cabinets in a large L.A. home. Complex scribes, custom-fitting each piece—a headache, but a chance to explore the practicalities of using hand tools when convention dictates power tools. Each morning, to appease the contractor, I uncoiled the extension cords, but then I played and experimented.

> EACH TIME YOU GIVE A MACHINE A JOB TO DO YOU CAN DO YOURSELF, YOU GIVE AWAY A PART OF YOURSELF TO THE MACHINE. THAT'S NOT PRACTICAL. IF YOU DRIVE INSTEAD OF WALK, IF YOU USE A CALCULATOR INSTEAD OF YOUR MIND, YOU HAVE DISABLED A PORTION OF YOURSELF. ON THE OTHER HAND, EVERY TIME YOU REMOVE A TECHNOLOGY FROM YOUR LIFE, YOU DISCOVER A GIFT.
>
> —Bill Henderson, from *Minutes of the Lead Pencil Club*

Building with wood is predominantly a process of cutting pieces of material to length and attaching them in place. All things being equal I am quicker with a power saw and nail gun than with a handsaw and a hammer. But I found that all things are not equal. It takes me a minute to buckle on my tool belt, which holds my hammer and saw. It takes twenty minutes to unravel the extension cords, wrestle with the table saw and miter saw, drag out the compressor, attach the hoses, and get electricity flowing to where I want it.

It takes me half a minute to cut a 1 × 4 with a Japanese kataba handsaw, ten seconds with a power saw. But if I want to still listen to Beethoven and read the *Funny Times* when I'm eighty, I need to put on goggles and ear protectors before I click that switch, so add a few seconds. The handsaw's whisper requires no protection.

A handsaw makes coarse dust that quickly settles to the floor, but power saws throw up a dust so fine it hangs suspended in the air until we breathe it in, blocking sinuses, causing allergies and asthma. In a profession that uses more and more toxic glues and chemicals in laminates and particleboard, we would do well to keep airborne dust to a minimum if we still want to breathe deep and smell the roses in our golden years.

The handsaw weighs half a pound, the power saw ten. I used as much effort—calories—to lift and maneuver the heavy power saw into place as I did positioning and making the cut with the kataba. And even if I did have to take a little more air into my lungs, the pleasure of using a saw that has evolved over seven hundred years more than makes up for the extra time and effort.

I have several interchangeable saw blades that clip into one rattan-covered wooden saw handle: a rip blade, three cross-cut blades with teeth fine enough for dovetails and coarse enough to cut 12-inch logs, a curved blade for starting a cut in the middle of a board, a narrow keyhole blade to cut curves, and a metal-cutting blade. All these blades and one handle I wrap up in a canvas pouch. The total cost, maybe $120. I once thought to be a real carpenter I had to spend thousands on power saws. Now no more.

So I take the board I cut and go nail it into place. Actual nailing with a nail gun takes a second. Using a hammer takes me five. But a hammer always hangs from my tool belt. It's an easy split-second motion to grasp the handle in my palm, the head swinging. The nail gun I've got to lift and drag around like a dead albatross. The hose is too short, the compressor needs moving. With all that hauling around, and donning the ear protectors again, I get pretty close to making up the four seconds lost in actual nailing. Besides, I enjoy the practice of swinging a hammer with grace. Any dimwit can pull a trigger.

Consider economics again. My hammer, a Hart Decker, cost $25 seven years ago. Hasn't broken down once. A compressor and nail gun will cost $500. You work out the cost of repairs and downtime over seven years. Nails for the machine cost five times what ordinary nails cost. But what the hell, the homeowner pays (borrowed from the bank, so multiply the price by three if you include 9 percent interest on a 30-year mortgage).

I can go on unearthing hidden costs. Injuries, for instance. I've never heard of anyone cutting off a finger with a handsaw, but thanks to power blades spun with incredible force by impartial engines, there are plenty of fingerless and toeless carpenters. As for a little job-site acupuncture with a nail gun, or punctured eyeballs, or deaf ears . . . but hey, that's what worker's comp is for. This is not to say that injuries never happen with hand tools, but the severity and frequency are far less.

I focus here on economics and speed because I'm part of a culture that values productivity over process, getting it done over just doing it, completion over creation. But to get to the heart of this questioning I need to look deeper at the unquantifiable.

Carpenters were once craftsmen who knew how to make, adapt, and tune their tools to reflect their individual needs and quirks. Carpenters are now machine operators, factory workers without the factory, assembling modular units. The pride in craft is lost. No longer do we use tools of individual character, but mass-produced tools designed and marketed to the lowest common denominator. Tools that are unadaptable and too complex to repair oneself. The life cycle of a power tool is but a few years, with the years diminishing due to built-in obsolescence. My unborn children or grandchildren will not inherit my circular saw, drill, and orbital sander. But my draw knife, block plane, froe, chisel, brace—already a generation or two old—my offspring will have the pleasure of using.

No doubt about it, power tools make some work easier. Ripping a half inch off a 4 × 4 with a table saw takes a lot less time than doing it by hand. But I noticed a strange difference in my body on days where I predominantly used hand tools compared with days spent directing power tools. I can work far longer with focus, joy, and grace using hand tools: at the end of a nine- or ten-hour day I may be tired but never drained, while after five or six hours in front a machine I am exhausted; although I spent fewer of my own calories, the juice of vitality has been sucked from me.

Why? The power these tools have to do me harm depletes me. My body—afraid, tense—on full alert turns subtle flexibility into rigid, tense muscles. Reflexes slow, the mind falters, mistakes happen, blood flows. With tense bodies, the chances of strains and wrenched backs are far greater than with a body that all day is being given a *gentle* aerobic and stretching workout by using hand tools. Maybe that's where the extra vitality comes from. When my cells are regularly flooded with fresh blood-carrying oxygen and nutrients, my body responds with more life to give.

Then there is decibel fatigue from the loud screeching noise that permeates every building site. More and more, this is the predominant reason I choose hand tools over machines. Our ears, attuned to lover's sighs, falling rain, friend's laughter, wind whispers, are not adapted to cope with frequent loud noises. We withdraw into a shell of numbness, deaf to the world. I want to work in an environment where my timid senses emerge in the silence to partake of creation, where the flow of conversation or thought remains free to meander, explore, and fall again to silence, not censored, interrupted, broken by machines.

Much of the bad rap hand tools have gotten is justified. Without the stern vigilance of craftsmen demanding only the best, the modern tool manufacturer sells a quality of hand tool that is shameful. It is no surprise that the tool buyer turns away in disgust and resorts to electrical force to get the job done. It is a rare store, staffed by knowledgeable sales people, that stocks a wide selection of quality hand tools. But what can compare to the serendipitous pleasure of finding a quality tool at a garage sale?

With the diminishing availability of quality tools, the wisdom of how to use them is also being lost, and needs to be rediscovered if we are going to use hand tools to their full potential. How best to clamp, fasten, hold material as I cut, chisel, or plane? How do I use the strength of my body in an efficient, graceful way so that I don't fight the tool, the wood, but turn the work into a dance instead? This is a study, a search worthy of my attention.

Working by hand allows time to ponder: Is faster better? What have we gained with excess power? Building by hand encourages us to build more deliberately, ponderously, aware of our actions that ripple beyond us. With only direct sweat labor, would human dignity allow the building of strip malls, tract homes, McMansions, and superhighways? What happens to our souls encased by machine-made objects of dull perfection? To know we exist as humans, we need to see the touch of another in the creations that surround us.

I am no purist. My power tools, well used, cared for, will continue to be used, although with less frequency as I discover again the joy of using just my body to propel tools to do their magic. For there *is* a magic there, a mystery. I eat oats and honey, bread and cheese and red bell peppers. I breathe in air laced with oxygen transpired from trees. And miraculously my body converts all these into motion, strength, finesse. I lift a plane, sharpened and tuned, and lay it to the wood. Then somewhere in the infinite realm between my hand and the tool, alchemy happens. Flesh, steel, wood combine in motion, and I am graced with translucent ribbons of shavings curling through my fingers, setting free the scent, revealing beauty. A gift.

Site Respect and Preparation 9

Many of the most serious problems that natural builders experience come not from constructing the building but while making or not making the connections between house and site: choosing a site and preparing their site for building. It is hard to overemphasize the importance of both processes. So be prepared to put a lot of slow careful work into site preparation.

Architects working in England in the 1960s used the 30-30-30 rule. Of all the time, skill, and money spent putting up a house, 30 percent would be below the walls, 30 percent would get up walls, floors, and roof, and 30 percent was for everything else. The floating 10 percent was tucked in wherever the particular job demanded. Houses in the United States today are not that much different, so the formula is still useful. If you wanted to avoid long-term problems, you would put the extra tenth into site preparation, access, drainage, foundations, pipes, wires, and water management. With really responsible natural building, perhaps the proportions should be more like 40-30-20.

❧ You are on a stage where you can act out your ecological conscience, minimize disorder, conserve life, and create beauty. ❧

This chapter will help you prepare your building site so as to make the rest of the building process enjoyable, safer, easier, and more efficient. There is a logical sequence to this work that makes each step of the process flow naturally into the next. Overlooking a step or postponing it until after construction has begun almost always makes more work in the long run and can lead to serious problems with the building or its surroundings. The sequence below works well in most cases, but adapt it as necessary to your individual site and circumstances. Be sure, also, that you have read chapter 5, "The Site You Build On."

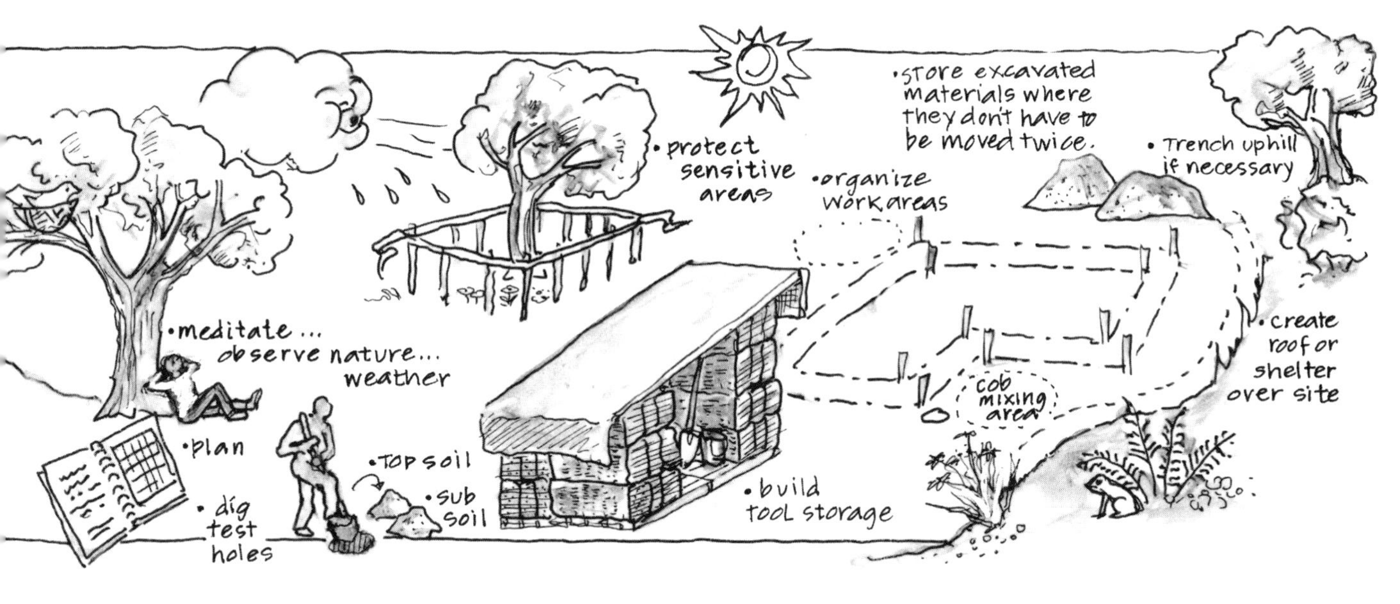

RESPECT THE DEEP ECOLOGY OF THE SITE

Go meditate at your site, in comfort, for a long time, on the Deep Ecology of the site, the inherent right of that place to be undisturbed by human activity. Think about the life you will take, from the unhatched songbirds whose parents will have no thicket to nest in to all the tiny earthworms and mites and sowbugs and bacteria. Meditate especially on the life that can't escape—the trees you'll cut, bulbs you'll crush, all the plants of the place. They can't fly away; you have them trapped. Consider the soil life, all carefully stratified at differing depths, and meditate on each spadeful you turn, which will be a death knell for millions. Ask their forgiveness. In the natural order their relatives will in time eat you. Treat them kindly, and restore as much of their habitat as possible.

It's good to remember also that every nail you bring to build with is made of iron dug out of the ground where it had lain peacefully for millions of years, heated to melting by gigantic furnaces blowing out sulfur into the clean air, hauled by big, noisy, smelly diesel trucks on a concrete highway laid over farms and forests. In driving that nail you will contribute to all those processes which destroy the world we love. Yet we persist in building too big, on too much land, creating new roads, hauling materials long distances, then barely living in our house because we want to be hiking in the undisturbed forest far away.

By care and skill you can at least minimize the damage. A restorative approach can mitigate damage, and building to *enhance* ecological richness will provide new habitat. You are on a stage where you can act out your ecological conscience, minimize disorder, conserve life, and create beauty by the choices

Preparing the site, be thankful for each spadeful you turn. Talk to the decomposers, whose descendents in time will eat you.

The 30-30-30 rule.

CONVENTIONAL construction

site protection, access, drainage, water management foundation, subfloor, services.

structure of walls, floors, roof, roof sheathing, bracing, siding

windows, doors, insulation, wiring, sheet rock, plastering, painting, plumbing, counters, fixtures, furnace, exterior sheathing, etc

COB construction

site preparation, drainage, water management, foundation, pipes & wires access.

roof structure, cob walls, niches, plumbing, wiring, windows, door frames, floor rafters

Floor, sculpted benches, counters, doors, final plasters, insulation, back-up heat source,

you make and how respectfully you treat your site.

Don't Clear Any Trees Yet

Don't cut any trees or clear any views until it's unavoidable. Once those trees are gone, they are gone forever. Remember in summer that deciduous trees won't block winter sun or views; remember in winter that you may need that maple on the west side for August shade. Consider whether trees will be affected by the decisions you make about access, services, and drainage.

You may need to cut or trim trees in order to maintain good solar access, but before any cutting, wait as long as you can. Try snipping twigs, pruning branches; wait a few weeks, try a bit more. The fir trees here in Oregon grow four feet a year. Suddenly a mountain disappears, the lake vanishes, we can't see the vegetable garden. Even so, we're very conservative in taking out whole trees. Every winter, we prune out dangling branches, snip twigs on the peach tree by the window, reveal new dimensions to our space, with pruning shears or a pole saw. Very different from the developer's bulldozer/chain saw one-shot approach.

Work Slowly, Carefully, By Hand

There's a basic law of ecology which says you always pay extra for emergencies. Get help from someone with more experience in setting up a time budget. Assess progress regularly and reprogram constantly to prevent emergencies from forcing your pace. Remember that most steps take twice as long as you expect, and cost three times as much. If you stay alert, a job that takes longer than planned need not be a reason to rush out and buy a machine, unforeseen costs won't make you anxious about the budget, and the rainy season won't arrive before you're ready for it.

Plan for slow rhythmic work that adjusts to season, weather, and workers. This way

Create a view with pruning shears and pole saw, not with bulldozer.

you can enjoy the rhythm of hand labor. When you work by hand, slowly, you pay attention. There's time for your perceptions to sink into consciousness. You're much less likely to take actions you'll regret later. It is a reward in itself to get a real feel for your place and how it works, by the slow turning of the earth, the shovel revealing how the soil was laid down, ax biting into stump. Savor each task, you will only undo this particular work of Nature once.

We have a cultural attitude that values speed and mechanization over beauty, ecological health, and the satisfaction of physical work. Ditch digging has acquired a reputation as onerous and degrading, and many people recoil at the idea of clearing a building site or digging foundations and drainage ditches by hand. The fact is that for a small building site, it generally isn't such a lot of work and will be both educational and enjoyable.

When preparing sites for cottages and other small buildings, we recommend avoiding the use of heavy machinery in site preparation. With a pick or shovel in your hand and your feet on the ground, you will be

sensitive to subtleties that machines generally ignore. For example, you may notice special rocks and trees that can be left in place within or around the building. Digging by hand, you will gain valuable knowledge about the hardness and solidity of the soil at various depths and therefore its ability to support the weight of the building. You may decide to create a split level along the contours of a sloped site, which not only reduces the amount of excavation but also makes the building appear to grow out of its site rather than being imposed on it. You'll also develop a sympathy for all the little creatures whose homes you're demolishing in order to make your own—the lizards, beetles, earthworms, and songbirds that are destroyed willy-nilly by machines, but which sometimes you can save when working by hand.

Our advice against using heavy machinery is based not so much on neo-Luddism as on practical experience. Despite the most careful plans and explanations, and even under constant supervision, we have repeatedly seen machinery, including pickup trucks, tractors, and brush mowers, severely damage trees and other vegetation, compact soil, interrupt natural drainage, and leave building sites so scarred that they may take decades or longer to recover their natural functions and beauty. This is due not to ill intentions on the part of their operators, but rather to carelessness, inexperience, and the sheer speed and destructive capacity of these machines.

Backhoes break tree roots and weaken their structure. Five years later, the tree falls on your house—perhaps the very tree that was the whole reason for building there. That ready-mix concrete truck you ordered, now stuck in a mud hole, has to dump its load in order to escape. In two hours you have a twelve-ton block of concrete set up for all of geological time in a place you don't want it. However careful the driver, in tight situations, a backing truck runs over your wild orchids, crushes the rabbit's hole (with the rabbit), compacts the ground. Next year you wonder why nothing will grow there. "Gee, sorry, I guess we . . ." (choose one of the following: ". . . cut the wrong tree, took the wrong turn, creamed up your garden a little, smashed down your peach tree" . . .).

An added bonus of not using mechanized equipment is that your construction site will be quiet. Snoopy public officials will be much less inclined to interfere, and your neighbors won't be likely to get irritated. To many people construction means bulldozers, concrete trucks, nail guns, and chain saws. If they don't hear any of those, there must be nothing

BACKHOE MADNESS IN CALIFORNIA

BY IANTO

A FEW YEARS BACK The Cob Cottage Company was asked to help design and build a small cob house in Northern California, in gorgeous rolling hills covered with golden grasses and scattered oak trees. Native groups had hunted there for ten thousand years leaving only a few arrowheads and the stone querns where they ground acorns for flour. Nobody had ever built in this place. In spring there are wildflowers so thick you can't walk without crushing them, all across those lovely hills.

The clients were nice, sensitive people; they understood the fragility of the place and appreciated what they had. They had chosen a spot in a little basin facing south, sheltered from the north wind, with ancient, crooked oaks to east and west, and views to the south across valleys and ridges for miles and miles.

The intention was to hand dig foundation trenches and to hand level a building site. We set an April date for a construction workshop, knowing that April is usually warm and sunny. Winter came early, with lots of rain, before the clients had a chance to start, and by March they were concerned the site would not be ready for the workshop. A neighbor offered to help with his tractor, just to level a building pad. The tractor had rubber tires and they knew he was skilled in using it, so damage to the meadow would be minimal. Just to be sure, the woman would watch and direct as he worked.

The big tractor crept in gently, just chewing up the sod ever so slightly, and got to work. The neighbor piled up sod, topsoil, subsoil, carefully, in a heap as big as a house. It looked like a terribly big hole he was digging, just to put up a 20 × 30-foot building, but she wasn't experienced in construction, so she asked why he was going so deep. He said "I took it down to 'grade.' You need to be sure you are on subsoil. If you build on topsoil, it will slide."

The neighbor was almost finished when the woman was called to the phone. When she returned twenty minutes later, oh horrors! "Well, I didn't want you to be stuck with that big pile of dirt, so I just spread it around. It looks a little raw but you can plant it up after the building's done." A quarter acre of mountain meadow encased in a foot of heavy wet clay, well compacted so "the construction equipment won't get bogged down." Bye-bye wildflowers, bye-bye ground squirrels. And a perfectly flat, rectilinear slice out of a hillside, dug in four or five feet at the back, hard and dense and impervious to the drizzle, little muddy puddles building up all across it.

After the pad was made, the owners brought in a small Kubota backhoe and dug a four foot curtain drain around the north and west sides to drain the site, which was now turning into a pond with the heavy winter rains. Once you start using machinery there's a commitment formed: machinery tends to require more work and more machinery.

happening. We have built discreetly within yards of adjacent homeowners who hardly noticed there was anything going on.

Work Out the Best Routes for Deliveries

Plan in advance where your building materials will come from. If you need wooden poles, can you use trees from the site clearance? Can clay excavation leave a pond in the best possible place? Opening a way for your access road might yield rock for the foundation.

Plan for delivery of heavy building components. How will drain rock, sand, straw bales, and roof beams get to the building site? What options are there? Could you wheelbarrow them the last fifty yards? Or would moving the building site really be the best option? If delivery is to be by truck, is there an existing road and is it passable year-round? Does it dead-end at your site and, if so, how do delivery vehicles turn around? You can attempt to limit vehicle size, but be prepared for much bigger vehicles than are reasonable. Expect them to be very heavy, very wide, and sometimes very long. They have wide turning circles, and drivers are not always skillful. Will you need to build a road specially for deliveries? Will it be temporary, or could it double as the permanent access? Or could you let a truck drive over the unprepared ground in dry weather? If so, how will you protect against ground compaction?

Protect Sensitive Areas

If there are ecologically sensitive areas on your site, be determined in keeping vehicles out of them. Polite notices or explanations to drivers don't always work. You'll need to *blockade* them out, for the duration of construction at least, with a robust and visible fence. Bright ribbons attached to the fence are indispensable. If a builder is contracted to do any part of the work, a penalty clause in the contract may help, specifying in writing and on a drawn-out plan exactly what you want protected (trees, fragile areas, etc.), laying out financial penalties for damage.

When you order sand or gravel, ask the exact height of the delivery truck and make sure any preparatory work such as lifting power cables, pruning low branches, routing surface water through a culvert is done in advance. Don't get caught having to work in a hurry with a nine-yard dump truck idling with diesel fumes and the driver waiting.

Walk-in Buildings

Occasionally owner-builders will decide not to take vehicles to the building at all. "We want to keep cars well back from the house. They're uncivilized and dangerous." This seems like common sense, but be prepared for difficulties. If you apply for a building permit, the fire department may require that they can get access with a full-sized fire truck, and turn it around. And unless it is made completely impossible, at some stage a well-meaning driver will attempt a doorstep delivery. If there's no clearly laid out access, you may get to the site one day to find your prime wildflower patch has been ground into the earth.

If you intend to build a temporary construction road for delivering materials, then take it out after the building is finished, keeping in mind that inertia will fight for it to become permanent. "Well, we'll just drive in a couple of times, in really dry weather. It will only mean cutting five trees, and what's five trees to a whole forest, and we won't lay a roadbed, I mean we won't even put down gravel." Ianto's granny used to say the road to hell is paved with good intentions. Maybe this is what she meant. Grim reality is that creating a road is usually a one-way street. It only takes once when the ground is wetter than you thought, a loaded truck gets bogged-down, requiring the neighbor's tractor to drag it out, and you have a situation that can only be fixed by more heavy machinery and a bed of gravel.

If you plan on having a walk-in site, think ahead to other users, after you've gone. How will they feel about carrying paper sacks of groceries a hundred yards in the rain? Would access be difficult with two toddlers and a baby? Will they decide to bulldoze a road, in frustration, across the only wetland for miles? Should you instead create an access road, now, in the place you know to be most suitable?

Ultimately, if you're determined and realistic about a building that has no vehicle access, here are a few don'ts:

- Don't build where an emergency vehicle can't get access, unless you're prepared for the consequences.
- On the other hand, don't leave any gap where an unwanted vehicle can get physical access, or it will.
- Don't build up a steep hill from the roadhead. You'll get frustrated hauling sand uphill in a wheelbarrow and yearn for a road.
- Likewise, don't build uphill from your main building materials, firewood, or water supply.
- If you have fears about vehicle security, consider how parking can be arranged within sight of your house.
- Don't tell the Building Department or Fire Department you're working there—they may demand a regulation road that could be expensive, destructive, overbuilt, and completely out of scale.

SITE PREPARATION

Up to this point, you may have barely set a spade to the ground or taken a saw to a branch. Your building site should appear much as it did when you first saw it. All that is about to change. Soon you will have a busy building site covered with trenches, piles of earth, and other materials. There will be no return to the undisturbed site you know. As you begin to excavate and clear the site, following the order of the steps listed below, retain your sensitivity to the Deep Ecology of the site. You are about to become an active component of the local ecosystem. You have a responsibility to the site and to its present inhabitants to build with great awareness and care.

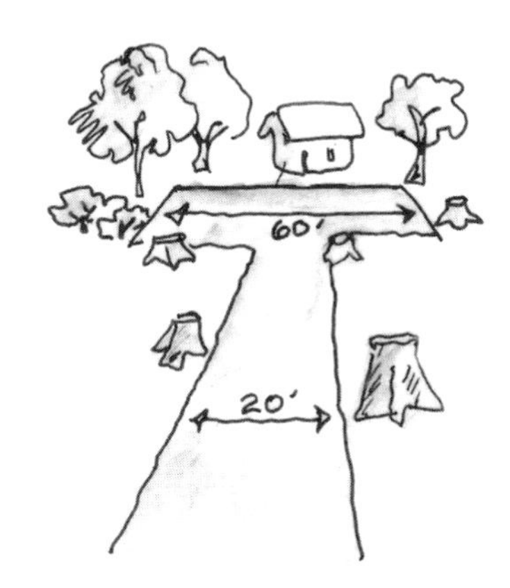

Access for firetrucks may require clearing, leveling, providing a hard surface.

Dig Test Holes

As we discussed in chapter 5, dig test holes all around your building site—at least half a dozen, inside and outside the proposed wall line, uphill and downhill. Dig until it becomes difficult, until you hit rock, or a minimum of 3 feet deep. These holes will tell you how suitable the soil is for building, how solid the subsoil is, whether you're going to have problems with hidden rock ledges, buried garbage, swampy substrate, and so on. Carefully lay out what comes out of each hole, in sequence, so you can see it clearly, about a foot from the hole so none of it falls back in. Cover the holes so nobody falls in them.

To get a rough idea of how good your drainage is, pour in five gallons of water, wait ten minutes, then pour in another five. Watch carefully. If the water stays in the hole

Dig at least half a dozen test holes around the site, at least 3 feet deep. Note how various horizons of different material are laid out sequentially.

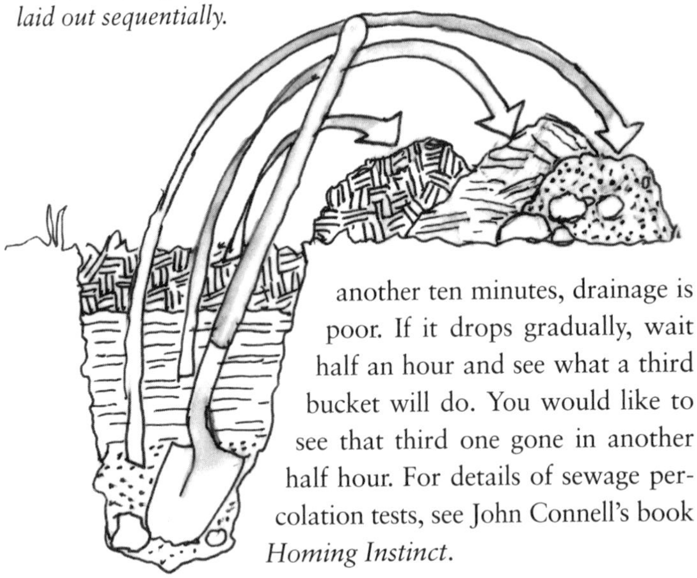

another ten minutes, drainage is poor. If it drops gradually, wait half an hour and see what a third bucket will do. You would like to see that third one gone in another half hour. For details of sewage percolation tests, see John Connell's book *Homing Instinct*.

Transfer Your Design onto the Ground

By now, using the procedures described in chapter 5, you should have your building site precisely located. You should also have a finalized design, a scale model, and/or a plan drawn to scale on a durable paper with a plastic rainproof cover. Part of your design process was preparation of a scale drawing on squared paper. Now you are ready to transfer that design onto the actual site. With an irregularly shaped building this can be complicated. You will need a clear head, a good helper, and lots of patience. It's helpful to have a couple of extra copies of your plan, all to the same scale, in case you make too many mistakes on one of them to erase. Collect a large number of straight wooden stakes, about 2 to 3 feet long. You will need a way to label them permanently, either with paint or indelible marker. To be completely accurate with your layout, you will also need a compass adjusted to the local declination (the difference in angle between true, solar north and compass north).

To start, mark on your plan a point near the center of the building. This will be called point A. Now draw a line through point A running precisely north–south (according to solar north, not magnetic north; see illustration on page 149) and another one running east–west. These will be called your N–S and E–W baselines. On your plan, follow the E–W baseline east and label a point at a fixed distance (say 6 feet) beyond the outside of the east wall. Label this point B. Do the same to the south, west, and north, marking those points C, D, and E.

Now transfer these points onto the site. Start by driving in a stake where you want the center of the building to be, and label this "A." Measure the distance on your plan from point A to point B, and record it on the plan. Then, using your compass and a tape measure, locate point B on the ground. Pound in a tall, very sturdy stake and label it in a durable manner. Also make sure it goes in vertically. This will become one of your permanent site stakes, from which everything else will be derived, so you need to be certain it won't fall over or get moved. If you can't drive it in just at that point because of a stone or tree root or because it's in the middle of a pathway, move it a measured distance one way or the other along the baseline. Be sure to move the corresponding point drawn on your plan. Repeat this process with points C, D, and E.

Double-check that your two baselines are straight. If you squat behind stake B and close one eye, the stakes at points A and D should disappear. If they don't all line up, move one of the peripheral stakes. Always leave the point A stake in place, unless you decide that the siting is wrong, in which case you will have to move all your stakes and start the whole process over again.

To check that your baselines are perpendicular, use the most convenient of geometry tricks, the 3–4–5 triangle. From point A, measure along your baseline a multiple of 4 measuring units toward point B, and mark the point B1 with a temporary stake. Then measure 3 units toward point C and mark point C1. If the distance from B1 to C1 is exactly 5 units, your lines are at right angles. Right on! If not, you will have to adjust two of your stakes (B and D or C and E) to get it right. The larger the triangle, the more accurate the results, so choose a measuring unit that makes the triangle take up most of the space between the stakes. When you achieve a perfect 3–4–5 triangle (usually a tolerance of ½ an inch is close enough), remove stakes B1 and C1.

You have now established a permanent quadrant system that can be used to accurately locate any point in the building. To continue the layout, you will use a process called triangulation. Let's say, for example, you want to find the exact location of the kitchen sink. Mark the center of the sink on your plan with a point called X. If it happens to be in the southeast quadrant of the building, measure on the plan the distances to points A, B, and C. Have your helper hold the end of a tape measure over stake A and walk out in the approximate direction of the future sink until you reach the exact distance. With a sharp stick held vertically alongside the proper distance mark on the tape, scribe an arc along the ground by walking back and forth a few steps. (If the mark doesn't show because the ground is covered with vegetation, try using lime, chalk, or flour). Now, repeat the process twice more, measuring from points B and C. All three arcs should intersect at a single point, which is point X, the future kitchen sink. Mark it with a stake.

You can use this procedure to transfer the whole design onto the ground precisely, but the process is fairly tedious. If you are still playing with the design and don't need perfect accuracy, just locate and mark a few key points precisely and then sketch in the rest by

LAYING OUT THE BUILDING.

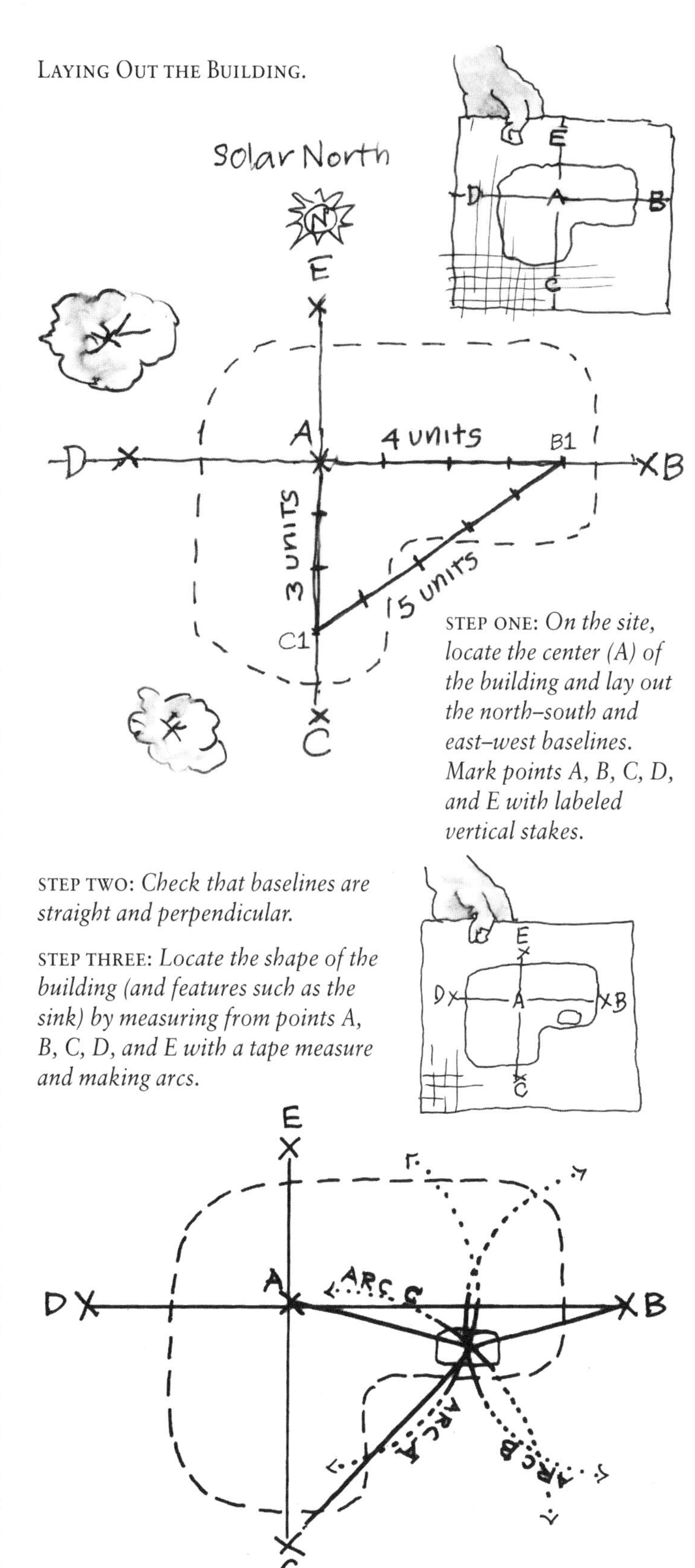

STEP ONE: *On the site, locate the center (A) of the building and lay out the north–south and east–west baselines. Mark points A, B, C, D, and E with labeled vertical stakes.*

STEP TWO: *Check that baselines are straight and perpendicular.*

STEP THREE: *Locate the shape of the building (and features such as the sink) by measuring from points A, B, C, D, and E with a tape measure and making arcs.*

eye. Also note that the farther the site is from flat and level, the less accurate this process will be. If the site is very sloped, wait until you have leveled it to lay out the foundation trenches permanently.

Build a Full-Scale Mock-up

Using the stakes you have driven as a guide, you can now mark out the location of the walls onto the ground. Because cob walls are so thick, to keep reality in view, mark both the inside and outside of walls. "Sticks and Stones" is a technique Linda developed, using whatever marking materials the site offers. You can use short sticks, rocks, or seashells to sketch the walls on the ground.

Then put in long stakes (maybe bamboo, cane, thin saplings, or even rebar or PVC pipe), as high as the roof will be. String bedsheets or cloth strips between them to make "walls," leaving openings for windows. Set some poles up to be doorposts; it's important to know where entrances are. Add a folding chair, then set up a "desk," a "kitchen counter," a "bed." Playact your daily life at your new home. Make sure the spaces feel right, the way you imagined they would.

If something feels off—a passageway too narrow, a corner too square, not enough room to take off your boots inside the door—adjust the lines carefully. Be flexible, be prepared to rearrange many times. Get agreement between several people about the exact position of the walls. The curve and flow should feel just right to everyone.

The biggest danger in this process is that you will decide to expand the spaces, making the building much larger than you had originally planned. Try to avoid that temptation. The building is almost certain to feel smaller now than it will later on. If you move all your stakes to make each of the spaces "just a little bigger," the result could be an unnecessarily large building that will take too much time, money, and materials and cause too much frustration. For instance, spreading the walls of a 10-foot-diameter circle by only one foot in each direction seems like a 10 percent increase. In fact, the circle's area increases from 79 to 113 square feet (nearly a 50 percent in-

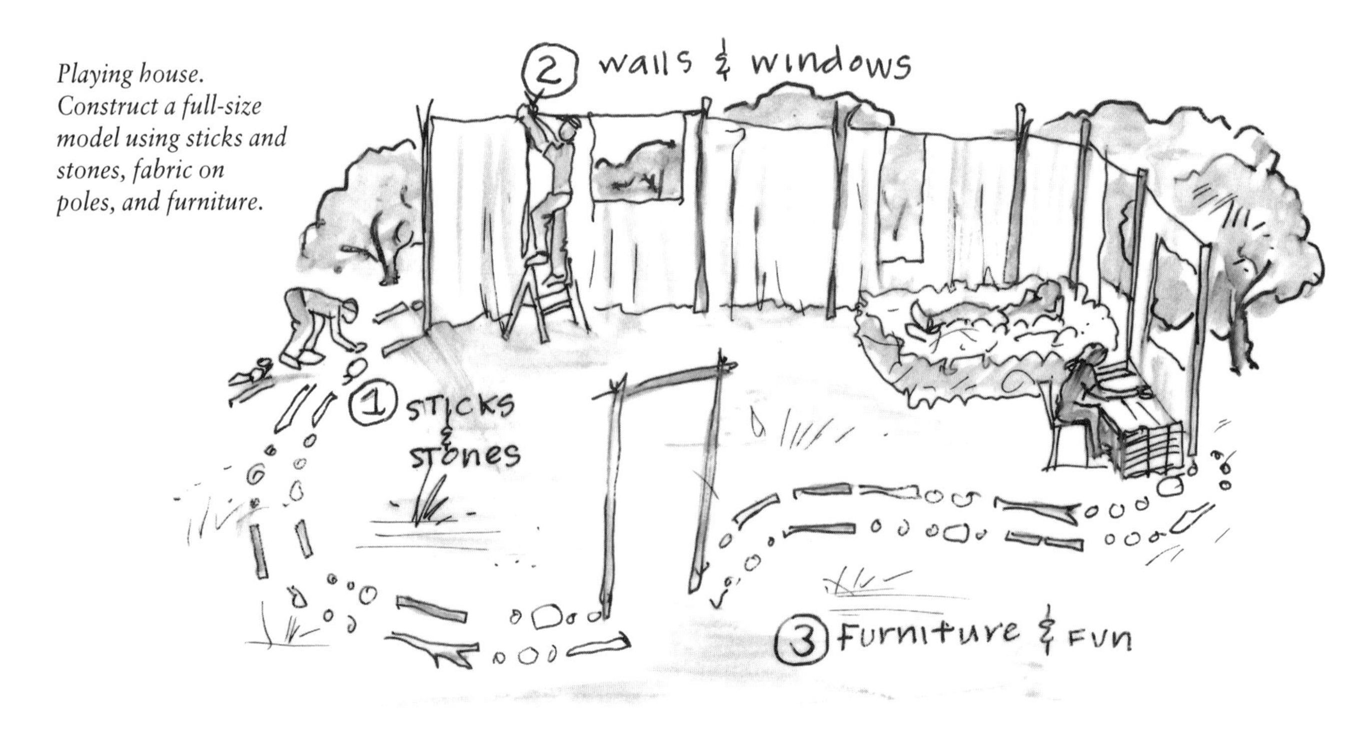

Playing house. Construct a full-size model using sticks and stones, fabric on poles, and furniture.

crease), and the wall length grows from 31 to 38 feet. Worse, a beam to span 12 feet needs to be 44 percent stronger than one to span 10 feet, which makes roof and suspended floors heavier, uses much more wood, and keeps you busy paying for all of the extra lumber.

If possible, go back to your on-site model in different extremes of weather—very hot afternoons, pouring rain, high winds, snow. Go at night, in every season, at dawn or sunset. Celebrate the cosmic calendar there—equinoxes and solstices—so you can watch sun and moon at their extremes of height and lowness. What ideas does your site visit give you for window location, access pathways, outside rooms? Are there any adjustments you need to make to the design before it's too late?

Drive in Foundation and Datum Stakes

When you're no longer moving the stakes back and forth, bang them in really well—you don't want the dog to pull them up, or the next-door kids to rearrange them for fun. Make sure they're large enough to be seen, or site workers, frustrated at having fallen over a half-hidden projection several times, are likely to pull them up, or accidentally remove them while digging. Drive in enough stakes to clearly mark the inside and outside lines of the foundation, which should be at least as thick as the base of the wall (see next chapter).

Also, if your staked-out design differs from your original plan, draw a new accurate plan. Work in reverse from the layout process described above, making sure everything is to scale. That way, even if your stakes get moved, you can locate them again.

This is a good time to establish some permanent "datum stakes" to mark final floor levels. It's handy to have a datum stake in each major room, especially if the floor level changes, and more datum stakes outside, if any grading needs to be done. Use a transit or water level (a clear plastic hose filled with water) to mark the same reference height on each datum stake. Call this height "0" and use it to draw a cross-section of your building, to scale. To avoid confusion, reference all of your final floor levels and ceiling heights by measurement up or down from that zero point.

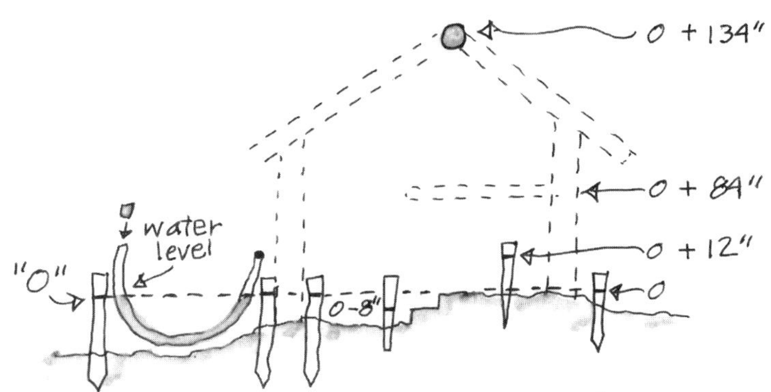

Locate permanent datum stakes to mark the finished floor level for each major room and in the surrounding site.

Because they need to remain securely in place throughout the entire process of construction, datum stakes need to be visible, very firmly driven into the ground, short, and out of the way of workers, wheelbarrows, and paths used at night.

Finally, Begin Clearing the Site

As the final design solidifies and you become committed to the exact building size and location, clear vegetation from the footprint of your building by hand. Clear the area between your foundation stakes until you can see the naked earth beneath the vegetation. Don't yet extend clearing beyond the footprint.

Remove all the vegetation, loose topsoil, and as many roots as possible from inside the footprint of the building. Dig down to solid mineral subsoil anywhere there will be weight bearing, such as a mass floor and especially foundations. Depending on the type of floor you will install and how much depth it needs, cut away the ground to create a level platform on which the floor will be built.

By now you should have decided where materials will be stored during construction, including those to be brought in. Mark some

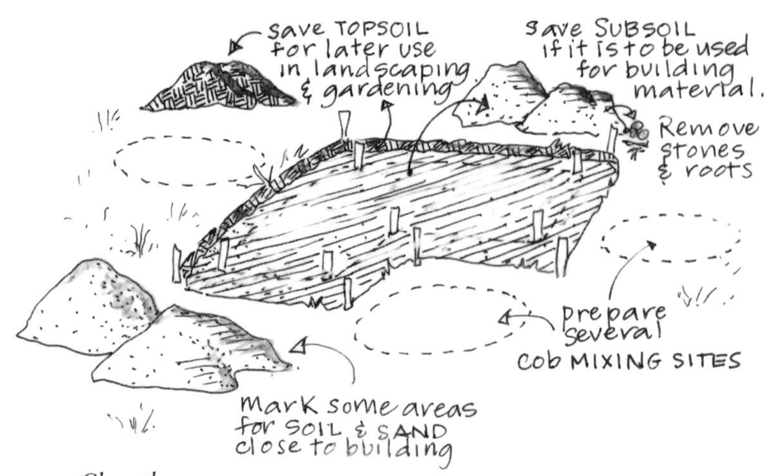

Clear the building's footprint down to solid mineral subsoil, and save the valuable topsoil.

areas for piles of soil and sand, close to the building perimeter. If you intend to use any subsoil as building material, clean it of roots and stones and pile it in the designated storage area. Store excavated materials where you won't have to move them twice, where they won't asphyxiate local ecology or be in the way of the builders. It's convenient to have several piles of soil excavated from foundation and drainage trenches, each within a shovel's toss of a different part of the trench.

You will also need several cob mixing sites, either inside or just outside the building.

Two ways of pre-erecting the roof before construction of the walls.

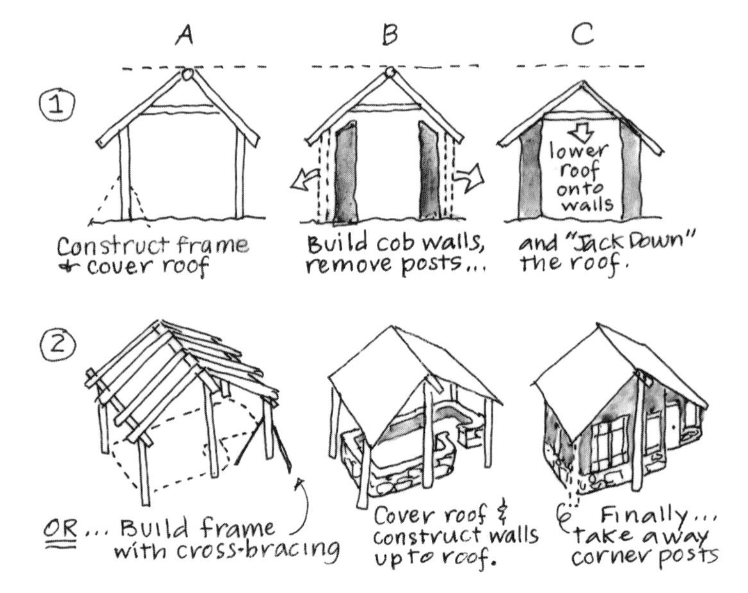

Mixing areas should be clean, level or slightly dished, at least 10 feet in diameter. You need unimpeded access to them from both the future walls and to the piles of earth and sand, as well as water and straw.

Erect a Temporary or Permanent Roof

A large tent or tarp over the site will protect you and your building from rain and sun. Make sure that the tarp is high enough not to interfere with construction and that stakes and ropes aren't in the way. Cob builders often erect the permanent roof structure at this stage, setting it up on temporary posts and covering the rafters with tarps, plastic sheeting, or even the permanent roof sheathing. Sometimes the posts are left about a foot too long, to allow space under the ceiling for finishing the tops of the walls. The roof can be lowered into place on jacks when the walls are finished (see chapter 15 for more details).

INTERVIEW: SITE AS SANCTUARY

Christopher Martin took a cob workshop led by Janine Björnson and me in 1997. The thick, solid walls reminded him of the houses he grew up in England, and he resolved to build his own home. With his partner, Lucy Sharp, he bought land in northern California and they began to design their dream house.

MICHAEL: We're sitting here on a beautiful, sunny January day, looking down the hill to the south with the gnarled oaks below us, a few old growth firs beyond them, and the ridges fading to blue in the distance. We're sitting next to your half-finished cob house. I wonder if you could tell me about how you came to put a cob house here on this particular site.

CHRISTOPHER: Well, there are two possible explanations. I'll give mine, which is that it's one of the few places around here which is actually level enough [laughs]. So that was a pretty good recommendation. We had spent some time here over a year and a half or so. We started looking around, tried various other spots, but they didn't seem right. And then I came upon this site, which was a little bit out of the way. One of the things which drew me here was that it's very protected. Just a short way up there on the ridge and you can get howling winds that will knock a building down. And yet here, we've left our tent up over two winters, and it's not a sturdy tent. But Lucy has a different sort of explanation.

LUCY: I feel like we were invited into this circle of trees, so we built to that. We had a choice of cutting into the hill, to make our foundation easier, which would have meant taking out trees, changing the look of the place, putting in a little rock garden behind the house. It was a pretty significant choice to just try to build to the spot instead, and leave everything as it was. I think that will have a big effect on the house when it's all done, because it will just be where it was supposed to be. It will look like it was just set in the trees.

CHRISTOPHER: I like the way the inside of the house follows the slope of the land, rather than just ignoring it or riding over it. The fancy of it, really, is that we just lifted the top layer of the earth off and slipped our home in between.

LUCY: We're not here as owners of the land, but more or less as protectors, just to live in harmony, and it seemed like a spot where that could work, and that we'd get help from the elemental kingdom around us.

MICHAEL: And do you think that has, in fact, happened, during the building process?

LUCY: I believe so. And it's confirmed all the time by people coming to say, "It's just so beautiful." It's changed, but it hasn't ruined anything.

CHRISTOPHER: It seems like that to me, too. Just things like, when I made the path, it wasn't until I got to nearly the end that I realized how much rock there was around and how I could easily have run into a lot of rock, and I didn't. It was relatively easy, and it could have been very difficult. And when we dug the foundations. We're fairly close to the trees around the house, and yet we didn't have to cut any significant roots. That's pretty remarkable.

MICHAEL: So what's your end of the bargain? If you've been getting this help from the site or from the spirits of this place, what have you been doing in return?

LUCY: Trying to treat them and the land itself with the utmost respect that we know how to do. When people came to help us, we tried not to show them, but have them feel what was around us so that they would show the same respect and in return get the same respect. If you are building a house like this, and you're trying to do it cocreatively, I think it's important to set up guidelines for living there if you have more than a few people. Not really rules, but guidelines, like "stay on the path." It's really nice to sit down and think about it before you invite a group of people in.

CHRISTOPHER: It's definitely something you want to keep on top of. When you get a group of people together, they can very quickly overwhelm the space that they're in. Certainly one of our major focuses has been to minimize the impact that the building will have on the place. Because that so often happens: you find a place you love, build a house and destroy the place. So we've restricted the activity to certain areas as much as possible. This area, here, down in front, we've kept clear because we don't want it to be messed up. An important thing for me was to stay on top of keeping wastes cleaned up, as a way of respecting the land. And for safety purposes. But our biggest decision in that respect was not to make a road capable of taking vehicles. Because as soon as you get cars or machines, it affects the site dramatically.

MICHAEL: What were some of the impacts of that decision?

CHRISTOPHER: The biggest impact is in the quality of the site now. We're in the middle stage of construction and it's almost like it's undisturbed. I'm really pleased with how well it's surviving. Thriving, in fact. And there was a big impact on the process, because it entailed a lot of extra work.

MICHAEL: How so?

CHRISTOPHER: Just transporting materials in. Sounds easy to wheelbarrow it, but it's not that easy. We learned tricks along the way that helped. After quite a lot of wheelbarrowing, we got a little ATV with a trailer, and used that, filled with buckets. That worked pretty well. But however well it worked, you're still double-loading the stuff. It's never as convenient as having a truck come right up and dump it at your doorstep.

MICHAEL: And in your case, since you were hiring a lot of the labor . . .

CHRISTOPHER: Yes, that made it very expensive.

MICHAEL: Are there any recommendations you would make to other people who choose to do the same, to build on a site that is not accessible to vehicles?

CHRISTOPHER: Make sure you do have a decent sort of access. At least the path in here is not too steep at any point, and that really helps.

MICHAEL: You don't seem to have any major regrets about it.

CHRISTOPHER: No. To have built a road in here would have dramatically changed the place. It would be a pity to sacrifice the place just for the convenience of the process.

MICHAEL: I notice that there's an area roped off. What's that about?

LUCY: It's a nature sanctuary. Since we are doing so much, and there's so much commotion—more impact than has been here. Because usually it's been just Christopher and me, and that's pretty low key.

CHRISTOPHER: Before that there was nobody.

LUCY: Yeah. So what we did was, we roped off an area, a nice beautiful section, that's just for the nature spirits themselves. No humans go in there, no kids, dogs, nothing. It's just a place for them to hang out, undisturbed.

MICHAEL: How do you imagine this level of care and sensitivity to the site will affect what it feels like to live in the building when it's finished?

CHRISTOPHER: I'm really interested to see that. I think it's going to be great. It will be hard to distinguish which part is the result of that and which part is the result of the history of how the building came to be and the people involved, and the materials. It's pretty good so far.

LUCY: I think it will keep a really nice harmony to the area. We won't have to go through years of trying to get it back to how it was. I think we'll be able to just get right back into the flow.

Drainage and Foundations 10

THIS CHAPTER IS ABOUT TWO OF THE least visible but most important parts of your building: drainage and foundations. We talk about the importance of good drainage and how to achieve it—including redundant systems on very rainy or wet sites. Then we discuss what a foundation needs to do, and several different ways to lay your own, even without previous experience. Don't be disheartened if we emphasize what can go wrong, for it is our job to keep you out of difficulties, as much as possible. We ourselves have individually and collectively made or seen enough mistakes that we can save you some heartache.

> THE BUILDING NEEDS TO STAND ABOVE ANY SURFACE FLOODWATER THAT MAY OCCUR IN ITS LIFETIME, WHICH COULD BE A MILLENNIUM . . . OR WE COULD, LIKE OUR ANCESTORS ON MOST PARTS OF THE EARTH, ALLOW OUR DWELLINGS TO DECAY AND RETURN GRACEFULLY TO THE EARTH EVERY COUPLE OF GENERATIONS.

SITE DRAINAGE

Cob can absorb large amounts of water without damage if it is able to dry quickly. But if cob is flooded or saturated for long periods of time, the embedded straw will begin to rot away and the wall will lose its tensile strength. A cob wall that is saturated all the way through its base will lose compressive strength and may collapse (see appendix 3 for a fuller treatment of this problem). To prevent this from ever happening, you need strategies for removing water from the vicinity of the walls. Plan in advance what forms of drains you will need to divert surface runoff and roof runoff from rain and snowmelt, as well as subsurface flow and seasonal springs.

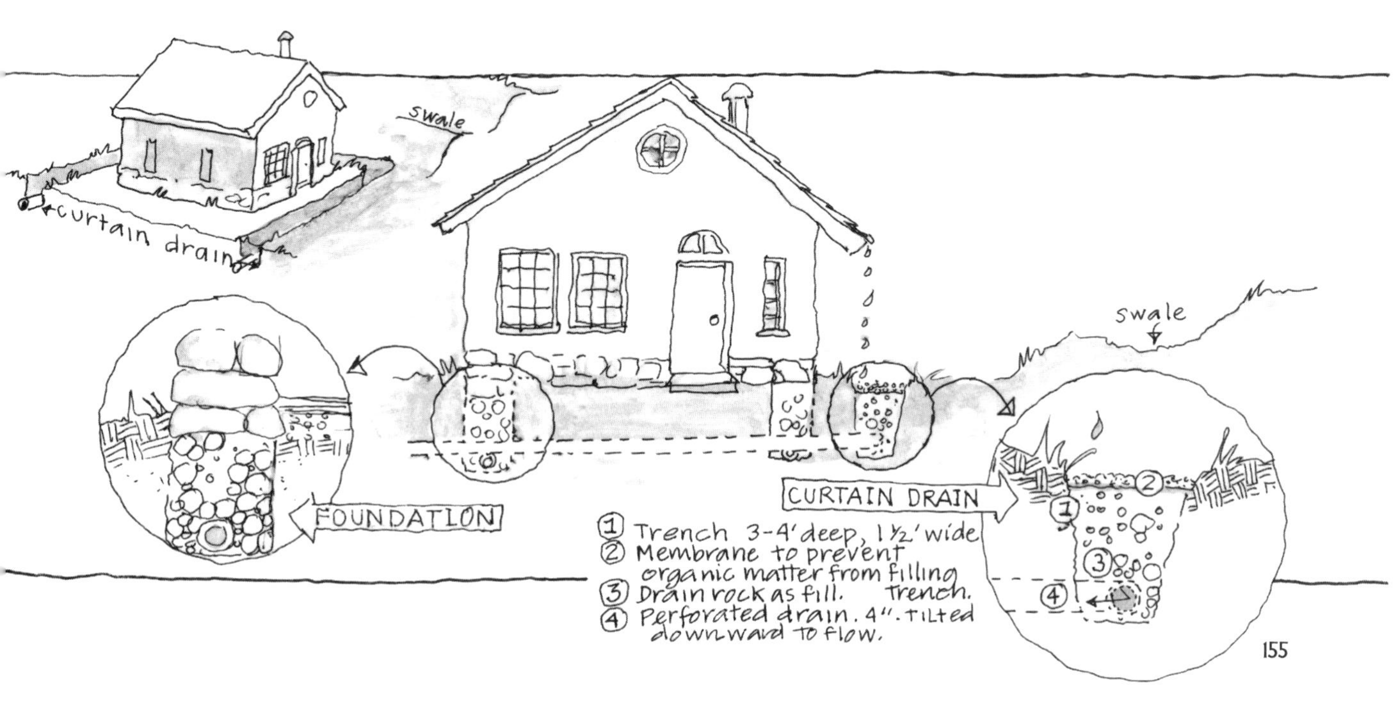

First, consider your foundation. The building needs to stand above any surface floodwater that might occur in its lifetime, which could be a millennium. Also the ground beneath the building should stay dry, to prevent the shrinking and swelling of underlying clay from moving the foundation, and to keep the interior dry and free of mold.

Rubble Trenches

Unless you are building directly on bedrock or on super well-drained soils (coarse gravel or extremely sandy) we recommend a rubble trench all around the perimeter of the building. The rubble trench is a traditional drainage system from the Middle East, introduced to the United States by Frank Lloyd Wright, beginning in 1902. He spoke very favorably of it in contrast with conventional concrete foundations but commented also on the obduracy of building inspectors.

A rubble trench is merely a trench located directly beneath a stemwall and backfilled with drain rock (round gravel or small stones). We usually lay a 4-inch perforated polyethylene drainpipe in the bottom of the trench to improve drainage. Traditionally a rock, brick, or sometimes wooden tunnel was constructed for the same purpose, and until recently terracotta drain tiles in 1-foot lengths were common and inexpensive. The bottom of the trench should be sloped, so that water will run out of it "to daylight," meaning that it eventually comes to the ground's surface downslope from the building. If your site is not sloped, a much inferior option is to drain the trench into a "dry well," a deep-hole dug some distance away from the building and filled with stones.

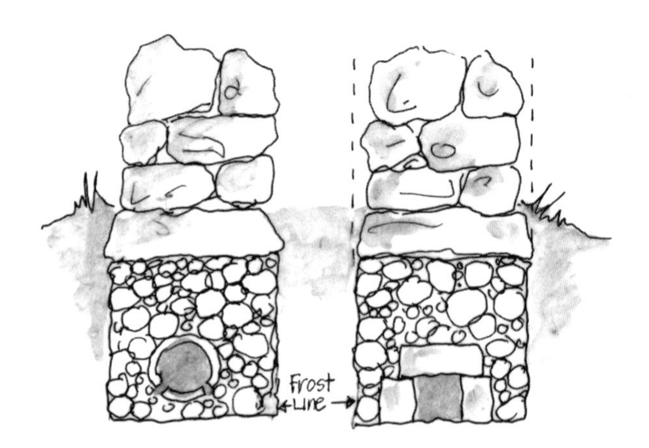

Rubble-trench foundation. Use 4-inch polyethylene drainpipe or perforated ceramic drain tile, or construct a rock or brick drainage channel. Note: Foundation flares beneath ground level, so trench should be slightly wider than the wall above.

Other Drainage Techniques

In dry climates the rubble trench may be sufficient in itself, but in other cases you will want to create additional drainage outside the building. This is particularly true if the ground is marshy, poorly drained, or composed of heavy clay, or if there is ground sloping down toward the building that may carry runoff during heavy rain.

To divert runoff, dig shallow ditches or swales upslope of the building. A swale is a broad, shallow ditch, running slightly off contour, often with a slight berm or bank on the downslope side that may be planted with moisture-loving vegetation. Swales slow down surface runoff, giving it a chance to percolate into the ground, and redirect it away from the area you wish to keep dry. You can integrate swales into orchards and gardens, or use them to fill ponds or cisterns.

As a second line of defense, we often recommend the installation of a "curtain drain" or "French drain," particularly on sloped sites where you have cut into the bank to create a level building site, and on marshy sites where groundwater rises close to the surface. A curtain drain is a deep trench filled to the top with porous material (we usually use 1½- or 3-inch drain rock above a 4-inch perforated flexible plastic pipe). The bottom of the trench should at all points be lower than the interior floor of the building and fall continuously to an outlet downslope from the site. It should be located approximately under the edges of the roof, around at least three sides of the building (the uphill sides).

Don't guess whether your site drainage works—test it! Fill several 50-gallon drums with water just above the building site. Tip them all over at the same time to mimic a heavy downpour. Or be there during a heavy rain. Watch exactly where the water goes. You want storm water and roof runoff to disperse and penetrate slowly into the ground rather than concentrating where it can cause erosion.

Digging Trenches

All your trenches should be at least deep enough to reach solid, compact subsoil, beneath any organic matter or roots. In areas where the ground freezes, it is customary and prudent to dig foundation and drainage trenches down below "frost line," the lowest level that freezes in winter. Ask locally how deep that would be. In very cold winter climates such as the northeastern United States, frost line can be as much as 6 feet deep, creating a prodigious challenge for manual ditch diggers. If you're building in such a place, consider that theoretically the rubble trench should work even if it isn't that deep, because when the weather is warm enough for water to be liquid, it should flow out of the trench before it freezes. Even if it should freeze in the trench, it can expand into the air spaces between the drain rock, rather than heaving up a solid foundation. For extra security, in deep-frost locations you can line the outside of the trench with rigid foam insulation. If your building is in an area of mild winters where the ground doesn't freeze, a depth of 12 to 18 inches is usually adequate for a rubble trench.

The walls of the trench should be vertical, and the trench floor fairly flat and clean. Remember that water needs to drain out of the trench, so slope the bottom a minimum of ½ an inch in every 4 feet. Test it to make sure by dumping in a few buckets of water. Keep digging as necessary until all the water flows out the open end of the trench. Tamp the bottom

THE IMPORTANCE OF GOOD DRAINAGE

THE IMPORTANCE of good drainage cannot be overemphasized. When I built my first cob cottage, I didn't understand this well enough. The site we chose was near the bottom of a hill in rainy western Oregon, so there was a large area upslope of the house that collected rainwater. The groundfloor was designed with a split level to follow the contours of the ground, so that the floor in the sitting area was about two feet lower than the kitchen. The stone foundations were built hastily, in shallow trenches with only a few inches of gravel at the bottom. The foundation trenches beneath the uphill walls weren't as deep as the level of the lower part of the floor. There was no curtain drain, because the site was too rocky to excavate by hand. To make matters worse, I installed a poured adobe floor in the lower section without any gravel beneath it for drainage.

During the heaviest rains of the first winter, the predictable happened. I woke up one morning after an all-night downpour to find the sitting area flooded, with almost two inches of water on top of the adobe floor. In a panic, I scooped the water up with a dustpan, threw it out the door, and cranked up the stove. The floor took many days to dry out, during which I couldn't use the sitting area for fear of damaging the floor. The miracle was that the earthen floor, coated with several layers of linseed oil and beeswax, survived without permanent damage. Luckily, the cob was raised on its stone foundation above the level of the floor, so it didn't get wet. I tried to deal with the drainage problem by digging a diversion ditch uphill from the building, but it was too late to do a good job. Each of the three winters I lived there, I experienced a flood, generally after the heaviest rain of the season.

Trenching tools.

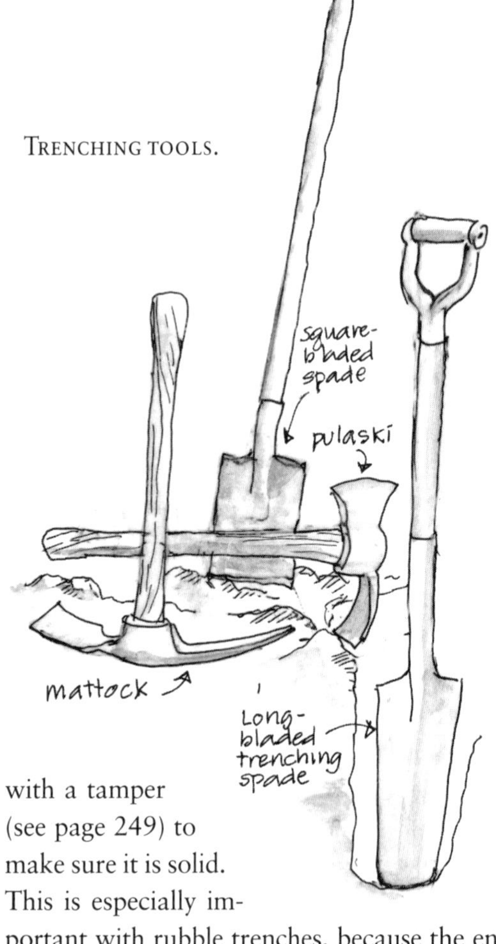

with a tamper (see page 249) to make sure it is solid. This is especially important with rubble trenches, because the entire weight of the building will be bearing on the bottom of that trench.

The best tools for digging trenches are sharp, square-bladed spades or long-bladed "trenching spades." To loosen up hard subsoil, use a mattock or pulaski.

Most of what you dig out of a trench is potential building material, so plan in advance where to store it. Separate topsoil for gardening, landscaping, and earthen roofs into one pile, subsoil for building into another, and rocks into a third. Earth is heavy, so avoid moving it twice.

On wet and rainy sites, give thought to what will happen to the water flowing out of your drains. You don't want to create an erosion gully or mudslide downhill from the building. It's good to drain your ditches into an on-contour swale or a small pond. Either of these will slow down the water and give it a chance to drop any silt it may be carrying before percolating into or spreading out over the ground.

Filling Trenches

Before filling your trenches, install utility ducts for the building, running underneath the foundation. These are short sections of pipe through which you can later run electrical wires, water pipes, and wastewater drains.

Drain rock should be fairly round, between 1 and 3 inches in diameter, and clean of fine particles such as silt and clay. Some fist-sized rocks, broken brick, old concrete, and so forth can be incorporated, to clean up your site and economize. If you're really lucky, you'll have a source of clean gravel on or near your site. For example, on the land where I live there is a seasonal creek that has deposited a large bed of gravel a hundred yards from my house site. To fill the rubble trench

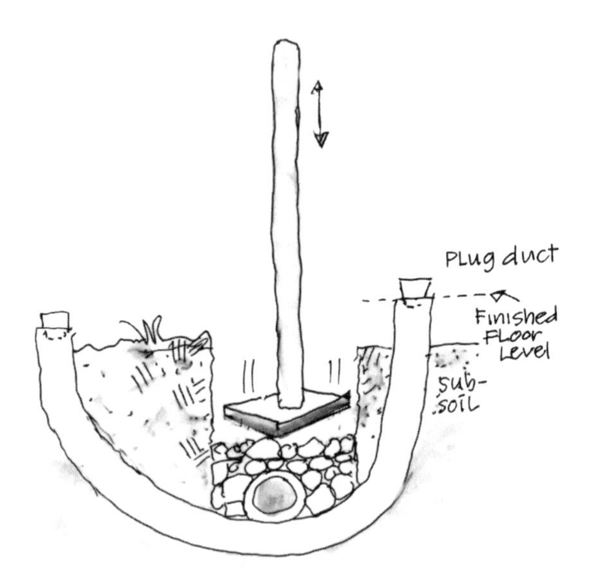

Utility ducts are curved sections of pipe laid beneath the foundation and plugged at either end for use later running electric and water line and wastewater drain. First install the utility duct, then fill trench with 1–3-inch rocks, and tamp every six inches.

and drain the floor under the building, my friends and I had to screen around a hundred wheelbarrow loads of gravel and haul them uphill to the building. It was a lot of work but rewarding, considering the environmental impact of most commercial gravel operations.

The less purchased rock you use the better. Most drain rock is dredged out of rivers by large mining operations that release huge amounts of silt into the river, increasing turbidity and diminishing habitat for many kinds of aquatic life, including endangered fish. The only convenient alternative in many places is crushed rock from a quarry. This is also an environmentally destructive and energy-intensive process, and it produces an inferior product for drainage purposes. If you have a choice, purchase drain rock harvested from old (not current) riverbeds or glacial deposits.

Fill in the trench with layers of no more than 6 inches at a time, compacting each layer with a tamper before adding the next. For this purpose the broad, square tampers you can buy at the building supply store work well. You can also make your own by welding a heavy steel plate onto the bottom of a rod, or bolting a square of thick plywood onto a wooden handle.

To prevent your lovingly dug and filled trench from clogging up over time with silt and soil, you should protect the top. Rubble trenches will be mostly covered by the stemwall and floor, but they may be a little wider on the outside. In this case, and always with uncovered curtain drains, use a horizontal layer of open-weave fabric, about six inches below ground level. There is a commercial product available for this purpose, called "landscaping cloth," but you can also use woven polypropylene feed sacks or woven poly tarps. Ever wonder what to do with those unsightly decomposing blue tarps that seem to plague building sites everywhere? Now you know.

FOUNDATIONS

A foundation is the durable masonry substructure upon which a building stands. It needs to raise the base of cob walls above any chance of getting soaked through, to hold the building together in case of ground movement or earthquakes, and sometimes to spread the weight of point loads such as posts on soft or inconsistent ground.

Most conventional foundations include both a "footing" or "footer"—a wide, solid base that spreads the load of the building over a larger area—and a stemwall, which raises the bottom of the wall away from contact with the earth. Because cob walls are thick and monolithic, they impose a very even load on the ground, so spreading their weight with an additional footing is normally unnecessary. In place of a footing, we usually build a rubble trench, which combines load-bearing and drainage functions. A foolproof drainage system simultaneously eliminates frost-heave problems, rising damp, damp floors, "shrinkable clay" expansion pushing your building apart, and all sorts of rot, mold, fungus, and mildew problems, not to mention minor flooding.

A stemwall or plinth is the above-ground portion of a foundation. Stemwalls can be built of any long-lasting geological material, including stone, fired brick, concrete, and perhaps soil-cement. Most continuous perimeter foundations for conventional houses are of poured concrete, often with a complete concrete floor slab. The owner-builder would probably prefer on both economic and ecological grounds to avoid some of the sixty cubic yards of concrete that go into the average new, American, wooden house. However, on unstable slopes and seismically active areas, it is advisable to pour a continuous reinforced concrete bond beam on top of the rubble trench, as a base for the stemwall.

For cob construction, the foundation stemwall should be at least knee-high (more than

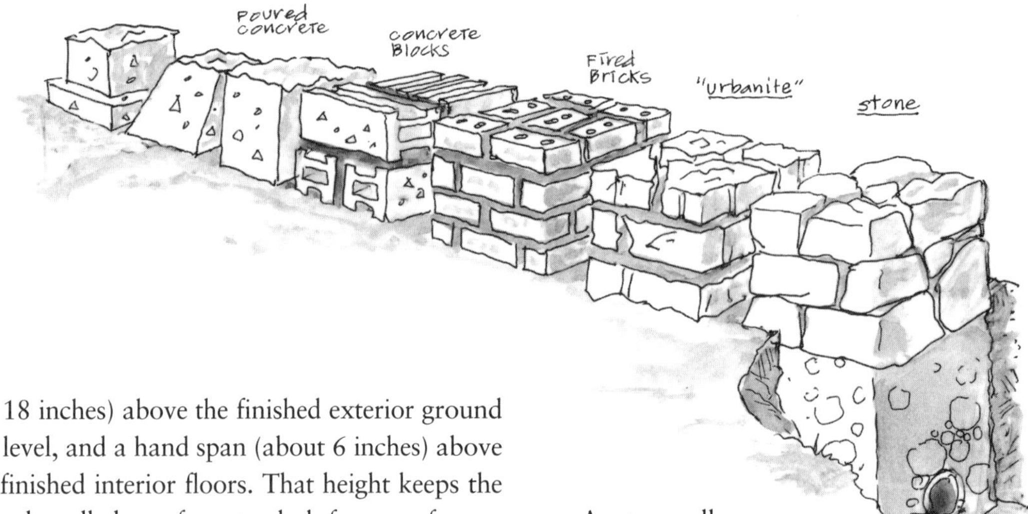

Stemwall options. A variety of materials can be used. A rough top surface ensures good contact with the cob wall.

18 inches) above the finished exterior ground level, and a hand span (about 6 inches) above finished interior floors. That height keeps the cob wall clear of most splash from roof eaves and leaky gutters outdoors, and relatively safe from flooding caused by plumbing failures indoors. Some English cob houses have a much higher stone plinth, up to three or four or in some cases even eight feet. This may have been to protect the cob from water and rubbing animals, or it may simply have been because people preferred the look and durability of stone but didn't have enough to construct an entire building.

An unusually high stone plinth in Devon. The unplastered cob shows a lot of exposed shillet, or sharp gravel. Note also reed-thatch roof.

HOWARD HOUSEKNECHT

A stemwall should be as wide as the cob wall, perhaps a little wider at the bottom for added stability and to help with spreading the load. Don't let the top of the foundation project beyond the base of the cob on the exterior side, as driven rain could accumulate on top of the stemwall and penetrate the cob, weakening the base of the wall.

In earthquake zones or mountain areas where ground movement is at all likely, a concrete bond beam should be poured above the rubble trench and below the stemwall. This bond beam should be a minimum of 6 inches deep beneath the full width of the stemwall. Pour the bond beam all at once with plenty of tensile reinforcement such as rebar, old wire fencing, barbed wire, or polypropylene baling twine. You can also incorporate the same kinds of tensile members into the mortar between courses of masonry in the stemwall. If you're building the stemwall of stone or concrete chunks, set the bottom course directly into the bond beam while it's still wet. Note the need to maintain a continuous bond beam beneath doorways. If you don't, the cob may crack severely as the two sides of the foundation settle differently.

Our favorite stemwall material is stone, to which we devote an entire section below. If you live in town or anywhere that stone is

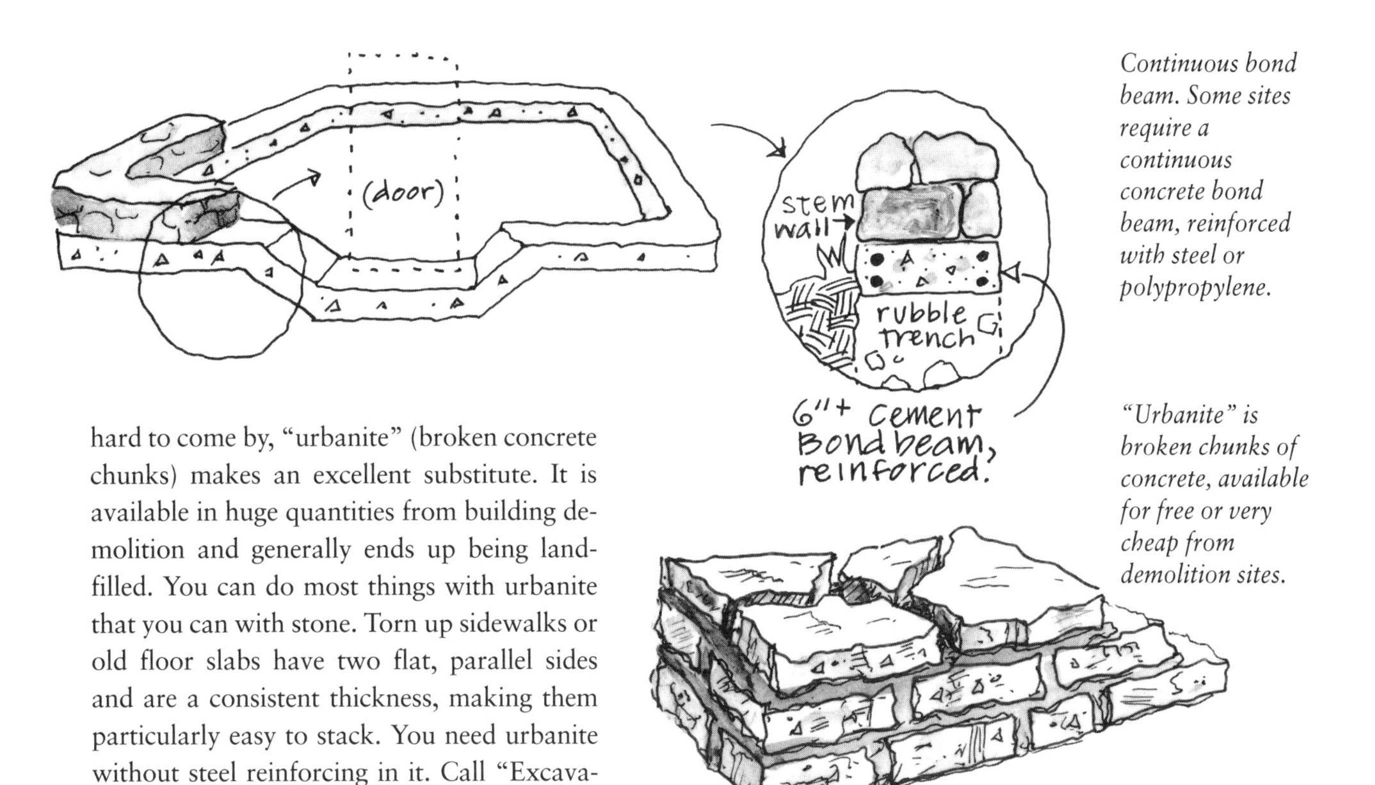

Continuous bond beam. Some sites require a continuous concrete bond beam, reinforced with steel or polypropylene.

"Urbanite" is broken chunks of concrete, available for free or very cheap from demolition sites.

hard to come by, "urbanite" (broken concrete chunks) makes an excellent substitute. It is available in huge quantities from building demolition and generally ends up being landfilled. You can do most things with urbanite that you can with stone. Torn up sidewalks or old floor slabs have two flat, parallel sides and are a consistent thickness, making them particularly easy to stack. You need urbanite without steel reinforcing in it. Call "Excavation Contractors" in the Yellow Pages; they are often happy to deliver used concrete to your site by the truckload for free. To build an urbanite stemwall, follow the instructions below, as for stone.

We try to avoid all-concrete, poured foundations, the universal norm in the building industry. Despite (or rather because of) its many useful qualities, concrete can be seen as an environmental disaster of epic proportions. Concrete is second only to water as the world's most consumed substance: slightly more than a ton of concrete every year for each human being on the planet. Cement kilns release enormous quantities of carbon dioxide and are major contributors to the greenhouse effect. The manufacture of cement is highly energy-consumptive, as is its transportation. Because toxic wastes are often used to fire the kilns where it is made, cement may contain PCBs, dioxins, and heavy metals, some of which could be released into your house. Many parts of the planet appear to be in the grips of a concrete monster, rapidly converting all that is green and alive to a lifeless gray wasteland. Since concrete is effectively indestructible, how future generations will deal with the enormous quantities we are bequeathing to them is a drama that will continue to unfold during many millennia. Perhaps the future will be full of urbanite foundations and retaining walls.

Building a Stone Stemwall

Stone is a natural for building foundations. Strong and water resistant, and in many places locally available and cheap, a well-built stone foundation is very attractive and can last for centuries. Many so-called stone walls are in fact largely concrete, with stones inserted for show and to reduce the amount of cement. But given good building rock, you can make a stemwall that relies on the strength and fit of the stones themselves, with or without mortar, to bear the load of the walls and roof. Mortar keeps air, moisture,

A high stemwall of rounded glacial rock and cement mortar.

weight carefully on each stone in turn, and not a single stone should shift.

As you place your base course, focus your attention on both the inside and outside surfaces of the wall, trying to make them smooth, attractive, and close to plumb. Often, you are actually building two walls, one facing out and the other facing in. The space between can be filled later with irregularly shaped stones, clean rubble, and mortar. To tie the two walls together, use frequent "tie stones" bridging the whole width of the wall and visible on both the inside and the outside surfaces. You may want to insulate between the two faces of the wall with vermiculite, perlite, pumice, or mortar with insulant aggregate such as polystyrene packing peanuts, vermiculite, or sawdust. Or you can attach lightweight plaster or rigid foam insulation board to the outside of the stemwall, later.

Although most stones are not really shaped like bricks, they should be stacked in a similar manner. Almost every rock in the wall should rest firmly on at least two in the course beneath it, and in turn support two more in the course above. To achieve this, it's necessary to create level courses in which the tops of adjacent stones are at the same height. Make these level courses as long as is easy, but periodically it's fine to step up or down as you proceed along the wall. At the boundary between two levels, first build up the lower course to the height of the higher, then place a large stone to bridge the crack. Each stone you set should be completely stable, even before you put other stones on top of it. This is where small, wedge-shaped rocks—"chinkers"—come in handy, to stabilize any stone that rocks on its base. For maximum "tooth" between the foundation and the cob, it's good to leave the very top of the foundation as irregular as possible.

If you use mortar, construction can still proceed exactly as above. After each course of stones is in place, wet it down and fill in all

and small animals from passing through the wall. Although a sand/clay mortar may be adequate (particularly in seismically stable areas), for novices we generally recommend cement/sand mortar, which stabilizes the wall during construction and requires less perfection in selecting and placing each stone.

Before you begin constructing your stemwall, gather as many stones as you think you will need and spread them out on the ground around your building site so you can see them all. Place the biggest ones nearest the foundation trench, but leave a three-foot pathway to move and work in. Spend some time familiarizing yourself with your stones, turning them over to see what they look like underneath. Then begin by placing the largest stones first, forming a solid base course. Each stone should touch its neighbors, and should be bedded into the gravel of the foundation trench until it is completely stable. At any time during the process of building a stone wall you should be able to walk along the top of it, putting your

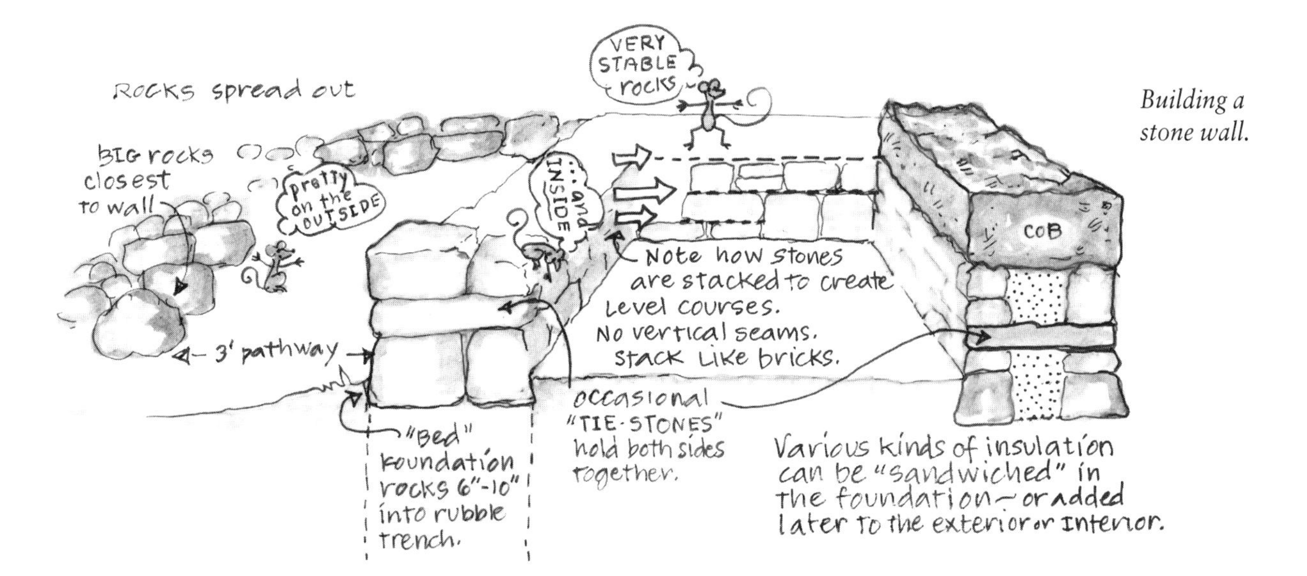

Building a stone wall.

the cracks and cavities with mortar and small stones, working them in with a mason's trowel, putty knife, or stick. Before the mortar dries, wash off any surfaces that will be visible. A strong and inexpensive mix for this sort of work is one part lime to three parts clean sand, if you want to avoid cement, or one part lime to two parts Portland cement to nine parts sand. The lime is not strictly necessary in a Portland/sand mix, but it makes the mortar a little more flexible and slower to set. Lime mortar sets very slowly, so allow at least two weeks before building on top of it. Portland cement mortar is firm a day later. For best setting, keep lime mortar shaded and dry, and cement mortar shaded and wet, for several days after building. Always remember to wash thoroughly any tools that have been in contact with the mortar.

The major drawback to stone foundations is their instability during earthquakes and ground movement. Between every two stones is a joint that can crack and separate when stressed. In earthquake areas, lots of cement mortar can help hold a wall together. Taken to an extreme, the result is a concrete wall with stone faces for visual effect. For details on building a "slipform" stone-faced concrete wall, read Helen and Scott Nearing's *Living the Good Life* or Karl and Sue Schwenke's *Build Your Own Stone House Using the Easy Slipform Method*.

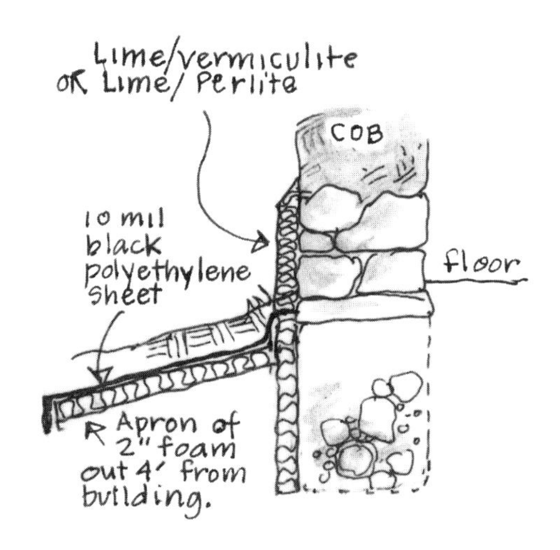

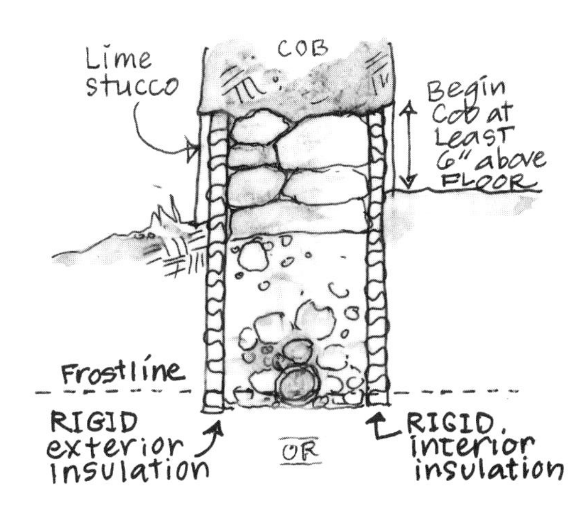

Insulation options for interiors and exteriors of foundations.

Adobe on massive tapered stemwall in seismic zone, Tlaxcala, Mexico.

Experimental Foundations

Stone foundations and reinforced concrete both have their drawbacks, as noted above. So far, foundations have received less attention from the natural building movement than wall systems, but opportunities are ripe for experimentation and research. Two innovative foundation types that may not be "natural," yet seem suitable for cob in some circumstances, include rammed tires and earth bags.

Rammed Tires: There has been quite a lot of experimentation in recent years with recycled tires as a building material. New Mexico architect Michael Reynolds has developed a housing style called "Earthship," which uses tires rammed full of earth for most of the exterior walls. Earthships are generally recessed or bermed into the ground, which makes them best suited to dry climates.

To make a foundation for cob, use just two or three courses of tires stacked on a rubble trench or sealed on the bottom with some sort of waterproof material to prevent moisture from wicking up into the tire fill. The tires are laid out touching each other, then each is rammed full of slightly damp earth using a sledgehammer. The average car tire will absorb around 300 pounds of earth. The second course, laid in a "running bond" like bricks, conforms itself to the shape of the course below. Rammed tire foundations are probably stable and durable. They are thought to have excellent seismic resistance because they can bounce around a bit in earthquakes but will find their way back to where they belong.

Unfortunately, rammed tire foundations seem to have as many disadvantages as advantages. Foremost is the extremely taxing labor of filling them. A strong person can fill one tire in perhaps an hour of constant ramming, and most of us are exhausted and disheartened after ramming a single tire. Tires are unnecessarily wide for a stemwall, and it can be difficult to deal elegantly with the extra width. Their size and shape limit your control over the form and flow of your whole building. It can also be hard to get clay-based materials such as cob and earth plasters to stick to rubber. Many people are concerned, too, about the health effects of possible offgassing from rubber tires.

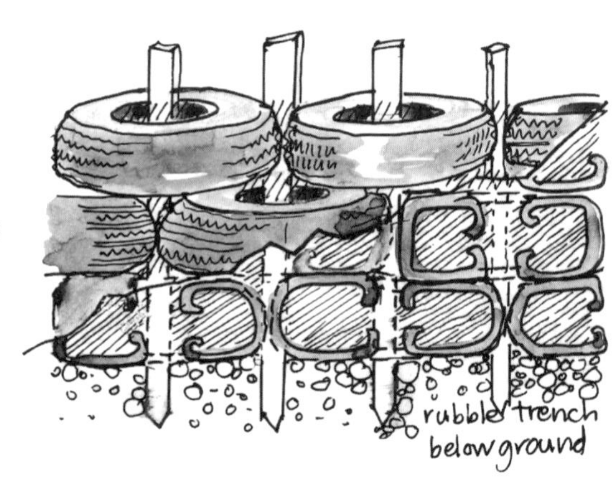

Rammed tires filled with earth and tamped solid. Stakes serve as anchor for cob wall above.

Earth Bags: Rammed earth bags have most of the advantage of tires, with fewer problems. Many sorts of bags can be reused this way, from natural fiber coffee sacks to woven polypropylene feed sacks. The latter are also available new, fairly cheaply from flood-control supply centers. Nader Khalili, who pioneered this technique at the California Institute of Earth Art and Architecture (Cal-Earth) buys rolls of poly tubing before it is cut and sewn into feed sacks. Using tubes up to 30 feet long, he builds earth bag domes and arches that have proven resistant to rain, floods, and earthquakes. Because the bags can be filled with anything from sand and gravel to heavy clay soil, this is a remarkably versatile and inexpensive technique.

Whether you use short or long bags, fill them in place with slightly moist earth, then sew the ends closed with nails or simply place them end to end against each other to prevent opening. As each course is completed, stomp it down with your feet, then tamp from above with a heavy tamper, a process much easier than ramming tires. For additional earthquake resistance, place continuous strands of barbed-wire "mortar" between each course of bags. This not only prevents the bags from slipping, but provides tensile reinforcement through the whole foundation, helping to hold the building together as a single unit.

The disadvantages of this system mostly have to do with durability. Natural fiber bags (while nontoxic and made entirely of renewable materials) decay when exposed to water. Polypropylene bags, while immune to rot, break down rapidly in ultraviolet light; fortunately, polypropylene is a relatively "clean" plastic that degrades into benign components. Protecting the bags with a mud "sunscreen" immediately after construction and then covering them with earth or lime plaster should prevent solar degradation.

Another concern is the possibility of moisture migrating through the fabric of the bag and either causing the clay soil inside to expand and heave, or traveling upward to saturate the base of the cob. In rainy northwestern California, where earth bags are appealing because of their earthquake resistance, I have adopted the following approach. First I build a rubble trench, making sure it drains well. On top of the filled trench, I place one or preferably two courses of bags filled with gravel (round drain rock 1½ inches or smaller is preferred, although crushed rock also works), creating an aboveground extension of the rubble trench footing. If more height is desired, additional courses of soil-filled bags can be added. To improve the "keying" between the slick bags and cob, stout wooden stakes are pounded into the top course. Although we feel enthusiastic about this low-cost, quick, and easy foundation system, its long-term durability and water resistance remain unproven.

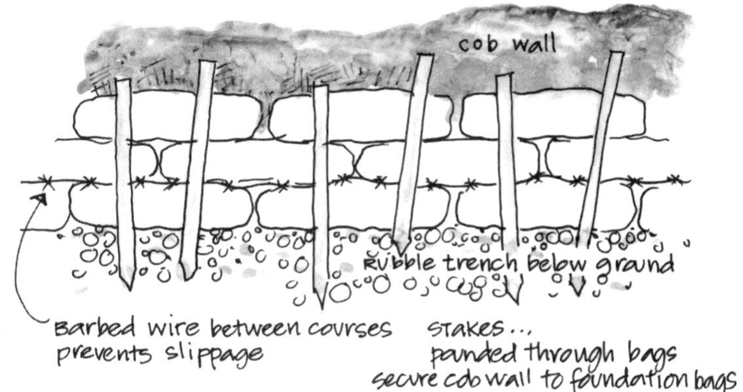

Earth bag or sandbag foundation.

Each of these foundation systems, while promising under some circumstances, has serious drawbacks. We are apparently still far from finding a durable, nontoxic, earthquake-resistant foundation that can be assembled easily from locally available materials. One approach to this problem is to adopt an active and rigorous research program, trying new systems and new combinations of materials in the hopes of developing a foundation system as strong and healthful as stone, as earthquake resistant as reinforced concrete, and as easy as earth bags. Another option would be to emulate our ancestors on most parts of the planet and abandon the goal of permanent housing, allowing our dwellings to decay and return gracefully to the earth every couple of generations, leaving no unsightly and toxic detritus behind.

11 Making the Best Cob

❧ MUD TRANSFORMS WORK AND MAKES IT PLEASURE. . . . BUILDING IS PARTICIPATION IN CREATION, IT IS COMMUNION, IT IS SHARED LABOR AND A SHARED GIFT, BOTH. ❧

THIS CHAPTER DESCRIBES THE CORE OF the process of cob building. Here you will learn how easily you can turn the ground beneath your feet into a building material, without machinery or chemicals, and at almost no cost.

Traditional cob, made entirely of ingredients dug from the building site and using only waste straw or wild grasses, often survived for centuries, despite being of less than ideal composition. With the modern advantages of cheap transport, inexpensive straw, and possibly mechanized mixing, Oregon cob can be expected to be stronger in both tension and compression. Carefully measured components can be varied to create a wide spectrum of cob qualities—a different mix tailor-made to each unique application.

When making Oregon cob you will find it worthwhile to strive for very complete mixing, yet constantly adjust your technique for speed, rhythm, ease, and quality. This is a new science, not yet ten years old, and complete novices often hit on innovative techniques.

Mixing is the most time- and energy-intensive part of the whole cob building process. While developing Oregon cob, we have experimented with many methods for mixing, both manual and mechanical, and have gradually increased our speed and efficiency to a level that at first seemed impossible, without losing the basic joy of the work.

HOW COB WAS MIXED HISTORICALLY

Traditionally, mixing was done by people's or animals' feet. There were, of course, local vari-

ations in mixing technique, as in every other aspect of cob building. Following are two accounts of traditional British cob mixing:

> The old method of mixing cob by hand is as follows: A "bed" of clay-shale is formed close to the wall where it is to be used, sufficient to do one perch. (A perch is a superficial measurement described as 16½ ft. long, 1 ft. high, and the amount of material will vary according to the thickness of wall required.) Four men usually work together. The big stones are picked out. The material is arranged in a circular heap about 5 or 6 ft. in diameter, and starting at the edge the men turn over the material with cob picks, standing and treading on the material all the time. One man sprinkles on water and another sprinkles on barley straw, from a wisp held under his left arm. The heap is then turned over again in the other direction, treading continuing all the time. "Twice turning" is usually considered sufficient.
>
> —from Williams-Ellis and Eastwick-Field, *Building in Cob, Pisé, and Stabilized Earth.*

> The soil is first broken down to a fairly fine tilth, all large stones greater than about 50 mm dia. [2 ins.] being removed in the process. It is then spread out in a bed some 100 mm [4 ins.] in depth on a hard, pre-wetted surface on top of a thin layer of straw. Water is then added and a second, thicker layer of straw is spread evenly on top (about 25 kg of straw per cubic metre of soil—1.5 to 2.0 percent by weight [2 lbs. per cubic foot]—is considered adequate). The straw is then trodden into the soil which is turned several times, more water being added as required . . .
>
> —from "The Cob Buildings of Devon 1: History, Building Methods and Conservation" by Devon Historic Buildings Trust

In parts of France, rammed earth and sometimes cob were known as *torchis.*

Traditional English cob construction.

> *Torchis* was prepared in the hole from which it was dug, and the hole afterwards would become the farm pond. Water was added to the clay, and the mixture was puddled for several hours. In the Brière region of Brittany until the 1940s puddling was done by men who linked arms to tread the clay barefoot, dancing and chanting as they did so. Then gravel was added to stabilize the material, short lengths of straw to bind it and often cow dung to make it more adhesive. After this the dance was repeated and the resulting *torchis* was left to dry for several days before use.
>
> —from Paul Walshe and John Miller, *French Farmhouses and Cottages*

PREPARING CLAY SOILS FOR MIXING

To make good cob, the water, sand, clay, and straw need to be intimately combined. Every grain of sand and every stem of straw needs to be smeared forcefully with clay to ensure good adhesion. To best achieve this, in most cases, the clay (or clay soil) should be thoroughly soaked before adding sand and straw. In nature clay is sometimes deposited almost pure, but usually the clay contains an admixture of

Presoaking dry clay soil, either by adding water to a hollowed-out pile or making a soaking pit with straw bales and a tarp, or in buckets or bins.

other things—sand, silt, organic material—in short, it is a *clay soil.* For convenience in this section we will generally refer to clay soil as clay, though that doesn't imply it contains nothing else.

Dry clay soils can be hard and lumpy, and difficult to work with. Mixing will be much easier if you presoak the soil. Clay can be soaked by hollowing out a big pile like a volcanic crater, then filling with water, or you can soak it inside buckets or barrels. If you're digging out a pond, loosen compacted soil in the hole every evening, then add water when you leave so that the soil on the bottom can soak all night. Aboveground, you can prepare a clay-soaking pit by staking straw bales to the ground in an open rectangle, then draping a waterproof tarp over and between them to create a lined trough. Tie back or weight down the tarp's edges so they don't fall into the pit. Make it big enough to soak a whole truckload, or sufficient clay soil for several days' building at a time, and locate it where you can shovel into it directly from the truck or out of the ground. Add water to cover everything, and leave to soak. This clay-soaking pit can later be used for a cob-mixing pit (described later). Try various methods.

TESTING YOUR MIX

For an ideal cob mix, the critical proportion is the ratio of clay to sand. You need enough clay to make a plastic, cohesive, workable mix, but not so much that the mix shrinks a lot and cracks. Depending on the coarseness of your sand and the quality of the clay and other components in your soil, your final mix should end up between 5 percent and 25 percent clay. Based on the percentage of clay in the soil you estimated from the shake test (see chapter 8), you can guess at a workable proportion of soil to sand. Perform the "snowball test" and "crunch test," described below, to help refine your estimated proportions of sand to soil. Then make test batches and test bricks to confirm that the proportions you have selected will work for building.

The Snowball Test

To test for a good building mix, combine measured cupfuls of clay and sand in different proportions: 3:1, 2:1, 3:2, 1:1, 2:3, 1:2, 1:3. When the samples are thoroughly mixed, add just enough water to make them stick together when you squeeze a double handful really tight. They should be quite dry—drier than pie crust dough. Make a compact sphere about 2½ inches in diameter out of each mix, like squeezing a snowball really hard. Hold your snowball up between thumb and index finger of one hand, then squeeze gently between index and thumb of the other, at right angles to the first squeeze. A good ball should be dry enough not to deform more than a quarter-inch and hard enough not to break—not pliable, not wet, not crumbly.

Next, take each sphere in turn, hold it a meter above soft ground (a grass lawn, for instance), and let it fall. If the ball shatters on impact, it is too dry or it contains too much sand. If it flattens or deflects, it contains too much clay or too much water. The ideal mix should hold its shape on impact and look just like it did before you dropped it.

The Crunch Test

The goal with Oregon cob is to build a wall of sand grains mortared together with clay. Sand

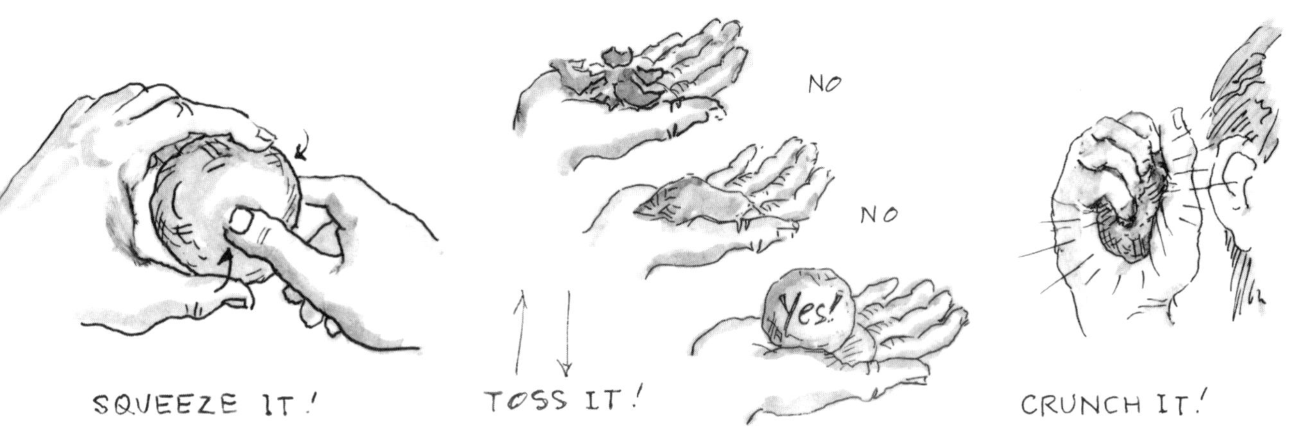

Snowball tests, to see which sample cob mix is best: squeeze, toss, and crunch the cob to see (and hear) how it behaves.

grains should be touching one another, so that the mix can't shrink much as it dries. One very simple way to determine this is what we call the crunch test. Take a small handful of each mixture. Hold it up to your ear. Squeeze and listen. If the mix has enough sand, you will hear the sharp grating sound of sand grains rubbing against each other—Crunch! CRUNCH!—like scraping sandpaper. There is usually a marked difference in sound between mixes with enough sand and those without. Try many different samples; get used to the *sound* of a good building mix.

Test Batches and Test Bricks

For comparative purposes, it's useful to make a few test bricks, which you can dry and examine for cracking, crumbling, and strength. Using the tarp method below, make a few small (4 to 6 gallon) batches without straw, using slightly different proportions of clay and sand, keeping track of which is which. Form part of each mix into test blocks approximately the size and shape of bricks (about 4 × 8 × 2 inches). Number them and keep a reference list so you know what proportions of sand to soil were used. Then add straw to the rest of each mix, tread it in thoroughly, and form it into similar-sized blocks.

Leave the test bricks in the sun or in a slightly warm oven until they are dry all the way through. Complete drying may take several days. You might have to break a brick in half to be certain. Surface cracking indicates too much clay. Try scratching the bricks with a nail or a knife to test for hardness: they shouldn't score deeply or crumble easily. Try breaking one of the bricks containing straw by twisting it between your two hands. If you have a good mix and the brick is dry all the way through, it should be nearly impossible to break this way. You may be able to break one over your knee, but be careful! You might get bruised in the process.

Make test bricks of different mix proportions. When completely dry, examine them for cracking, crumbling, and strength.

IANTO EVANS

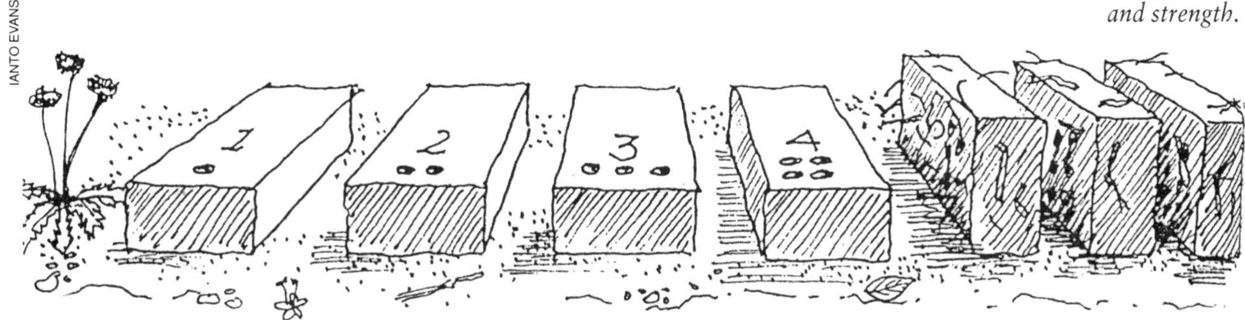

MIXING COB ON TARPS

A great breakthrough in manual cob mixing occurred in 1994, when Becky Bee developed a system for mixing on tarps. Before that, we had been mixing with shovels on a level platform made of tamped earth, concrete, or plywood. The tarp method is quicker, easier on the lower back, and requires fewer tools. It has now diversified into several quite different techniques—"different folks, different strokes!"—and it pays to change technique as circumstances demand. Experiment!

You will need a squarish piece of durable, slick, and water-resistant material, six to eight feet on a side, larger than your armspread by about a foot. Some people prefer a tarp a little longer in one dimension, 7 × 8 feet for instance, or 8 × 10.

Lay the tarp out on clean and level ground, close to your cob ingredients and your building site. We have found it saves work to dig out a shallow dish for the tarp to lie on: about 8 feet in diameter and 6 inches deep in the center works well. Spread the ingredients out on the tarp, alternating buckets of sand and clay to accelerate mixing. We normally use from three to five 5-gallon buckets of ingredients for a single batch of cob. This is the largest amount most people can handle easily and repeatedly.

Solo mixing, using a tarp.

When mixing by yourself, you can simply grasp one or two corners of the tarp and walk backward over the tarp until the mix is folded back upon itself. Do this repeatedly, rotating to a different corner each time, until the dry materials are mixed.

Though it is quite easy to mix alone, the initial stages of mixing are faster with a partner, so call over a helper if one is nearby. Each person should firmly grasp two adjacent corners of the tarp, then lean back slightly. Keeping their spines straight, both partners slowly rock *side to side* from one leg to the other, using the greater strength of their legs to roll the material on the tarp back and forth. Part of the tarp and most of the weight should remain solidly on the ground as you roll the dry materials across the tarp. After a few long rolls, in which the mix travels all the way from one side of the tarp to the other, stop and rotate positions 90 degrees. Then rock and roll in the other direction to make the mixing more thorough.

The dry mixing stage should take less than a dozen rocking motions. Once you can no longer see pockets of different colors and textures in the dry mix, it's time to add water. Using your hands or feet or a shovel, make a crater in the middle of the mound of mixed earth, and pour in water. Always add a little less water than you think you'll ultimately need. It's easier to add more than to compensate if you add too much.

With the water added, repeat the mixing process described above, rocking the mix back and forth a few times. Then tread a little, roll the mix, add more water if necessary, tread, roll, and so on. Don't get stuck on any one motion, and stay alert to what will have the most effect for the least effort.

For treading, take your shoes off. Barefoot mixing keeps you in contact with changes in the mix and is really enjoyable and therapeutic. It is advisable to have toughened your feet in advance by going barefoot for a few days. If you have to wear boots because of cold weather, skin problems, or lots of rocks in the mix, try leather workboots or other flat-soled footwear. But each day, at some point, take off your shoes and dance barefoot. It's exhilarating, even in freezing weather, and the sensors in

LINDA SMILEY

Steen *Møller, mixing cob at a workshop in Denmark .*

your feet, adapted by evolution to fine-tuned alertness, will tell you instantly how well your mixes are developing. There is no substitute for the sensitivity of barefoot mixing.

Learn to dance a twist on the cob mix, using a smearing motion to break up lumps and distribute clay particles. Either raise your knees high and jump on the mix or use a heel-first tread to maximize pressure. Try different methods.

If you are working with a partner, approach the task like ballroom dancing, formally. One treads, the other lifts tarp corners. Treader retreats from turner, turner lifts, rolls, replaces tarp (note positionof the turner's right foot in diagram). Turner moves clockwise, to the left. Treader, facing turner, *without breaking the rhythm,* moves to his or her *right,* also clockwise, onto the most unmixed part of the mass. With rhythm, you can work very fast, very smoothly. When either tires, she or he shouts "change!" and trades positions. Two people, energetic and rhythmic, can make and build a cubic meter a day (35 cubic feet, about a ton and three-quarters wet).

Once the lumps are all broken up and the water evenly distributed through the mix, begin to add straw. Hold a flake off a bale (2 to 3 inches thick) under your elbow and allow it to sift down onto the mix as you dance. Keep treading, using your heels now, until all the straw is dirty and worked into the mud, then turn the mix by pulling one corner or edge of the tarp toward you until the mix folds over on itself. Do this repeatedly as you add more straw, pulling from a different corner each time, making certain to pull the tarp far enough to turn the center of the mix, so you don't end up with an unmixed mass there.

How do you know when you have added enough straw and your mix is done? After practicing with the process, you will know

"Ballroom dancing": One person holds a corner of the tarp while the other treads the mixture. As the tarp holder lifts the corner to turn the cob, the treader retreats. Then the treader changes direction and the tarp holder takes a different corner.

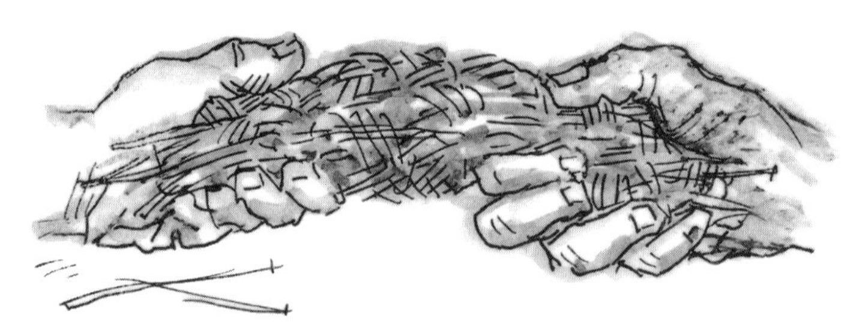

Take a double handful of cob. The straw in the mixture should resist pulling apart.

intuitively, but here are several tests to get you started. As you add straw, you should be able to feel the consistency of the mix changing under your feet. Suddenly it will reach a point where it feels like a tough, cohesive substance rather than like loose mud. It will get harder and harder to sink your feet in. If the straw is long enough, you will reach a point where as you turn the mix it rolls up like a burrito and holds together in a single mass instead of cracking along the line where it is folded.

When working in a little more straw seems to be taking a lot of effort, stop adding it. Your material should be a good consistency for building. Reach down and seize a double handful of cob. Is it difficult to pull away from the rest of the mass? Does it hold its form and pack readily into a ball that you can throw and catch without breaking? Have your partner hold on to one end of a cob loaf and take the other, your fingertips touching theirs. Try to pull it in two. The straw in the mix should resist your pull, making it difficult to break.

Another way of mixing solo. Dance backward down a slope, pulling the upper corners of the tarp to roll the mixture from side to side.

Another way to tarp mix solo is to work on a sloped site, carefully spreading all the ingredients out across the top of an 8- × 10-foot tarp, for instance. Add water carefully. Jump onto the pile and tread, then grasp one top corner and pull back diagonally over the whole pile, flipping the mix toward yourself as you tread. When the mix gets to the side of the tarp, grasp the second top corner (don't let go of the first). Dance backwards, pulling diagonally until in turn this side can't go any further. Work your way backward down the tarp, pulling up more tarp with each hand as you go. By the time you get to the bottom of the tarp, the mix may be finished. If not, lay the tarp flat and haul both the bottom corners uphill, rolling the mix up with you (you may need help for this maneuver). Then repeat, until the cob is well mixed.

Yet another technique involves a large tarp; 10 × 14 feet (or up to 30 feet) is ideal, allowing two people to make larger mixes if the materials are dry. Spread the materials across one end of the tarp, then each grasp a corner closest to the pile. Walk back across the length of the tarp, both of you, rolling the material all the way to the far end. Repeat in the opposite direction. Often only three or four rolls will have thoroughly mixed sand, clay soil, and straw. Then add water and tread, rolling the heap as needed.

A basic goal of any heavy manual work is that it be rhythmic and effortless, a smooth and comfortable dance. The joy comes from the dance itself, not from anticipation of finishing the job.

The key to making mixing easy and enjoyable is a consistent rhythm. Keep trying until

With two people, you can mix a bigger load of cob on a larger tarp.

you find it. Work to reduce the time and effort it takes for every phase of the process. You'll be amazed at how efficient you can get!

Notes on Tarps

The tarp has become one of the most important tools in cob construction. Finding tarps that suit you is critical to making the work enjoyable and efficient. Below is a partial list of kinds of tarps we have tried, with some of the pluses and minuses for each. Many thanks go to Misha Rauchwerger for his research in this department.

Used Lumber Wrappers: First choice! Lightweight and easy to hold while mixing, often available in large quantities for free from your local mill or lumber store, these woven polypropylene tarps (sometimes with a paper backing) are used to keep plywood and kiln-dried lumber dry during transportation. They aren't very durable, but the price is right, and if you can get some productive use out of them on their way to the landfill, so much the better. Ask at your lumberyard, or check their rubbish pile.

"Blue Tarps": These are made of woven polypropylene and are also available in other colors. They are inexpensive and readily available. They shed the cob mix easily. Unfortunately, they are very susceptible to damage by ultraviolet light in sunshine. Store them in the dark if you want them to outlast a single season and don't drag them across rough ground when loaded, as they tear and puncture easily. Be aware that like many products of the construction industry, they are usually smaller than the advertised size.

Plastic Sheeting: Available in many thicknesses, usually black or clear, plastic sheeting is ubiquitous. Rolls of 8 or 10 mil black polyethylene are inexpensive, but hard to find in small dimensions. Most kinds tear and puncture easily and are slick and difficult to hold on to, so look for plastic with fiberglass reinforcement. Clear plastic photodegrades rapidly, except for UV-resistant types made for greenhouses.

Housewrap: The woven polyester sheeting used to wrap new construction (also known as Tyvek or Typar) is extremely durable, but is hard to get in small quantities unless you find an offcut discarded on a building site, or beg one from a contractor. If you are using salvaged housewrap, it may be brittle due to exposure.

Canvas: Expensive and heavy, especially when wet, canvas wears out rapidly and tears if exposed to lots of water and rough treatment. It takes a lot of work to keep canvas tarps clean, since the mix tends to stick to them. Oiled canvas, available as expensive trucking tarps or from army surplus, is much more durable but heavier and awkward to use. Canvas is not recommended though it seems ironic to build natural houses with synthetic throw away tarps. Is there a better natural fabric?

RV Awnings. Recently Misha Rauchwerger made a discovery at his local recreational-vehicle shop, where he was allowed to cut damaged awnings from their aluminum housings. Vinyl reinforced with nylon fibers, they are waterproof, tear and puncture resistant, and much more UV resistant than polypropylene. One 7'6" by 16' awning cut in half makes two perfect-sized cobbing tarps. These and housewrap are the most durable tarps we have used, often lasting through two cobbing seasons. But recycled awnings are heavier and stiffer, therefore more difficult to grip than woven polypropylene.

THE PIT METHOD

After they have mastered the tarp method, which provides the most consistent and homogenous mixes, some cobbers move on to mixing in a pit. Pit mixing has several advantages: one person can easily make a mix far larger than the usual tarp-mixed batch; big, wet mixes can be kept on hand until needed; and the clay can be soaked first in the same pit, eliminating the need to shovel and move cumbersome soaked clay.

One way to make a mixing pit is just to dig an unlined saucer-shaped depression in the ground, perhaps 6 feet in diameter by 1 foot deep. It's helpful to have a solid, well-compacted floor so you can tell where your cob mix stops and the ground starts. The advantage of unlined pits (at least in well-drained soils) is that they dry out the mix faster, and you can scoop the cob out of them with a fork without tearing up a tarp.

Using a mixing pit, one person can mix a larger batch than with a tarp, and a big, wet batch can be presoaked or stored in the pit.

Alternatively, you can use something like the aboveground clay-soaking pit described earlier. Place four or more straw bales with corners touching. Drape a tarp across them, but don't tie it to the bales or pin down the corners. The advantages of using a tarp are that it's easier to keep the cob mix separate from the underlying soil and the mix can be kept moist by folding the tarp over it. Bales can be pulled aside for convenient removal of the finished cob.

To mix cob in a pit, combine clay soil and water in the pit and mix and stomp vigorously until all the lumps are dissolved. Make the mixture much wetter than for tarp mixing—about the consistency of pancake batter. After the clay is completely suspended in the water with a smooth consistency like a milkshake, add the sand slowly, mixing as you go. In a pit, most of the mixing happens by sloshing about with your feet, which can be a lot of fun, and is easier for children and small adults than rolling a heavy tarp. A large hoe or similar implement helps the initial mixing. If you are using a tarp-lined pit, pull back on each of the corners of the tarp in turn to move unmixed material into the middle. Make sure the clay, sand, and water are thoroughly combined before adding straw and treading it in. Note that the wetter mud absorbs proportionally much more straw than a dryer mix. So watch out that you don't add *too much* straw, making the mix hard to build with.

The main disadvantage of pit mixing is the difficulty of getting the ingredients mixed evenly, particularly if you're using an unlined pit. It can also be hard to judge when there's enough, or too much, straw. If you make too large and wet a mix, it may sit for so long before being applied to the wall that the straw will begin to rot, so the wall will lose strength. If your mix sits for more than a couple of days before being used, mix in more fresh, strong straw just before you use the cob.

USING MACHINES FOR MIXING

There has been a perhaps inevitable interest in mechanizing the labor-intensive process of mixing cob. Machinery can play a valuable role in excavating and delivering raw materials, and we are slowly refining our techniques for mixing with machines. In England, today's renaissance cob builders use a tractor to mix, often on a concrete pad. In the United States, several machines have been tried for mixing: the tractor with front bucket or backhoe, the Bobcat, and both concrete and mortar mixers. Mechanized systems have certain advantages, but also involve problems.

A Tractor with a Bucket: A tractor can make enormous batches fairly quickly, but you need a hard base to work on. Soil, sand, and water are mixed together with the bucket or backhoe, then the mix is driven over to compact it. These two steps are repeated until the cob is thoroughly mixed, while a helper on foot periodically sprinkles on an even layer of straw. Frequently check the progress of mixing from the ground, as it is very difficult to see what is happening from atop the machine. Unless the operator is unusually patient and methodical, the resulting tractor cob tends to lack uniformity and is often short on straw. Compared to the silence and safety of an unmechanized worksite, the noise, smell, and hazards of heavy equipment can be stressful and exhausting.

Another problem with mixing huge quantities at one time is that the straw can begin to rot if the cob is not used within a few days. Rotten straw not only weakens the cob, it can also produce an almost unbearable stench. On one site in northwestern Washington, the cob was mixed by tractor in the late fall, as the weather was beginning to turn gray and wet. The cob already smelled foul by the time most of it was built into the walls, and the unpleasant odor lingered in the finished building for a year.

Tractor mixing, Denmark.

Bobcob. A Bobcat is a small, 4-wheel-drive tractor capable of spinning its front and back wheels in opposite directions. Several people we know have had good results using a Bobcat to mix cob. In the summer of 1999, in an urban lot in Eugene, Oregon, Mark Lamberth built a small cob cottage in nearly record time. Using a Bobcat dramatically increased his building speed. In six hours, he and a friend were able to mix more than 20 tons, approximately 17 cubic yards. In Mark's own words, this is how he did it:

> I ordered a full dumptruck (10 yards) of sand and had it placed in the street, just beside the piles of clay. I then began moving loads of clay into the street with the tractor, while a helper was wetting down the mix with a hose. I drove back and forth over the mix, using the tractor tires and the front loader to grind and homogenize the mixture. After a couple of hours, and several hundred gallons of water, the mix began to coalesce and feel like cob.

MUD, LABOR, PLEASURE

BY KIKO DENZER

MUD (AND THE SHARING OF IT) transforms work and makes it pleasure. It took a while before I really understood this. The pleasures of ploshing around making huge batches of wet cob were apparent when I had help (about a third of the time), but doing it alone seemed like toil.

One weekend I was cobbing with a friend to funk music. My knee-jerk reaction was that tape-recorded funk is wrong for natural building. Fortunately, when you're mixing cob, you can't jerk your knee without landing your other foot in a pile of mud. So I got into the rhythm. One rap group, Arrested Development (now defunct, sadly), has a great song with the refrain: "Put your hands in the dirt. Children play with earth." Their rhythms were complex, and invited either fast or slow dancing. It occurred to me that we need rap cobbers and cob rapping. We need to cross the urgent nowness of urban rap with the stable foreverness of earthen building: dynamism with depth. Society, like biology, requires diversity to evolve.

After that weekend, I started taking my little boom box out with me when I went to work alone. Now this seems obvious, but still I was amazed at the effect music had on my work. The very hard physical labor of moving and mixing and forming tons of earth, sand, straw, and water suddenly became a dance that I could do all day without tiring. Indeed, at the end of a day working to music, even by myself, I would be elated and energized. I think the surprise may have been due to achieving a *physical* understanding of that relationship that defines who we are—a relationship that words simply do not describe—but that action does: rhythm, pattern, breath, and sound; steps, touch, texture, feeling, weight. And it is not just a temporary physical "high." Building is participation in creation, it is communion, it is shared labor and a shared gift, both. And it does not end when you stop working—the building acquires a life of its own, and the maker moves on.

We eventually added several bales of straw. By late afternoon, we had a huge mess and a load of cob the size of a pickup truck. It was great! I delivered the cob to the building site from the street.

I returned the tractor to the rental yard the next morning. I had to rent a pressure-washer to clean up the street. The tractor rental was $100 for one day and the pressure washer cost $35.

Mark found it difficult to get the mix just right, to keep the proportions, and to incorporate enough straw. But he had no trouble keeping the cob wet enough to use for a whole month by covering it with plastic. Surprisingly, he had no problems with rotting straw.

Concrete Mixers: Standard drum concrete mixers tumble dry material well but don't smear the clay forcefully onto the sand. They work okay for mixing soil, sand, and water, but not for incorporating straw. If you're determined to use one, put a couple of 20-pound rounded rocks in to tumble with the cob. Each time they fall, the rocks will smash some of the clay and sand together. Better yet, pour the mix out onto a tarp and mix in the straw by foot. The extra work seems hardly worthwhile.

Mortar or Plaster Mixers: Unlike a concrete mixer, which has a fixed paddle and *tumbles,* a mortar mixer is a large drum with an independently rotating paddle and *smears*. The paddle moves the mix in a figure-8 pattern that seems quite effective for cob. Different people have had different results with mortar mixers. Some claim they can make good stiff cob with plenty of straw as long as the batches are kept quite small. Other people prefer to make large, wet batches in the mixer and then tread the straw in by foot on a tarp. Otherwise there can be problems with straw tangling around the paddle and overstraining the motor. Choose a fairly powerful motor—at least 8 or 10 horsepower. You may be able to rent a mixer one day a week and mix up enough cob for several days' building, keeping the site free of machines the rest of the time.

The prospect of mixing enough cob to build a sizable structure may seem like an overwhelming task. Before cob can become a viable alternative for even mainstream building contractors, it may be necessary to develop better mechanization. For now, even setting aside the issues of mix quality, noise, risk of injury, fuel use, and embodied energy, buying a mixing machine is outside the budget of many cob owner-builders, though renting is a possibility. Mixing by foot on a tarp or in a pit is still the cheapest and easiest way for most of us to make good-quality cob.

If you do decide to mix by machine, keep in mind that the quality of the cob will most likely be inferior. Make your walls thicker, and remix small batches by foot wherever high quality is essential.

TIPS FOR SPEED AND EFFICIENCY

We've found that by conscientiously working to increase efficiency and concentrating on logistical details in order to make things go more smoothly, people can learn to mix cob more rapidly and with less effort than they at first imagined possible. A fit person working alone can expect to mix and build half a cubic meter per day (about 15 cubic feet). If you are exceptionally strong and well organized, you might increase that up to a full cubic meter per day.

The following suggestions should help any cobber increase mixing speed and efficiency:

❧ ***Plan ahead to avoid moving heavy materials twice.*** Position piles of earth, sand, and straw as close as possible to where you'll be building, ideally within shoveling distance of the mixing area. This saves time and energy in moving cob ingredients to your tarp, and in moving the cob itself to the wall. Make sure you have enough buckets, tarps, and shovels on hand so that you aren't waiting around for tools.

❧ ***Soak the clay well in advance.***

❧ ***Spread out the sandy portion on the tarp first.*** This way, the clay soil doesn't get stuck to the tarp.

❧ ***Premix your ingredients.*** Alternate, sand-clay-sand-clay when dumping materials onto your tarp.

❧ ***Don't overmix.*** Pay attention to what you're doing and know when enough is enough. Mix without conversation until the work is completely automatic, as when learning to drive a car.

❧ ***Mix wet.*** If you can allow time to wait between mixing and building, mix wet. Water lubricates the cob and makes mixing much easier. Before building, you can leave your whole mix to dry uncovered for a few hours or overnight, or build while it is still muddy and let it dry out on the wall.

❧ ***Pay attention to weather.*** To avoid getting tired quickly, do most of your mixing

Leo Houck with an 8 h.p. mortar mixer, making cob.

COB MIX TROUBLESHOOTING GUIDE

Problem	Solution
Mix sticks to your feet or the tarp	Add more sand
Mix crumbles, won't hold together	Add clay and/or water (could also mean mix has too much straw)
Wet loaves pull apart easily	Add more straw, or longer straw (could also mean mix is too wet)
Test bricks crack while drying	Add more sand
Dry test bricks are soft, crumbly	Add more clay
Dry test bricks break in two easily	Add more straw, or longer straw

when it's cool (early mornings or evenings during the hot summer), or when there's a good breeze. Don't mix in the sun; create shade to work in.

- *Work mostly solo.* The most efficient (but not necessarily the most fun) way for a group of people to mix cob is for each person to have their own mix on their own tarp. They can pair up for the initial stages of mixing by rocking the tarp back and forth between them, then complete the stomping, rolling, and straw-adding stages on their own.
- *Develop your rhythm.* Treat tarp mixing as a kind of enjoyable dance. Music and drumming can keep your energy up and make mixing more fun.

CUSTOM MIXES

Once you have reached a level of expertise where you can consistently make strong, stiff mixes in a timely and efficient way, you may want to vary each mix according to its intended application. The relative proportion of the major ingredients—clay, sand, straw, and water—can be increased or decreased. Different sizes of sand or lengths of straw can be selected. Sometimes other ingredients can be added for special characteristics. Psyllium husk, horse dung, gravel, flour paste, white glue, chopped baling twine, all have been tried—the possibilities are endless. Use your imagination and your understanding of the functions and behaviors of the different ingredients. Below is a list of some of the reasons you might decide to alter your basic mix, with ideas on what you might change in each case.

For hot, dry weather, or slow building: Make the mix much wetter than usual—it will be much easier to mix. Then let it dry out a bit overnight or over lunchtime before building. The tricky part is knowing when you have enough straw in the mix, because your normal clues will be muddied by all that water, and you might go on adding straw until the cob becomes hard to work with.

For wet weather, or fast building: Make stiffer, dryer mixes. Mixes with more sand and less straw than normal will tend to contain less water and dry out faster.

For sculptural details and around windows: Use a finer mix, screening out rocks from soil and sand and chopping the straw, because long straws left sticking out of a sculptural detail or close to glass can be difficult to trim. We sometimes add a little extra clay for a more cohesive, workable mix, but be cautious; with too much clay, cob can crack.

❧*For extra insulation:* Increase the amount of straw or substitute perlite, vermiculite, or pumice for sand. Pumice cob is difficult to mix and hazardous to the skin because of all the tiny, glassy fragments. You might also try adding sawdust or wood chips, which increase insulation but reduce strength and slow the drying.

❧*For extra hardness or thermal mass:* Increase the amount of sand as much as you can, and reduce the proportion of straw. To improve thermal performance, you could build exterior walls with an inner layer of high-density, sandy cob with better-insulating, straw-rich cob on the outside of the wall.

❧*For straps, corbel cobs, arches, and shelves:* These all need extra long straw for tensile/shear strength. See chapter 13.

12 Building Cob Walls

❧ YOU WILL BE PROUD WHEN YOU REVIEW EACH DAY'S WORK, ENJOYING THE SATISFACTION OF A TIRED BODY, AND HAVING CREATED SOMETHING BEAUTIFUL. ❧

NOW THAT YOU KNOW HOW TO MANUFACTURE your own cob, you are ready to learn how easy it is to build with. This chapter covers methods of moving your mix to the wall, several ways to apply it, and how to connect it to previous layers. We will explain how to keep the wall vertical (or non-vertical if you prefer), how to build safely as the wall gets higher, how to trim and shape your work, and how to keep the recently built parts in workable condition—not too wet, not too dry.

Hopefully this chapter will answer most of your remaining questions about actually building walls with cob and save you from many problems. As with any manual craft, there is no substitute for learning directly from a master, so if possible attend a week-long workshop or join an experienced builder for a few weeks (see the resources listings in the back of the book). If that isn't possible, just start building; you will learn by doing it.

The basic intention in cob building is to build as durably as possible, with rhythm and enjoyment, at a comfortable rate and without frustration or mistakes. You will be proud when you stand back and review each day's work, enjoying the satisfaction of a tired body, and having created something beautiful.

COBS, GOBS, AND BLOBS

There are three main methods for building cob walls. Each uses the same basic mix, which you learned to make in chapter 11. Each technique is appropriate under different circumstances, although it is common to use all three in the same wall. You will develop your own preferences as you practice.

❧ ***Trodden cob*** (this has sometimes been called *pisé*) is fast, easy, and rather crude. It works well if you can mix right up against the wall you're working on. Its best use is for thick walls, below about eye level and without openings, though walls can be trodden as high as you can easily hoist the material. This is the most common way in which traditional English cob was built, and was sometimes combined with moveable wooden forms, a primitive style of rammed earth.

❧ ***Cob loaves*** or "cobs" take longer to prepare than trodden cob but can be placed more accurately and attached more firmly. The quality tends to be better, because you must feel every handful as you prepare each loaf, facilitating quality control. Cob loaves can be tossed to a builder quite high on the wall, so they are well suited for higher parts of the building. Since they can be gently and firmly sculpted in place, use them for parts of the building where there is a need for care and accuracy, where walls are thinner, and for special applications such as shelves and arches. This is a traditional system used in England, Africa, and Yemen.

❧ ***Gaab-cob:*** The technique we call Gaab-cob versatile and fast, using a wetter mix and combining the best features of trodden cob and cob loaves. Although it is a recent development, we use it increasingly for most general wall building. Its chief disadvantage is that it takes longer to dry, which can be an

LAURA AND DANNY GORDON

In Yemen, traditional cobbers catching cobs and building with them.

issue in cool damp weather or with thicker walls, or if attaining rapid height is a goal.

Trodden Cob

Trodden cob is applied by a team of two: a lifter on the ground and a treader on the wall. First, tread a fairly dry, firm cob mix out flat on a hard surface. If it was mixed on a tarp, roll it off the tarp onto solid ground alongside the wall, as you'll be using a digging fork that could tear the tarp fabric. A cob mix flattened out to about 2 to 3 inches thick is easiest to pick up and tread onto the wall. Using a long-handled digging fork (a garden fork with flat tines, not a pitchfork with sharp round tines), the lifter picks up flat plates of cob and places them quite precisely on the wall. The treader then walks over them, squishing them into place.

To make sure the material is well attached, it is customary to tread right to the edges of the wall, spreading the mix out over the edge where it will be beaten back or sliced off. Beating is done by the treader using a tool named with traditional English understatement, "the persuader." This is a heavy wooden paddle with a short, thick blade, with which you whack the living daylights out of the wall. You can easily carve a persuader from a 4-foot length of scrap 2 × 4. Under most circumstances you also may need to work the edges by hand. For trimming the edges, the treader can use a long-handled, flat, sharp-edged spade with the head in line with the handle, aimed downward from above. (See later in this chapter for instructions on trimming.)

Cob loaves should be sized for the particular use intended.

Trodden cob is fast and serves well for low thick walls without many openings, but has some shortcomings. It is difficult to ensure a good bond between layers, so test your adhesion regularly by trying to peel the new layer off the top of the wall. Excessive trimming requires more work, so try to avoid squishing too much of the mix beyond the line of the wall; all that trimmed material has to be picked up, reconsolidated, and forked back onto the wall again. And though a persuader can help with a thick, solid wall, be careful with walls less than about two feet thick. Excessive zeal in chastising irregularities can knock the opposite face of the wall out of line. Overpersuasion can also result in cracking later, when the wall dries out.

Cob Loaves

If you have ever kneaded bread, making "cobs" will be easy. First, roll the mass of finished mix to one side of the tarp, or break it loose from the ground beneath so you don't tear off your fingernails pulling it up. Kneel on the edge of the mixing tarp, maybe on a flake of straw, or wear kneepads. Plunge forward and seize a double handful of the mix, tearing off a chunk you can handle. Compress it under your hands as you roll it toward you. Using the weight of your whole upper body, bear down, rolling it into a crude loaf. With practice this whole action should take no more than five or ten seconds. For people who can't kneel, it may be possible to form loaves on a very solid table or platform, lower than your waist height when standing.

You are not making bricks. Cobs are made mostly to transport the material to the wall, so they don't need to be perfectly shaped, or smooth, or squared. Don't overwork them. The only criterion for a good cob is, "Will it hold together when caught, thrown from ten feet?"

Toss each finished cob loaf to someone building the wall, or stack it on the ground to your side. If it is fast drying weather, stack cobs in a close pile, then immediately cover them. If you made a wet mix for ease and need the cobs to dry out a little before building, stand them seprated from one another so the breeze can dry them more easily.

Moving cob loaves from the mixing tarp to the wall can be done as in a bucket brigade with a "cob toss."

Loaves should be sized to accommodate everyone who might be catching them and roughly shaped to fit the particular use you have for them. If somebody other than yourself is using them, check that the cobs you supply are the right size, texture, and shape; consistency of manufacture can streamline the building process, as will good communication.

Moving a pile of cobs from mixing tarps to the wall becomes fun and even humorous when the loaves are thrown from person to person in a "cob toss." Line your team up at comfortable intervals, and simply toss cobs along the line as in a bucket brigade. When they arrive at the wall, stack them near but not on top of the section to be built upon. With practice, you may be able to build them in as fast as they arrive! While it may not be strictly "efficient," most people love the cob toss. It may be the high point of their day, so don't forgo it if you have a big group of volunteer help!

Gaab-cob

Gaab-cob (pronounced "gobcob" and named after our colleague Dana Gaab, who invented this system in a 1994 workshop) is mixed very wet, then loaded onto the wall in *gobs,*

Ianto picking up big gobs, applying them to the wall, then sewing them in with a cobber's thumb.

DEANNE BEDNAR

as large as you can easily carry. It can be loaded by the forkful, or in big, loose, double handfuls or even armloads, then worked into the wall's surface by hand or with a "cobber's thumb," a smooth stick or other hard object that has a blunted tip like a thumb.

With both cob loaves and Gaab-cob, it is usually easiest to apply quite a wet mixture, a little stiffer than mashed potatoes. Be sure to leave the faces of the wall vertical (or shaped as desired) before moving to another wall section.

For speed of construction under good drying conditions, Gaab-cob is an easy winner. However, be sure to compensate for its wetness with a high sand content, or as the wall dries it may crack.

CREATING A MONOLITHIC STRUCTURE

Whichever delivery technique you use, it is important that there is good bonding between layers to ensure a solid durable wall. Gravity will hold a cob wall together even if the layers are not well attached, but good bonding will help your building survive earthquakes, ground settlement, hurricanes, and other shear forces. Until you have confidence that your technique is working well, *test regularly* by trying to pull new material off the top surface. The latest layer of cob should be really difficult to remove, even after only a few minutes. Drop-in volunteers can learn rapidly to make reliably good mixes, but check that they attach material to the wall securely.

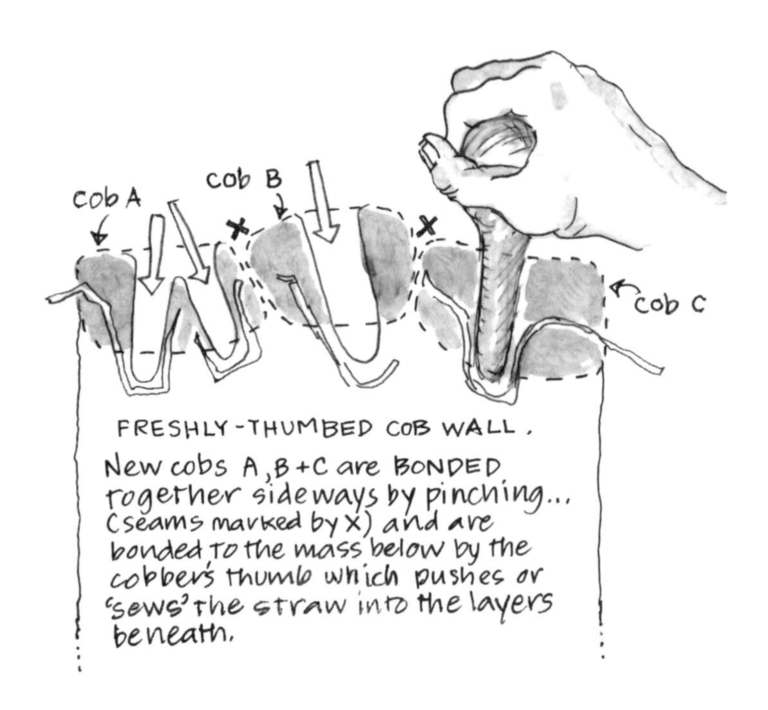

The most critical action is to marry each new application into the mass of the wall. The goal is a three-dimensional textile of interwoven straw, buried inside a strong and durable mass of sand and clay. This sewing action can be done either with your fingers or with a *cobber's thumb*. When you're done, it should be hard to tell where any one cob stops and another starts, and the top surface will be well perforated with finger-sized holes. These perforations help the next layer to bond, they help the wall to dry evenly, and they accelerate drying.

Work each cob into its neighbors. Pinch them together, side to side. Smear with the heel of your hand down the face of the wall, along the seams between new cobs and old, to eliminate any cracks into which air could penetrate, causing two sections of the wall to pull apart as they dry. Check by looking up at the top of the wall from below. A shadow along the edge where you just added material indicates a poor bond.

Cob loaves and Gaab-cob can be applied to the wall in any pattern you choose, but we often use one called "spine and ribs." This pattern leaves a really lumpy working surface to improve bonding between successive layers and helps dry the wall's center more equally with the sides, frequently an issue when many people are working together on a small building. The spine and ribs pattern also helps to equalize the rate of drying between the wall's center and its faces. Make a high ridge of cobs down the middle of the wall, all attached well

LEFT: *edges have not been smeared down tight, and the cobs are poorly bonded, as the straw isn't sewn between layers. Note top and slumping shoulders.* RIGHT: *A monolithic mass is formed by sewing straw between layers with cobber's thumb. Note rough surface on top, a good base for next layer to grab.*

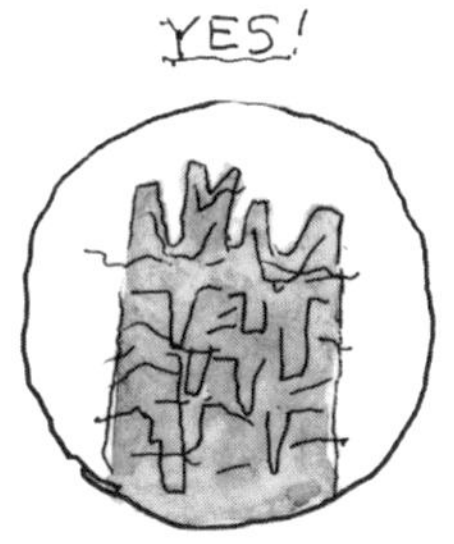

Spine and ribs.

to one another, and to the wall beneath. On each side of the central spine attach cobs at right angles to it, running from the spine to the inside and outside surfaces of the wall. Leave a space between consecutive "ribs" about the width of a cob. These spaces will be filled in first when you come back to apply the next layer.

Work below your waist level, or you can strain your back, arms, and shoulders. The weight of your upper body is very helpful to bond the layers together.

Don't overwork the cob on the wall. Don't pat or slap it. Overworking can jellify it so that it slumps. To prevent wet cob from slumping, work each layer into place gently with your fingers, then go back over it with a cobber's thumb. While one hand pushes down to perforate the cob, the other can be used as a temporary form to hold the nearest edge of the wall vertical. Don't rub the surface of the cob smooth. A slick surface slows down evaporation, preventing the inner parts from drying quickly. Press the cob down quickly into place, perforate, then leave it alone. If parts spluge out, if the sides "mushroom" at all, wait until they stiffen up, then cut them off. Don't try to shove the wall back into line and above all don't whack it; you'll merely redistribute and exacerbate the problem.

Keep the working surface damp and soft. After the top surface has dried, you will never again achieve quite as good a connection. Unless that surface remains moist, there will be a line of weakness at the seam. As you work, be sure the top of the wall has not become crusty, dry enough to change to a lighter color. Wet cob will not stick as well to dry cob, so even tiny pieces of crust will prevent good adhesion.

Whenever you resume building, check not only for crusting but probe the top of the wall with a cobber's thumb. The material should be soft enough so you can push in the stick a couple of inches. If not, soak down the surface with water or clay slip. If you get delayed at all, recheck and rewet. If the wall gets really dry, you may have to soak it repeatedly over hours or days, perforating the surface with a sharp digging fork to let the water soak in.

Connecting Two Cob Walls

When you are constructing attached walls, try to build both walls up at the same rate, especially where they join. If one becomes much higher than the other, they will settle at different rates, imposing stress at the junction, with a potential for structural cracks. If for any reason one wall needs to be erected before the other, drive in wooden stakes, rebar, or some other kind of large pins as you build, to reinforce the connection between the walls. You can also chisel a vertical keyway out of the

Ideally, adjoining walls will be built at the same time, but if not, a "keyway" slot and embedded sticks or lumber will help lock the new wall into the older one.

first wall, a channel into which the later wall will lock. Then try to build the connecting wall slowly, allowing time for each few inches to settle before adding the next layer.

WALL THICKNESS

"How thick do you make cob walls?" There's a risk of answering, "It all depends," which really is a good approximation of how Nature does everything. There are some guidelines, though. One good one, also from Nature, is: *Put the thickest parts where you need the most strength.* (On the other hand, don't overbuild where you need less.) But where would the need for strength be greatest?

Tree trunks are wider at the base both to stabilize them and prevent them from falling over, and because the lower part of the trunk holds more weight than the upper sections. Their trunks taper, sharply at first, then more gradually. Similarly, masonry structures are more resistant to toppling if they taper.

Always consider what a specific section of wall needs to support. A wall that carries the

THREE CONTAGIOUS CONDITIONS

Splüging is the tendency of wet cob to bulge out horizontally when pressure is applied from above. Splüging can occur several feet below where you are currently building. It weakens the wall and necessitates lots of extra trimming. Remedy? Trim, and slow down.

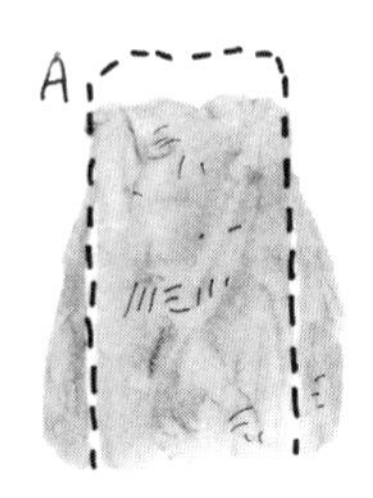

Mushrooming is the gradual widening of the top of the wall as each successive layer is laid a little wider. Trim off this bulge as soon as it is stiff enough, or your eye will tend to follow the wrong line, resulting in a wall that gets progressively wider as it goes up, exactly the opposite of what you want. By building onto these projecting sides, you add unnecessary weight, further accelerating the tendency to mushroom. In some cases, we have had to trim as much as 6 inches to bring a wall back to plumb. Check your vertical with a level each time you add new cob to a section of wall, whether you think you need to do so or not.

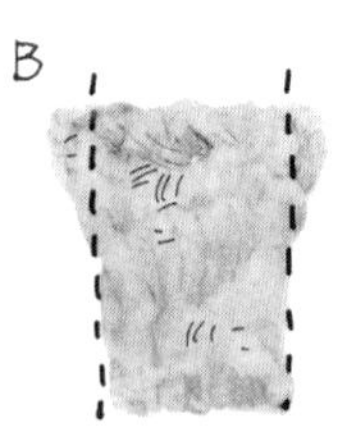

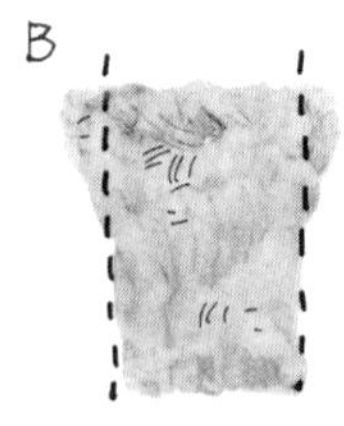

Shouldering often occurs when novice builders are not careful about forming the inside and outside surfaces of the wall as they build. The top of the wall slopes off giving the top a rounded profile. When you notice shouldering, fix it by laying long flat cob "straps" (like the "corbel cobs" shown on page 200) down from the top over the part to be filled out. Fix this problem as soon as possible or it becomes cumulative, and the wall develops a sharp spine.

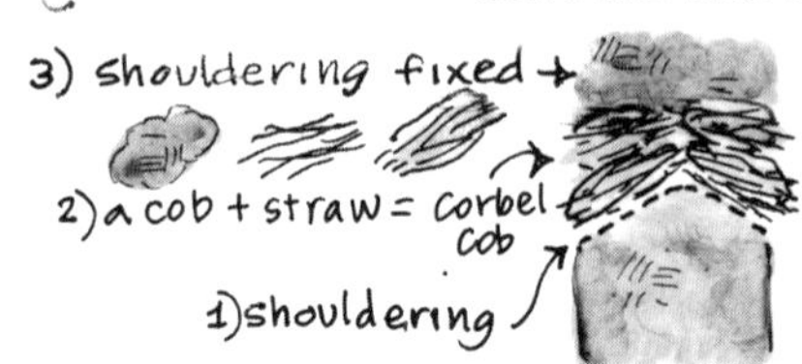

If the wall is "splüging" (A) or mushrooming (B), slow down and trim off the excess cob after checking for vertical plumb. If top of wall is "shouldering" (C) build up more even top surface with a corbel made of cob and loose straw, then continue with regular cob.

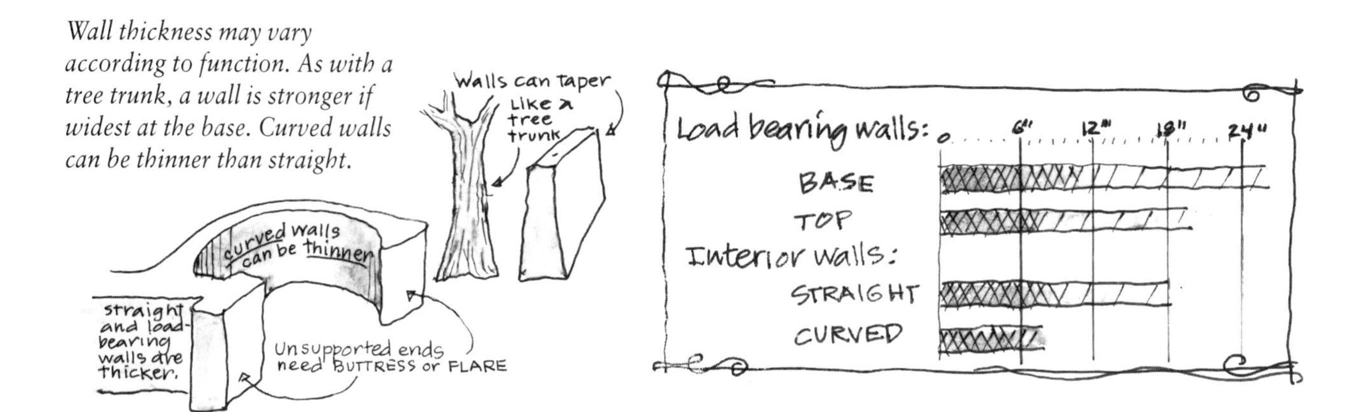

Wall thickness may vary according to function. As with a tree trunk, a wall is stronger if widest at the base. Curved walls can be thinner than straight.

weight of the roof, for instance, should be thicker than a non-load-bearing wall. Walls also need to be thicker and more tapered where they are straight; curves add strength, so curved walls can be thinner and untapered. The unsupported end of a wall needs extra strength, especially if the wall is straight, so flare it out a little wider at the end, or add deliberate buttresses.

Another variable with wall thickness relates to how much heat storage is needed in various parts of a building. For instance, walls close to a stove or fireplace may be thicker in order to store more heat. Similarly, interior walls reached by direct winter sun can be thicker, especially trombe walls directly inside south windows.

Interior walls can be thin, to provide more indoor space. "*How* thin?" In traditional English cob, interior walls are usually thinner than the exterior, between 14 and 24 inches thick. In New Zealand there are walls inside buildings as thin as 11 inches. But these walls are all more or less straight, and the cob is often of mediocre quality. Oregon cob, being stronger, can be built much thinner. Non-load-bearing interior walls are very robust at six inches, and if the length is short can, with a curve, be taken down below four inches. We have built curved partition walls as thin as two inches, and even then, they are surprisingly rigid.

In fact cob walls can be of variable thickness, changing from one section to another, reflecting the builder's sensitivity to the need for strength or extra depth for niches and shelves or other design elements. If you make the transition from thick to thin gradual, the wall will be stronger and it will feel better. Where walls meet, flare the connection out, in the way that a branch flares out from the trunk of a tree. Experience is your best teacher, but an approximate guide can be found in the chart above.

DRYING

With an efficient mixing team, the limit on wall-building speed is usually drying time. If you add too much new material before the cob underneath it has had a chance to solidify, for instance, in cool wet weather, your wall will spluge, requiring a lot of trimming later. The ideal is to be building onto a damp, sticky, soft surface supported by stable, solid, dry cob a few inches down.

The rate of drying depends on many factors, including air temperature and humidity, wind, the proportions of sand and clay in your mix, and how wet the mix is. Cob will dry only from surfaces exposed to dry moving air. It is not essential that the air be warm, just moving. The worst drying conditions are in sheltered places and in damp weather without wind. Thick walls dry more slowly than thin ones, because they have less surface area proportional to volume. A thick wall can look quite crusty on the surface, yet still be soft inside, creating "Brownie Syndrome," good in a chocolate brownie, bad in a wall. Fast building in cool, windless, damp weather presents problems—there's a danger of the wall splüging a couple of feet down, or mushrooming at the working surface, and gradually getting wider as you build up.

If the drying is going too slow for you, try these strategies:

- Create more surface area by using deep perforations or tall spines and ribs.
- Concentrate on getting the core of the wall to solidify before covering it—build a really tall central spine and leave it exposed overnight.
- Leave each new layer to solidify before adding fresh material.
- Use drier mixes, with more sand and less straw. Wet batches are easier to mix than dry ones, so you could make a number of wetter mixes and let them dry out substantially before building.
- Make sure you don't smooth the wall's exposed surfaces as you build, closing off the open pores that allow drying to happen.
- Fill the insides of the wall with solid materials. Rocks, scrapwood, chunks of concrete, plastic or glass containers filled with water or air, or any other hard, noncompostable materials make a good filler in the middle of walls. (We heard of a man who entombed his toaster oven, his TV, and his computer.) Filler reduces both the volume of cob that you have to mix and the amount of water that has to evaporate to make the wall solid. As long as the chunks are separated from each other by continuous sections of cob, they shouldn't weaken the wall significantly. To be safe, don't arrange embedded material in straight lines, and especially not in vertical stacks. Chunks of filler can, however, make it more difficult to punch holes through the wall later or to bury plumbing or wiring, so think ahead. Locate buried solids in the center of the wall, as trimming can be difficult if they are near to the surface.

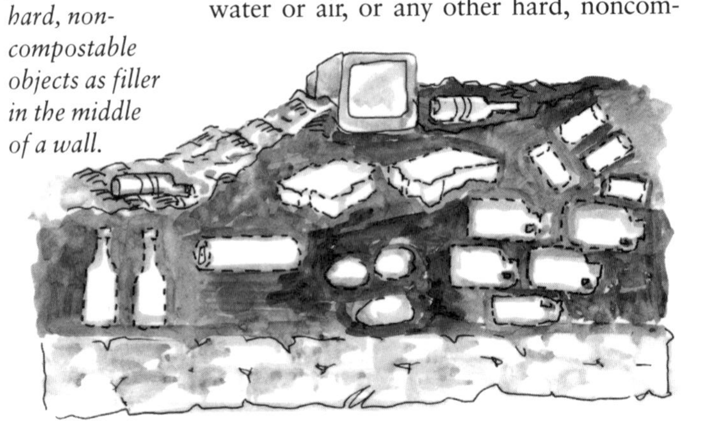

To reduce the amount of cob needed, you can add hard, non-compostable objects as filler in the middle of a wall.

Building slowly in dry or windy conditions will result in a crusty surface with poor adhesion.

If the wall is drying too rapidly, wet down the top surface regularly. In extreme conditions of desert, high mountains, or steady wind and sun, keep the wall top covered at all times, even for an hour or two between adding consecutive layers of cob. If you will be leaving the wall for a day or more, cover it with packs of soaked absorbent material: a few inches of straw, burlap, newspaper, scraps of cloth or carpet. Then tie plastic sheeting over the upper foot or two of wall, tight enough to prevent dry air from entering.

How long will a wall take to dry completely? Well, as with many questions in natural building, it all depends. Probably weeks, maybe months, possibly, with very thick walls in cool, damp climates, years. What matters is that the wall will be dry enough to continue to build on within a day or two, to receive plaster in a month or so, and to inhabit within the time it will take to complete the house. Linda and Ianto's first cob cottage, begun in early September, was dry enough to move into by mid-November. There were still damp patches on the walls—the last patch disappeared about Christmas—but the water evaporated in ways that created no problems. It seems probable that without dry weather or additional heat, the straw inside a thick wet wall stands a chance of rotting if the wall stays wet for months, so try to build early in the dry season rather than later, or be prepared to heat the building to dry it.

Building with PURE CLAY is a bad idea: Too much shrinkage...thus LARGE CRACKS OCCUR

original clay wall

BUT by adding sufficient SAND or GRAVEL & STRAW to the clay, the wall does not crack.

Clay shrinks and cracks; cob doesn't.

SETTLING

Cob shrinks in volume as it dries. As water evaporates, it vacates space, and most clays shrink on drying, usually by 6 to 15 percent in linear measure. If you were to build a freestanding wall of pure clay, it would contract a *lot* on drying. If all parts were free to move, the entire structure would shrink proportionately, but if the bottom part was attached to a foundation and was thus unable to move, it would need to crack; the cracks would probably be few but large, maybe running right through the wall. The upper portion of such a wall, free to move, might shrink but not crack deeply.

By adding sufficient sand or gravel to the clay, you can minimize shrinkage and avoid structural cracks. In fact, if the grains of sand are already touching before drying begins, shrinkage locks the whole structure together more tightly. A sandy mix will harden better, will be more stable, and will shrink less, overall. Both sand and straw help redistribute the forces of contraction during shrinking. Instead of a few large cracks, you will end up with many small discontinuous ones, which won't compromise the wall's strength.

Well-made cob shrinks mostly vertically. So long as the cob remains plastic, gravity is the major force that reshapes it, and almost all the shrinkage resolves into downward settlement. This is not a big problem, but you can minimize difficulties by understanding how and when settling occurs. Most settling happens in the first few days. Expect it to stop within a week, unless drying conditions are very slow.

New cob settles, but also slumps just slightly, spreading outward from the centerline of the wall. Slumping is particularly noticeable if you're building fast, or with a mix that's very wet or short of sand. As a result, you will have to trim more, slow down, or compensate by building the top section of the wall strongly tapered. As the cob slumps, solid objects such as windows will move, mostly by tilting toward the nearest face of the wall or, on a curved wall, toward the outside of the curve. You can compensate by tilting the top of the window toward the inside of the curve, or away from the closest face of the wall. Better yet, brace the window temporarily to the ground or to any solid object to prevent it from moving.

If you rapidly build the wall up the sides of an opening (window, door, or niche) be aware

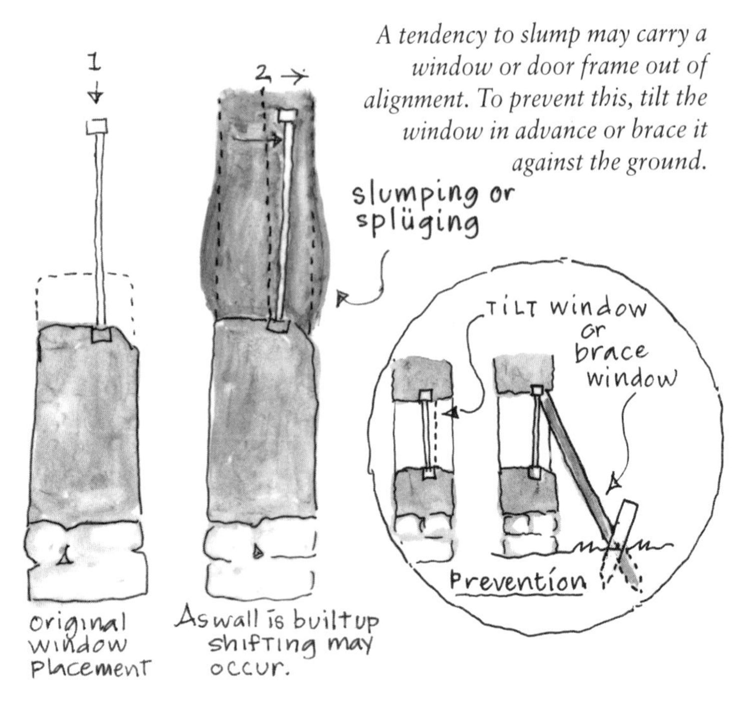

A tendency to slump may carry a window or door frame out of alignment. To prevent this, tilt the window in advance or brace it against the ground.

Cracking may occur at the corners of door and window frames, the result of building too quickly above.

that settling will continue for a few days, in cold wet weather possibly weeks. Be prepared for the cob at the top of a tall opening to settle up to 2 percent. In other words, if you're installing a frameless window 50 inches high by rapidly building up the sides, expect up to an inch of settling at the top. A lintel installed too early will settle with the wall, potentially crushing the window and leaving diagonal structural cracks up from the ends of the lintel. So either wait a few days before installing the lintel, or leave a gap between the top of the window and the lintel. If the cob ends up settling less than the gap, you can seal it with wooden molding. See chapter 14 for more detailed discussion of windows and doors.

Some references suggest that traditionally walls were built solid, then doors and windows were cut out later. That seems most unlikely. We strongly suggest that this would be a loony way to proceed; it would involve an immense amount of extra work. Build in door and window frames as the walls go up. Don't plan on coming back to cut them out, afterward.

GETTING IT STRAIGHT

As your walls rise higher, keep a level or plumb bob close at hand and use it frequently. Use a 4-foot level or duct tape a long, straight board onto a shorter level so you can read the straightness of the whole wall. Even if it's not important to you to have perfectly smooth and plumb walls, it's still advisable for structural reasons to keep them close to vertical, or decreasing in width with a predetermined taper. Consistent checking can save a lot of reshaping and trimming before plasters go on.

If the wall isn't vertical, immediately trim it back to where it should be, or strap on long, straw-rich cobs to fill it out, before continuing to build. Check new work frequently to avoid shouldering and mushrooming. The human eye is notoriously bad at gauging plumbness.

For tapered walls, create a taper level. For a 5° taper, duct-tape a 4-foot level to a 4-foot wooden wedge that is 3½ inches at the top and 1 inch at the bottom. For the vertical walls, use a regular 4-foot level.

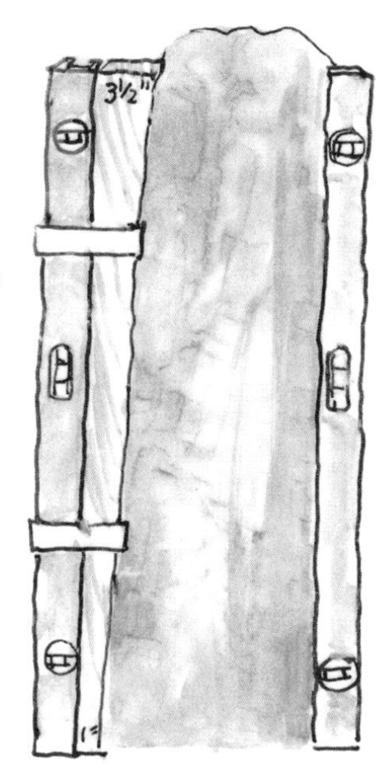

Tapering Walls

In many cases, we deliberately taper walls so that they are narrower at the top than the bottom. Why? The top of the wall has much less weight to bear than the bottom, which has to support all the cob above it. Tapering the wall significantly reduces the amount of cob you have to mix and build. Tapering also brings the center of gravity of the wall closer to the ground, making it less likely to fall over during an earthquake. And not least, a slight taper helps a building look securely based on teh ground, growing out of the landscape.

We usually taper walls at about 5 percent for every foot and a half of height, the wall loses about an inch of width. You can taper your wall more or not at all. Traditional English cob often tapers sharply for the first 2 to 4 feet then the taper decreases, like a tree's trunk. Yemeni earthen skyscrapers taper from about 3½ feet at the base to just 9 inches at the top, 60 to 100 feet up. Make sure load-bearing walls don't get too narrow at the top to support the weight of the roof. We recommend a minimum thickness at the top of 12 inches for exterior or load-bearing walls and 4 inches for interior partitions.

Often we build walls so that they are tapered on the exterior, plumb on the interior. Tapering the inside gives more psychological space, but is awkward with manufactured furniture. Or keep both sides of the wall plumb, but reduce its thickness in steps at the heights of shelves, counters, upper floors, and so on.

A taper level, a special tool you can easily construct, will make consistent tapering easy. Cut a 4-foot piece of 2 × 4 diagonally so that it has the desired taper (say, 5%) and strap it with duct tape onto one edge of your level.

Trimming Off Excess

Even if you build slowly and with great precision, it is almost always necessary to trim cob walls. The wall usually has bulges, lumps, and hollows that, if left, will make plastering more difficult, and there will be projecting straw that if not trimmed can poke through the plaster.

Linda trimming with a spud.

As trimmings fall, catch them on a tarp or board. They are valuable; you already invested a lot of energy in making that material. If you trim soon enough, trimmings may still be damp enough to reuse as they are. That saves work—adding water, letting it soak in, retreading. The act of trimming results in the straw being chopped short in the offcuts, so the recycled material has extra value for special jobs—underplaster or "window cob," for instance. They are very helpful for drying out overly wet mixes. And if trimmings make poor-quality material, bury them in the middle of the wall where structural quality is less important. (We've heard this called "Playing Doctor," because you bury your errors.)

Tools for Trimming

It is difficult to trim cob that is either too wet or too dry. Try to trim each section of wall a day or two after it is built, as it reaches a "leather hardness," firm and solid but still moist.

The most useful general purpose trimming tool is a *machete*. Keep the tip sharpened (both front and back) so that it will cut easily. You can use the machete by gripping the handle with one or both hands, swinging the blade to hack off unwanted cob, or by carefully grasping the point of the machete in one hand and the handle in the other, then scraping the blade against the wall. Because of their long blades, machetes are most useful for trimming relatively straight or convex surfaces. To true up bulges on the top of a wall, work your way forward a couple of inches at a time, chopping down all the way through the bulge before moving forward.

If trimming is necessary while the wall is still very wet, use an *old handsaw* dipped regularly in water. The sawing action has less impact on a wet wall than hacking with a machete. The saw works well for cutting into tight corners and sawing off unwanted bulges.

The most useful tool for trimming cob is a machete.

On curved surfaces, you can spring the blade to cut a curve.

For getting into tight corners and concave spaces, you need a shorter blade. A *hatchet* can work well. So can an *adze,* which is a speedy tool for hollowing out niches.

A flat-bladed, sawn-off sharpened shovel is called a *spud*. With its long handle, it is particularly useful for reaching high and for trimming dry cob. The spud most closely resembles the cob paring iron used by traditional English builders. Jan Stürmann, who has been

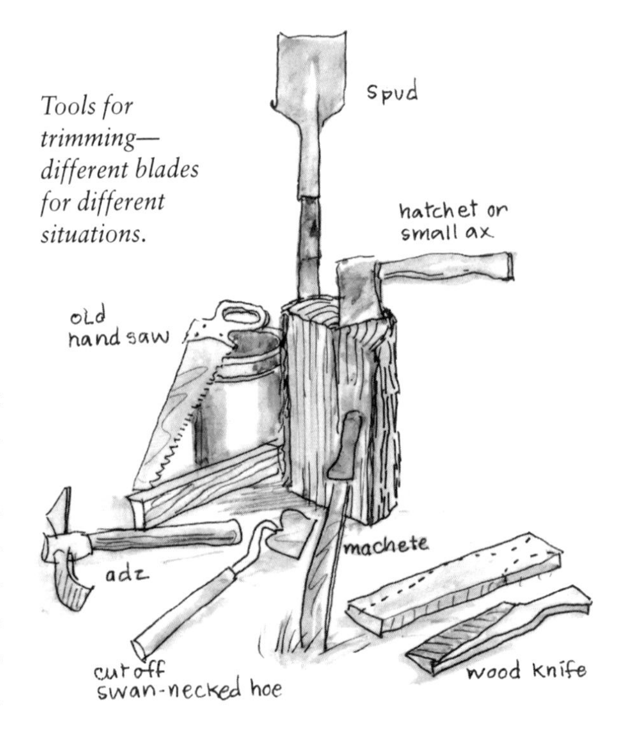

Tools for trimming—different blades for different situations.

building cob houses in South Africa and Massachusetts, swears by a sharpened *swan-necked hoe* for trimming walls. Cut the handle off at about two feet.

Any new craft rapidly develops new tools. Among those used on cob are a trimmer that Kiko Denzer made from a bicycle-chain wheel on a short wooden handle. *Wooden knives* are an innovation we use to trim along the edges of glass. To make them, split firewood into really fine kindling, in flat plates roughly 2 inches by a quarter inch, about a foot in length. Sharpen one end to knife shape, and whittle a comfortable handle.

A short, straight length of hard, thin board is also invaluable (a 1 × 3 about two feet long works well); grip it in both hands and systematically scrape across the face of the wall, first in one direction, then at right angles, to plane wet cob that will be left unplastered so as to create an even texture.

PIPES AND WIRES

Plumbing and electricity need to be carefully planned before you get to the building stage. Work out where you will want spigots and electrical outlets, then design a distribution system to supply them. Pass your plans by an electrician and a plumber. Pipes and wires can be built in to the cob as you go, before the walls get to counter height.

In walls without doors, it is safe and easy to run both pipes and wires along inside the wall, buried deep enough that subsequent excavation will not disturb them. For flexibility and maintenance, wires can go in conduit, though they should be very safe without it, as fire danger in a cob wall is nil and building around conduit is a drag. In zones with heavy frost, keep water pipes close to the inside face of the wall; in coldest regions, insulate pipes with straw or foam pipe insulation or run them through the floor. Try to keep plumbing joints in accessible places—a slow drip inside a solid wall can do enormous damage over time. Where services enter or leave the building beneath or through the foundation, be sure to plug with cob any cracks left around the service lines. Make a careful map of where service lines are buried, then make several photocopies and keep them in safe places.

Wires and pipes. For the electrical supply, wires should be deep enough to be safe from nails driven into the wall (1a). In a subfloor utilities strip (1b), wires should be covered by a board or other flooring. For the freshwater supply (2a), pipes should be deep in the wall but not more than halfway, or in a subfloor utilities strip (2b). The graywater drain should exit through the wall in mild climates (3a) or beneath the foundation (3b) in areas with frost.

For ease of maintenance, you may prefer to run main water lines and electrical conduit behind a moveable baseboard or through the floor. Even if your floor is earthen, you can leave a removable wooden utility strip along the base of the wall.

If for any reason wires or pipes need to be installed in a cob wall after the building is complete, use a squirt bottle and a hatchet to hack out a channel deep enough to avoid damage from nails or drills, insert your pipes or wires, then stuff cob around them.

INCORPORATING OTHER MATERIALS

Thick cob walls lend themselves to the insertion of other materials, for both decorative and practical purposes. Anything solid in the wall speeds construction up by reducing both cob mixing and drying time, but will change structural continuity and thermal qualities. Examples include:

❧ *Wood rounds.* Bolts of interesting wood projecting out of a cob wall add new textures, shapes, and colors. They can be convenient places to attach wooden shelves and counters, hooks to hang kitchen tools, coat racks and plant hangers, or sculptural details. Wood in a cob wall stays very dry unless exposed to rain, so rotting should not be an issue. But most wood will shrink over time, leaving a crack between cob and wood. Cedar seems to be very stable. Salt-impregnated driftwood shows a wonderful rounded softness, but covered up can cause discoloration of interior plaster as the salt leaches out. Redwood will stain any mud that touches it black. Entire walls and buildings are sometimes constructed using cordwood, whole or split, stacked tightly with mortar at each face and insulation packed sandwich-fashion in between. Our experiments with cob mortar for cordwood walls have shown promise.

❧ If the foundation rocks have dramatic or decorative faces, it is sometimes worthwhile to expose some of those rocks in the wall, giving visual reinforcement to openings and corners. Stone quoins can extend the stemwall up the sides and corners where abrasion is most likely to occur and where the stone will be most visible.

Cordwood embedded in cob.

❧ *Glass bottles* can be laid horizontally through the wall. Be sure the insides are clean and that they are well sealed. The shape of most bottles and the color of the glass prevent much light from coming through, so don't count on them for illumination. But located where the sun will hit them end-on, they can bring a sudden flash of light in at key moments, such as at dawn on the Equinox or noon on December 21st. Angle them up, tilted to receive the sunlight. Be very intentional in the patterns you create, or embedded bottles can look cheap and cheesy.

❧ *Glass blocks* were popular once, to let some light into city basements and through downtown walls where you'd be appalled at the view. They show up quite often during city demolition jobs, in several sizes, usually 4 inches thick and 4, 6, 8, or 12 inches square. They're a good privacy compromise for bathing or toilet rooms where there is nearby foot traffic, though they don't allow much light through, compared with a regular window.

❧ If extra thermal mass is needed, for instance, in a trombe wall, insert heavy rocks or water-filled containers to increase the heat storage capacity. Use glass or plastic containers, not metal, sealed very tightly.

REACHING HIGHER

As you build your cob walls, always try to be in a position where you are building just below waist level. Cob is heavy material, and unless you continually raise your working platform, you'll soon tire as the wall grows. Even at chest height the work is inefficient; you can't get help from gravity to press new cob into place, and you risk putting out your upper back wrestling with heavy awkward lumps of mud.

Scaffolding

Straw bales are invaluable to stand on when building at the three- to five-foot level. Use them flat, then on edge, alone or with short boards spanning two bales. As the wall climbs to about six or seven feet, stack the bales several high. Empty 55-gallon barrels are quite stable if you span two with heavy boards, and will even support bales on top, so that you can work in the seven- to nine-foot range, at the upper reaches of a single story.

Higher than about eight feet, you have several choices. You can use ladders, but they are somewhat hazardous and give little freedom of movement. Safest are fruitpicker's ladders with splayed rails, but not everybody has access to those. You can borrow or rent painter's scaffolds, collapsible steel platforms that can be stacked up to about thirty feet in height. When I'm sufficiently organized, I like to build-in scaffolding by laying logs or 4 × 4s through the wall, sticking out on both sides. I then wait for that part of the wall to stiffen, put boards on top, and voilà! Very simple, and stable. Saw them off later and you have ready-made deadmen for attaching counters and shelves. Or you could leave projecting wood in place permanently for use during maintenance, as is done in West Africa.

You can also easily build your own pair of lean-to trestles, which allow you to build to about twelve feet and to reach about fourteen feet. For each pair you will need four eight-foot 2 × 4s, an armful of shorter pieces of 1 × 4, screws or screw-nails or ringshank nails, and two strong pieces of 2 × 6, each about three feet long. Don't scrimp on screws and don't use unsound lumber, for your neck depends upon them. Well-made, these trestles will outlive you; they are a very good investment.

Lean-to trestles allow you to build to about twelve feet in height.

Second Stories and Up

Second (and subsequent) stories need special treatment and have special constraints. Building slows down when you work on a scaffolding, both because your movement is restricted and because it's harder to deliver the mix. These are minor problems and don't seem to deter Yemenis from adding another story to a seven-floor building.

Cob can be delivered to a high scaffolding by tossing loaves up from the ground; by a bucket and hoist, mechanical or hand-pulled; by a series of ramps up which you can push a wheelbarrow; or by the long-handled forks used in England to deliver mix up to about ten feet. Sometimes a combination of methods is most suitable, notably with more than two stories.

Remember that the floor of the upstairs, built in sequence with the walls, is immediately available as scaffolding to build the walls and roof of the second story.

HOW FAST CAN I BUILD?

All of this brings us to a common question: how much is it reasonable to expect to build each day? There are two limiting factors—what *volume* of cob can a person mix and apply, and how much *height* can be added?

A young, vigorous, and experienced cobber might mix and build as much as 15 to 25

Homemade roundwood trestles, 5' high.

cubic feet (0.4 to 0.7 cubic meters) in a day, mixing by foot. Use of a machine could increase that, but see chapter 11 for some drawbacks. Beginners, particularly those not in good physical shape, can expect far less—perhaps half that amount, though with organization and rhythm, their pace will pick up rapidly. Realize that this is not a race. You will want to build at a pace that you can sustain.

Below the need for scaffolding (about 4 feet in height), building usually takes less time than mixing, on a plain wall possibly half the time. As the wall grows and mix needs to be lifted and scaffolding moved, application can take longer than mixing. Sculpted windows and other details slow the process dramatically.

Suppose you are two people working together, without experience but energetic and in good health. You are working five-day weeks, on a cottage twice the size of Thoreau's cabin—that is, about 300 square feet—with walls averaging 16 inches thick, two doors, and quite a number of windows.

With materials already excavated, your combined rate of mix and build might begin at 20 cubic feet a day, quickly rising to about 40 cubic feet. The first week you would only build about 18 inches of cob. Don't be discouraged. Week two you speed up, getting another 30 inches built. By this time you're up to window level, so there's less volume in the walls, which are also tapering. However, the building will grow more slowly due to the extra work of lifting mix and using scaffolding and due to the detail work around windows. By the fourth week of steady building you'll need to install deadmen to hold the roof down. You may decide to stop cobbing and install the rafters before finishing the walls. But if there are no unexpected delays, you and your partner could mix and build the essential bare walls of a single-story cottage in less than a month.

In Devon, builder Kevin McKabe recently built the walls of a 3,000-square-foot, three-story cob house with only two helpers, in a season. The exterior walls are 3 feet thick and the entire structure, 30 feet high, weighs 300 tons. He used a tractor for mixing and raised much of the mud mechanically.

Sculpting with Cob 13

WHAT SETS COB APART FROM EVERY other building technique and attracts so many artists and creative people is its extreme fluidity of shape. Since cob requires no formwork and is not made in uniform blocks, it provides absolute liberation from the straight, flat walls and right angles that plague modern construction. Rectangular buildings are inherently less stable than curved ones, and we believe that square corners and straight walls contribute to the stress and anxiety of their inhabitants.

The generic box house is so boring that we must fill it with artfully crafted tables and chairs, potted plants, paintings, and sculpture to enliven its barrenness of form. By contrast the rounded houses of Africa have little need for furniture or incidental "sculpture." Their shapes are rich enough to be satisfying in themselves. A sensuously shaped wall carefully plastered needs no applied art. Framed pictures look superfluous; armchairs feel like an obstruction. They are unnecessary clutter, confusing to the satisfaction of true simplicity. When the house itself becomes the sculpture, your home is a living work of art.

❧ SCULPTING AN ENTIRE BUILDING REQUIRES NOT SO MUCH ARTISTIC EXPERIENCE AS A RELAXED AND OBSERVANT ATTITUDE. . . . LET INTUITION GUIDE YOU. ❧

Cob offers the opportunity to incorporate sculpted functional features and furniture, both within and surrounding your home. This chapter is about the nonstructural parts of a cob building—built-in sculpted benches, bas-reliefs, garden walls. Arches, niches, and

alcoves are also easily sculpted in cob. By building furniture and fixtures directly into the structure, a cob cottage can be made more space efficient, particularly as the rounded shapes of cob make it very inviting, with fewer sharp projections into the living space.

We will also discuss cob's special relationship with fire and give directions for several sculptural hearths—a cookstove, an oven, and a fireplace. We'll include information on how to make a wood-fired cooking/heating stove that cunningly uses waste heat to warm a cob couch.

SCULPTING A WHOLE HOUSE

Sculpting an entire building requires not so much artistic experience as a relaxed and observant attitude. From the very beginning be aware of the surrounding sculpture of the site—the shape of the ground, the enclosure made by buildings close by, the spaces created between nearby trees. Pay special attention to the volumes left vacant, rather than only the bulk of the solids. These emptinesses hold potential; objects block possibility. As you absorb the shape and scale of what is already there, your body memory will store those forms and prepare itself to match them gracefully as you build.

When you begin designing the building, let the aesthetic of the site flow into your sketches and models. Let intuition guide you, for example, to curve the foundation around a tree, to incorporate an existing boulder, or to mimic the line of the hills with a curved ridge beam.

Similarly, if you remain open to new possibilities as the walls go up, there will be times when your subconscious speaks. It may call out for a broad window seat where you never saw one before or a tiny peephole window onto a magical view. A sculpted dragon may grow above the doorway, a bust of Karl Marx may sprout from the mantel, or a curved piece of driftwood may unexpectedly become a towel rack.

Cob features don't always need names, functions, or explanations. Sometimes you want to sculpt a fin right out from a wall, for no other reason than it feels right. Some of the most endearing houses have unexplained lumps, strange protuberances, and curiously secret hollow places without obvious uses.

Perhaps we should urge restraint. Cob so naturally becomes three-dimensional art that it can be hard to contain your creativity. Many neophytes go overboard with niches and bas-reliefs, disparate window shapes, and a mish-mash of conflicting forms, designs, and details, rather than letting the beauty of the materials speak for themselves. In restful places, avoid shapes that contain right or acute angles, letters or words, or facial features. These shout for attention like a bumper sticker or a cartoon on a coffee cup.

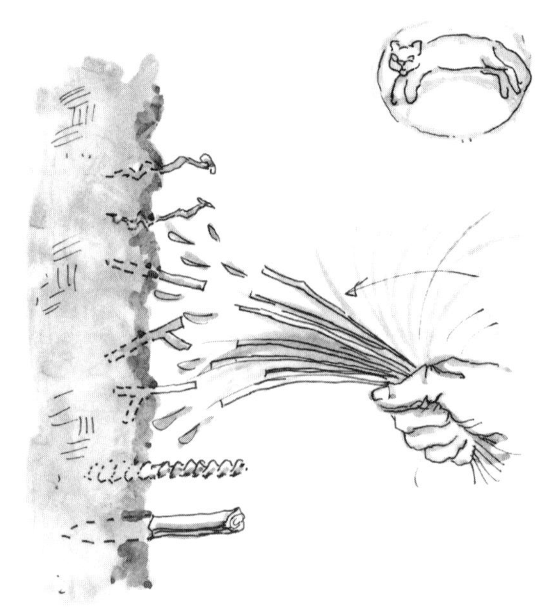

To provide an armature for attaching a sculpture to a dry cob wall, rough up the surface with a hatchet, then drive in nails or other metal or wooden protrusions. Wet the walls with "slip-dipped" straw. Upon this base build up the fresh cob. Use rebar or stronger stakes for larger sculptures.

Sculpting major features is usually easier done as an integral part of the initial construction. Cob shelves and large sculptures should be corbelled out of the wall as it is built, using a special technique described later in this chapter. Furniture that will be supported by the ground can be added later, because it doesn't require the same quality of adhesion. Where you know you will later be adding a cob bench or counter later, leave the surface of the wall lumpy and rough, possibly with sticks or metal stakes protruding as anchors.

Wall sculptures and bas-reliefs are best added after the wall is complete, or they will obstruct the process of trimming and smoothing the wall. If you want to add a small, nonstructural cob feature to a dry wall, first rough up the surface with a hatchet or claw hammer, then drive in long nails every 2 to 3 inches, with their heads sticking out an inch or two. Finally, wet the wall thoroughly and paint it with thick clay slip before adding fresh cob. To support a larger heavier sculpture, build a stronger armature by driving long rebar or wooden stakes into the wall and connecting these with wire or baling twine. For sculptural details we prefer a finer cob mix than usual, using sifted soil and chopped straw.

Remodeling

Most cob cottages are not only living sculptures, they are also works in progress, even years after their inhabitants move in. One of the joys of cob is that it can be altered, amended, carved away, or added to whenever the fancy or the need arises.

To excavate a new niche into a dry wall, all you need is a squirt bottle and a sharp tool such as a chisel, hatchet, adze, or machete. Soak the area thoroughly with water to soften it before you begin carving. Be patient: let the water soak in, scrape away the moist mud, then soak it again.

Snoozing pussycat, 4' long, by Kiko Denzer, in cob house in Oregon.

In the Heart House, the phone lives in its own cubby, a rounded niche three feet high with shelves for the phone book and Rol-o-dex. When Ianto and Linda got a fax machine, it was too wide for the niche, so Ianto took a hatchet and hacked out a hole just big enough for the fax, with a little extra space to reach one's hand in for the handset. This took about an hour, including applying a quick coat of gypsum plaster.

During the construction of *my* first cob cottage, overzealous novice cobbers left a huge bulge in the wall above the sleeping loft. In the rush to get the building enclosed before winter, the wall was never trimmed plumb. For nearly two years the bulges remained, becoming more and more of an eyesore as the rest of the cottage was finished and smoothly plastered.

One rainy day, I spread a tarp out over the bed, took my hatchet in hand, and began hacking with no particular plan in mind. As the cob around it was chipped away, a life-sized human head appeared at the highest point of the wall. Soon the whole upper body had taken form, with arms outspread and hands cupped to hold candles. Before building out the belly and legs, I drove rebar spikes into the wall, roughed up the surface with the hatchet, and wet the whole area with clay slip. One of the legs comes away from the wall to rest on a small cob seat at the foot of the bed. The freestanding lower leg has a hidden skeleton of sticks and wire. I plastered the finished figure with a smooth clay plaster, leaving her skin the rich brown color of the native soil and painting her hair with brilliant blue and white *alis*. With her solid stance and sensuous rounded shape, she seems to me to embody the spirit of cob.

BUILT-IN FURNITURE

Cob furniture has a solid, reassuring feel and appearance that complements cob walls and floors. When built in to the perimeter of the room, cob seats take up less space than chairs and couches scattered about. Seats can be shaped to fit curved walls, seeming to grow out of them just as the building itself grows from the ground. With no sharp corners or projecting legs, a sensuously sculpted cob seat or couch invites snuggling. It can even be heated by running the flue from a woodstove through its mass.

Manufactured furniture can be selected to fit well into small, curvilinear spaces. Choose round or oval tables and curved couches. Build a platform bed against the wall so that the wall itself becomes the headboard. Kitchen implements can be displayed decoratively: cast-iron pots hung from driftwood hooks over the stove, mason jars filled with colorful beans and grains lined up on cob shelves. If you intend to attach wooden shelving or cabinets to the cob wall, remember to leave exposed wooden anchors. See the discussion of "deadmen" and gringo blocks" in chapter 14.

A variety of arch shapes can be made using forms and templates which are removed once the cob is dry.

Bookshelves over sitting areas, kitchen shelves, bedside ledges and candle niches can all be made of monolithic cob, either recessed into a thick wall or projecting into the room, or a combination. Cob shelves can be plain or extravagantly sculpted, surfaced with plaster, wood, tile, or flat stone. Their inclusion gives an otherwise plain room a sculptural interest. Monolithic shelves can add to the apparent space in a room, rather than eating into the space as freestanding bookcases do.

In the Heart House, bookshelves were constructed by corbelling out the west wall above the heated bench, above seated head level. The books add a layer of insulation in an area where the wall is thin and never gets sun. The most essential reference books are there, where the spine titles are at eye level. To achieve an 8- to 9-inch shelf they built out 6 to 7 inches and cut back into the wall slightly. In a friend's cob house, I recently completed a corbelled cob mantelpiece that projects more than two feet into the room above a woodstove, providing a spectacular display area for pottery, sea shells, and other special objects.

CORBELLED SHELVES, ARCHES, AND NICHES

The key to many kinds of structural details is corbelling. Corbelling creates a horizontal projection from a wall, cantilevered in such a way that the weight of the projection and anything that rests on it is supported by that wall. We use corbels to make cob benches and shelves that project a foot or more from the wall, yet are strong enough to support heavy objects such as books and even people. Corbelling simultaneously from opposing directions produces arches, niches, vaults, and domes.

The most effective corbelling requires special cobs. Naturally enough, we call them "corbel cobs." Corbel cobs are like regular cob loaves except that they are small, long, and flat and have extra straw running in the long direction. Often they are about the size and shape of a small trout. To make corbel cobs, you'll need a rather sticky standard cob mix, with extra clay and long straw. First, spread out a mat of long, strong straws, laid parallel on the ground in front of you. Make oblong cobs and roll them in the bed of straw so that they pick up straws in the lengthwise direction. Knead until these straws are fully incorporated into the cob, then step on the loaf to flatten it into a straplike shape. Keep corbel cobs very moist until you use them.

To make a corbelled shelf, start by leveling the top of a section of wall, some distance below the finished height of the shelf. It's difficult to build out at more than a 45-degree angle, so if you wish your shelf to project 12 inches, start corbelling 12 inches below the eventual shelf level.

Attach the first course of corbel cobs at right angles to the wall so that their front ends project a couple of inches out from the wall. Sew the back of each corbel cob into the middle of the wall, taking advantage of the extra straw to get a really good bond. Work the fronts of the corbel cobs together sideways, pinching the seams between them and smearing the fresh cob into the wall below. Leave the top surface rough, and build up around and behind it, pinning down the back edge of each corbel cob with regular cob. Wait until the first course of corbel cobs is firm but not dry, then apply a second course in the same manner. Be patient! Project each course only an inch or two beyond the one below. Test your corbel to see how firm it is before applying more weight. Don't worry too much about the shelf's profile; if you don't like the shape, you can trim it later, when it's dry. *Warning:* Keep the middle of the wall dry while building corbels. A soggy interior can cause a corbelled shelf to sag and even fall completely off the wall while you watch, dismayed!

Corbelled bookshelf where the cantilevered projection is supported by cob with lots of long straw mixed in and sewn well into adjacent cob.

To make an arch, build up both sides until they are flat and level, then begin corbelling out from each side at the same rate just as you would for a corbelled shelf. Don't be concerned about finish; it's much easier to trim later. Build up the surrounding area with regular cob as you go, to support the arch, but keep the initial leading edge of the work as narrow as possible to reduce suspended weight. Be extra careful as you near the top. Allow the arch to dry substantially before inserting the "key cob" across the gap.

Both arches and vaults are easier to build to the desired shape if you use a template as you build. Cut a thin sheet of rigid material such as stiff cardboard or plywood to the shape of the arch. Hold this template up against the arch frequently as you build, in order to match its shape.

Another way to build an arch is by cobbing over a rigid form made of thin, bent

CAROL CREWS

LINDA SMILEY

MISHA RAUCHWERGER

COB COTTAGE COMPANY

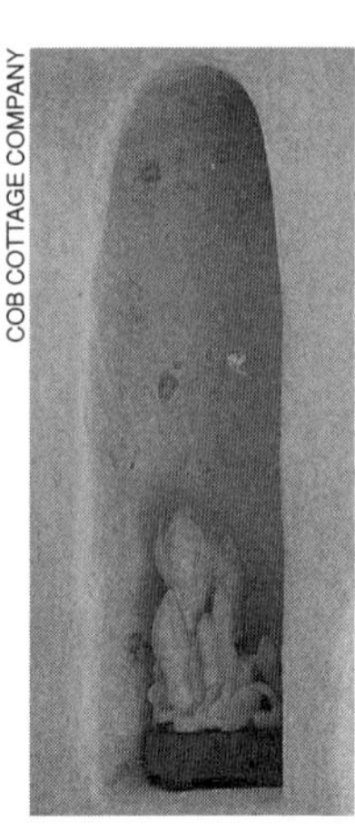

Arched alcoves and niches, from left to right: sleeping alcove, Carole Crews's house, Taos, New Mexico; Carole Crews's cob pantry; altar niche, Michael Smith's Oregon house; sculpture in a sculpture, Linda and Ianto's Heart House.

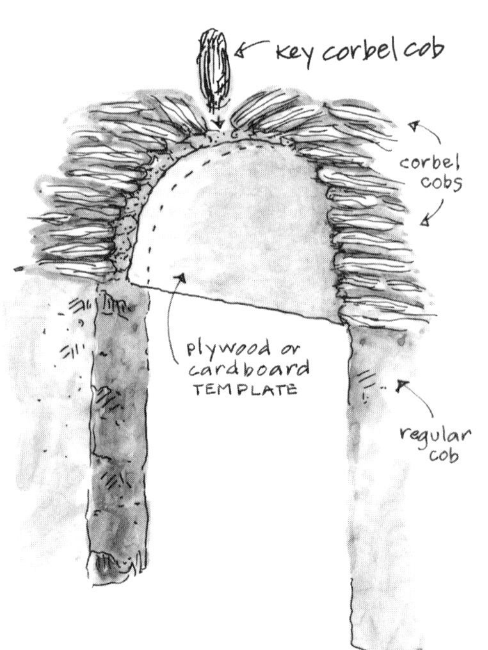

ABOVE: *Making an arch over a flexible form.*
RIGHT: *A corbelled arch built over a temporary form or template.*

plywood or boards, and propped very firmly in place so that it can not shift. Use great care to marry adjacent cobs to one another, which is more difficult because the form obstructs your access and will also impede drying. Build a thin layer and let it stiffen before applying the second layer, which should be stiff before you attach the third, and so on. Leave the form in place until the cob becomes quite solid, remembering that the face against the form will dry out very slowly by comparison with exposed surfaces. In this manner you can easily build archways 6 feet across, probably much bigger. It is faster than corbelling, but probably not quite as strong.

For its simple, pleasing shape we recommend you begin with a semicircular Roman arch. The template is easy to trace using a pencil on the end of a string attached at the center of the arch. With practice, more complex shapes will become easy—pointed Gothic arches, other segments of circles, or asymmetrical shapes that would be quite impossible in brick or rock masonry.

A thick wall not only encloses space, it is in itself a place, an occupier and redistributor of volume, which can be exploited by built-in niches, alcoves, and window seats, or short, vaulted tunnels between two rooms.

In a small building, building niches—sculpted cavities—into the wall is an efficient way to meet some of your storage needs and to expand the apparent space. In addition to being functional and ornamental, they lend themselves to ritual and symbolism.

Rectangular niches work all right structurally, but arched ones are more graceful. Sometimes it's effective to build in a pair, a trio, or a whole row of niches with slender walls between them. Completely hiding a secret hollow (behind a decorative plate, for instance) makes a good safe deposit box, fireproof and hidden but convenient of access.

There are two basic ways to build small niches. One is by corbelling to make a three-dimensional arch; special corbel cobs may not be necessary if your arch span is less than about 18 inches. The other way is to build a plain wall, then carve a hole in it after it solidifies. Both of these techniques work well, and it's mostly a matter of personal preference which you choose. If the niche is very small, it is usually faster to excavate it later. If the niche will be larger, when you factor in the time saved by not mixing the extra cob to build a solid wall, arching with corbel cobs may be just as fast. For *really* big arched alcoves, build over a removable form as described earlier.

A row of niches in a cob wall.

STARTER PROJECTS: GARDEN BENCHES AND WALLS

An ornamental outdoor bench is one of the most satisfying projects for novice cobbers. A little group of people can have a wonderful weekend party creating such a bench for a garden nook or a hilltop lookout. The bench can be as big or small as you want, curvaceous and complex or simple and temporary. It can be for one person or two, straight or curved, with or without a back. Don't be too ambitious, though—try to finish in a weekend. You'll feel a sense of achievement and your completed project will inspire other people.

The bench can be built directly on the ground, but a low foundation of brick, rock, or concrete scraps will speed the process and give it longer life. If you live in town, consider building it within the boundary of your yard, but against the public sidewalk so that other people can sit on it. At a school, use the bench for a bus stop. Or make a personal snuggle-nook in a sheltered corner. We've even seen a cob armchair, with a built-in stove for cold days. Many of the considerations for siting a house are equally relevant to a bench, so this can be a good practice project.

Benches need weather protection, or they will erode in the rain and be cold and soggy to

(continued on page 206)

A garden bench made of cob will need weather protection or will gradually erode away in the rain.

LEARNING AND GROWING WITH GARDEN WALLS

BY IANTO

BEFORE BUILDING THE FIRST Oregon cob cottage, Linda and I put up a test wall. It was three feet long, two feet high, and ten inches thick, built in the most exposed part of our site, facing due south. We chose not to give it any weather protection, no roof, plaster, or stucco, and it was made of the same mix we used for the cottage. For four years we monitored its performance and watched how quickly it decomposed under different weather conditions. We gained confidence in our material by submitting the newly built wall to various kinds of abuse—hacking into it, driving in stakes, and wetting it down to simulate downpours. Quite apart from being a wonderful exploration of how Nature reasserts her own kind of order, this wall gave us practice at refining our technique without suffering stage fright while committing to a building that could stand a thousand years.

KIKO DENZER

Bear in an Oregon garden. Kiko Denzer, sculptor.

Our second Oregon cob cottage (the Heart House) has its own set of experimental garden walls. We began by walling in a little courtyard in front of the cottage, then replaced part of the wire garden fence. Recently we have added a 500-square-foot cob greenhouse. All the walls are made of cob, and all are connected. There's a total of about 200 linear feet, roughly a foot thick and 5 to 8 feet high, containing three gateways, two of them arched over. Most of the wall has a little roof of cedar shakes, while a small portion is open to the weather. The wall contains fifteen windows of varying sizes, some with glass.

What began as a whim has taught us more and more about the advantages of freestanding cob walls, both as a teaching tool and as a way of raising the quality of home life. A walled courtyard can increase living space to help a compact cottage feel more expansive and can create a soothing outdoor retreat. The wall is a canvas on which to sculpt bas-reliefs and paint murals, a permanent trellis for colorful plants to grow on, with built-in planters for flowers. Our own courtyard contains a Rumford fireplace set into the wall, which makes a snug gathering place for cool evenings, with attached cob benches and niches carved into the wall for candles and flowers.

Outdoor spaces surrounding buildings can be amorphous, fading away into uncertainty with distance from the solidity of the house. Garden walls, by breaking up big spaces into linked, smaller, outdoor rooms, offer the mystery of part-enclosures leading one into another. And walls attached to a house anchor the building into the landscape.

Cob garden walls can enclose a courtyard and create outside alcoves and suntraps.

The Romans grew pomegranates and citrus in cold, wet England, two thousand years ago, by using masonry walls to store heat. Our own outdoor walls create dramatic microclimates for both plants and people. The thermal mass of the wall stores the heat of sunshine, keeping night temperatures higher in the surrounding area. The north side is very obviously cooler and damper in summer, a good place for peppermint and black currants, whereas the south side catches and reflects the day's sun, providing the perfect ripening ground for tender tropicals such as peppers and eggplant. We're planting figs, citrus, persimmons, and pineapple guavas, none of which are possible here in unsheltered conditions. With diversified microclimate comes diversity in life-forms. Dry-laid rockwork foundations arer homes for snakes, lizards, and small mammals, providing access to warm basking grounds and cool feeding areas. Solitary mason bees excavate east-facing sides and chipmunks live under the wooden shake roof.

Cob walls have great potential in urban areas, where they can provide privacy and protection from traffic noise. A curbside wall shelters pedestrians and is an opportunity for public education. Whereas a house needs privacy, a surrounding wall offers public display; it could be intensely sculptural and colorful, exhibiting a range of plasters. In America's all-wood neighborhoods, an earthen wall elicits plenty of interest, so this is a good place to set up an explanation board. Remember also that in most places the wall won't need a building permit ("it's just a fence," after all).

sit on. Pick a dry site, preferably under a long eave, in a greenhouse, or on a porch. If necessary, build a special roof, though on very dry sites you could get away with a water-resistant finish such as lime stucco or linseed oil and beeswax (see chapter 17).

A garden wall is another favorite cob project. It can be as short as you want, as high or low as you need, and quite thin; you will get building practice and plenty of effect for a minimum of mixing. You can give the wall a little roof, or leave it uncovered to find out how the local weather affects the material. A simple wall can also serve as a trial ground for stucco mixes and natural paints. You can practice building-in glass windows and open arches, bas-relief sculptures, planters, and nest boxes. Try different kinds of foundations, which allows you to experiment without danger of irremediable mistakes. You could also experiment with different mix ratios and cob application techniques on different wall sections, then try to demolish the wall to find out what worked best.

EARTH AND FIRE: COB HEARTHS

The great distinction of cob in natural building, according to our Danish colleague Flemming Abrahamsson, is that cob involves all four of the prime elements—Earth and Water to form it, Air that dries and breathes through it, and Fire, which it contains so well. Cob and fire, in fact, have a uniquely complementary relationship. Cob made without straw is completely fireproof and at high temperatures will be cooked to the consistency of low-fired brick. Cob combines with fire to make wonderful bread ovens, durable cookstoves, cozy heat-storage mass stoves.

A Heated Cob Bench

The most effective way to keep warm in a cool building is by conduction (contact heat), directly into your body. When your body is being heated directly by conduction, you can be perfectly comfortable in very cool air. In northern China and Korea, a horizontal flue system called a *kang* is built of earth. It is used by day as a cookstove/kitchen table, then at night as a heated bed, where often the entire family will sleep.

What follows is a description of an easy-to-build mass stove/heated bench, which will provide heat by radiation from its metal surfaces, by conduction from direct contact with the bench, and by slow release of stored heat into the building. A solid bench such as this can be made of cob, brick, stone masonry, or even tight-packed sand contained in a brick or wooden box. Let's assume you're using cob.

The bench needs a built-in flue along its length, the same diameter as the feed tube of the stove. Put in a liner of galvanized-steel stovepipe or ducting. Without it your bench could leak slightly, drawing in room air, and just possibly allowing carbon monoxide into the room. At each end, build in an inspection hatch with an airtight lid for cleaning the tube and for priming the vertical stack on occasions when the stove is difficult to start. The stack will never get very hot, so it can be kept inside the building as an exposed metal pipe, yielding a little additional heat, until it exits through the roof. This pipe should need no extra insulation or protection, as the whole aim is to offer it only gases that are no longer very warm. *Don't put a flue damper in that vertical stack.* A damper will be unnecessary, and in fact it is dangerous to back up emission gases in the system. Gas flow is instead regulated at the stove, by limiting the air going in.

Build cob tight around the horizontal flue, using a stiff mix with as much sand as possible (sand stores more heat than clay), and no straw at all in the layer surrounding the pipe. You could also pack the inside of the bench with dense rocks for more heat storage. Heat travels through cob at roughly an inch per hour, so if you anticipate firing the stove for three hours on an average winter day, three inches of cob above the horizontal pipe will allow the surface of the bench to heat up just when you close down the stove. The bench will absorb *conductive* heat from the duct equally in all directions, so make sure the cob is as well compacted below the pipe as above it, or sits on a bed of heavy rocks.

A stove works by the warm gas in a chimney stack rising, pulling fresh air through the fuel in the combustion chamber. A horizontal flue linking the stove and the chimney does not appreciably diminish the draw unless it is *smaller* in cross-section, or *very long,* or has *abrupt bends* in it or a rough surface. How long is too long? Well, we have built successful stoves with horizontal flues more than thirty feet long.

To be sure of a good draw of gases through the system and to prevent smoke leaking into the house, locate an airtight woodstove below the level of the bench. Before lighting the stove, it may be necessary to prime the draft by inserting a burning piece of paper in the primer hole, directly beneath the vertical stack. Later, when combustion in the stove is complete, and most of the wood has burned away completely, close off the whole system by sealing the air inlet to the stove. If you fail to seal it, the bench will continue to siphon warm air out of the room and discharge it up the chimney to be wasted.

Most kinds of airtight clean-burning stoves will do as a heat source. The critical thing is to burn a very hot fire using dry firewood, or you can get creosote build-up in the horizontal flue.

In Ianto and Linda's cob cottage, supplemental heating comes entirely from a gravity-feed experimental stove of Ianto's own design. This is a downdraft model, fed from above, made from brick, cob, and a steel barrel. Surplus hot exhaust gases, which otherwise would be lost up a flue, are routed through eleven feet of daybed/guest sleeping/lounging bench. The spare heat is absorbed slowly and is gently released into the room from four to forty-eight hours later. The exit temperature of the flue gases is down to 150° to 300°F where they leave the bench (compared to 400°–600°F in most woodstoves). With a long enough bench, gases can exit at below blood temperature. No smoky, slow combustion, no constant checking and stoking, no buildup of creosote in flues, no need to clean chimneys every few months. The stove burns much less fuel overall and therefore generates less greenhouse gas and air pollution. The half dozen prototypes have performed well over about a decade, though there are still some problems of durability to be worked out, as the internal temperatures reached are very high. More detailed instructions are available from The Cob Cottage Company.

MISHA RAUCHWERGER

Downdraft "Rocket" stove heats this 11'-long bench. Earthen floor with linseed oil/beeswax finish. Design and construction: Cob Cottage Company, 1994–95.

Rumford Fireplaces

Two hundred years ago, Count Rumford, a colorful character and friend of Benjamin Franklin, observed how pitifully inefficient the open hearths of the time were. Because Rumford was a Royalist sympathizer, he spent

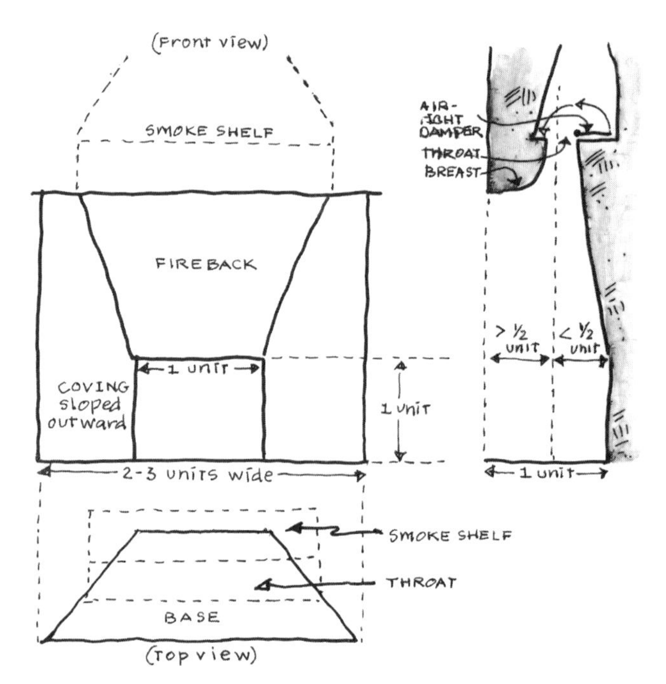

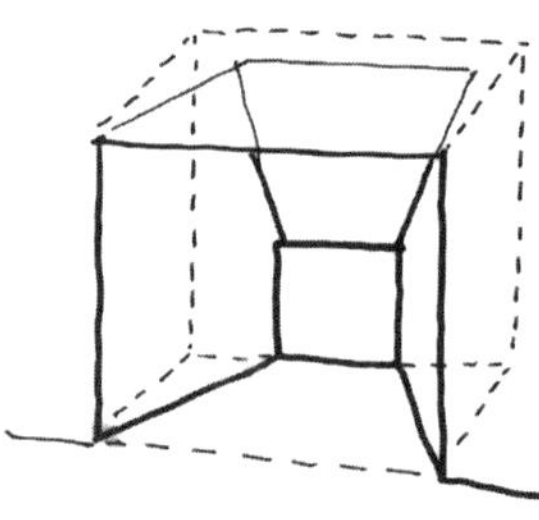

The Rumford fireplace is the most efficient and effective open hearth, and can be made partially or entirely of cob. Note how the sides and back of a Rumford fireplace are angled to reflect heat into the room. Its shallow depth also increases heating efficiency.

most of his life in exile from the U.S.A., but in Europe he was recognized as a genius. Without access to the information we now have about combustion or heat transfer, he designed a system for building the most efficient, clean-burning, and warming open fireplaces ever seen—so good in fact that they have never been surpassed. Recently, there has been a resurgence of interest in Rumford's designs, and tests have demonstrated that they burn efficiently enough to be classified as a "clean-burning stove" by federal standards. Rumford fireplaces are simple to construct, cheap to run, and can be quite beautiful.

The Rumford's efficiency lies entirely in its simple applied geometry, more or less independent of construction materials. The Cob Cottage Company has built many, partially or entirely of cob. They appear set to work well for a long time. To minimize maintenance, we recommend that the exposed fire-back be made of brick, though the rest of the fireplace can be made of cob, with or without rocks set into it. Parts that will get really hot, anywhere in direct contact with fire, should be made of a pure sand/clay mix, quite lean in clay, usually between 3:1 and 6:1, depending on quality of both sand and clay. Other parts should include long straw and more clay or other fireproof insulative materials such as pumice, perlite, or vermiculite. An industrious worker can finish a cob Rumford in a day, and in fact we have begun one at 9 A.M. and had a fire burning in it by suppertime.

A little above the smoke shelf, the exit gases can feed into a standard metal or concrete flue, or a flue can be fashioned from the cob itself during the initial construction. The inside of the flue should be made as smooth as possible for easy gas flow and to facilitate cleaning.

As with any open hearth, where the Rumford is placed in the building is critical to how satisfactory it will be. Plan in advance where a fireplace will go, before beginning the foundation wall. A good Rumford should sit down at floor level, and be built on an interior wall to limit heat loss to the outdoors. It will heat the space much more efficiently if it is supplied with its own airflow, directly from outdoors, so build in a 4-inch minimum diameter air supply with damper, which feeds into the *front* of the hearth. And don't forget to build a tight-sealing chimney damper at the smoke shelf to prevent warm air being lost from the house when the fire is not burning.

A Lorena Cooking Stove

"Lorena" is the generic name for a family of sand/clay cookstoves originally developed by Ianto in Guatemala in the 1970s, now used

The Lorena stove developed by Ianto is built by packing a solid block of sand-clay mixture into a form then carving while still damp.

worldwide. Lorena stoves burn wood, sticks, peat, coal, and agricultural wastes such as cornstalks. They provide long-term gentle cooking, protect the cook from radiant heat better than metal stoves, and burn fairly efficiently, with little smoke released inside the house. They are easy to make and satisfying to cook on. Lorenas are quite cheap to build, requiring only the expense of the sand and clay of which they are made.

A solid block of sand-clay "Lorena mix" is packed into a wooden mold or built freeform, then carved while still damp. A firebox sits under one or two pots that settle down into custom-fit holes in the surface, and the waste gases heat additional simmering pots as they pass through a curving tunnel leading to a vertical chimney. Lorena stoves are most common in rural areas in Africa, Latin America, and Southeast Asia but are also used in some homes in the United States, New Zealand, Canada, and Europe. They can be freestanding or built into the structure of a masonry house, adding mass and providing some warmth by gradual release of heat stored in the heavy mass. Their advantages include slow and even cooking and low fuel use. Their sculptural forms and colors add

Guatemalan woman who built her own Lorena stove.

life to a kitchen. Disadvantages are a certain amount of smoke escape, some wood ash and dust in the house, and the need to split fuel wood quite small.

For further details, consult *Lorena Owner-Built Stoves* (see the bibliography), or contact The Cob Cottage Company.

An Earthen Bread Oven

Wood-fired baking ovens have been used all over the world for thousands of years. The basic design is remarkably consistent: a hollow masonry dome, made of earth, brick, or stone, with a tight-fitting door. A fire is burned inside the oven, the oven absorbs the heat of combustion, the coals are removed, food is put inside to bake, and the door is closed. These ovens are extremely versatile. When fired to full heat, they make perfectly crisp pizzas. Then as they cool, they sequentially bake bread, pies, cookies, potatoes and other vegetables, granola, and finally yogurt. Even today, the finest bakers almost universally prefer wood-fired, retained-heat ovens. They claim that these ovens make the best-tasting, best-textured bread, and we believe them.

There are many ways to build an earthen bread oven. Perhaps the best known in the United States are the "hornos" of the Southwest, which are made of small adobe blocks. In Quebec there is a traditional design made by smearing mud over an upside-down basket

Earthen bread oven. This is a cob variation on a traditional design, easy to build and use.

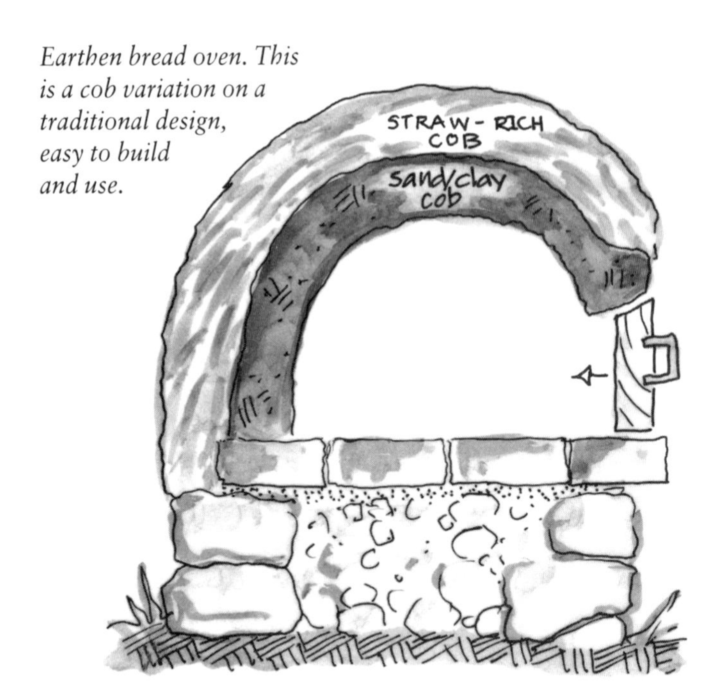

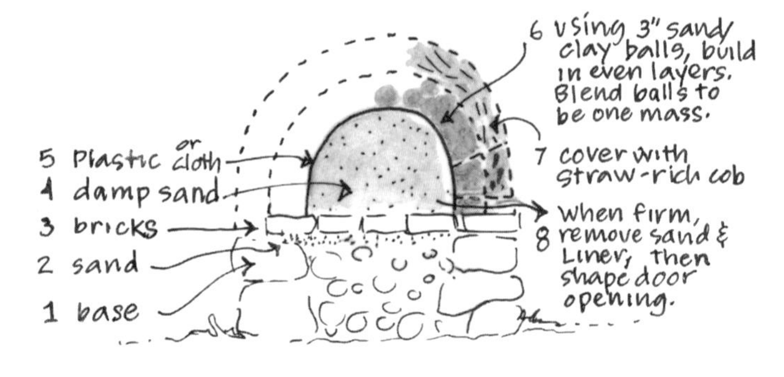

Kiko Denzer and friends on auxiliary bench of Phoenix Oven. New Mexico, 1995

of woven sticks. At The Cob Cottage Company, we have developed our own design, which is simple to build and easy to use.

We start with a stable platform of stones, bricks, cob, or even a tree stump—anything to elevate the oven to a height where it easy to use. On top of that base, set a floor of common red bricks into a layer of sand. Then build up a dome-shaped form by packing damp sand as if building a sand castle. This will serve as a mold for the inside of the oven. Make a mixture of clay and sand, just enough clay to hold the sand together firmly when dry. Roll this mixture into balls of a uniform size (around 3 inches in diameter is good). Cover the sand mold with a sheet of thin plastic (garbage bags work well), and begin to pack the clay-sand balls around it, building up in layers like a coil pot or an igloo. Be careful to keep the dome a uniform thickness as you go.

When the inner heat-retaining dome is complete, mix up some large batches of lightweight cob for an outer insulation layer. Use as much straw as you possibly can and little or no sand. Build a layer of light cob several inches thick over the clay-sand layer, concentrating on good adhesion between the two.

When the outer layer is complete, you can either leave it to firm up for a day or two, or immediately begin to excavate the door. Start by cutting a small hole through the dome with a sharp machete, and gently pull the loose sand and plastic sheet out through that hole. Then, if the dome feels stable enough, you can enlarge the door to its final size. The door should be semicircular, wide enough for your widest pizza pan or cookie sheet, and 60 percent as high as the inside of the dome. This last dimension is quite critical to make the oven burn well and cleanly. Make a door out of a heavy slab of wood that fits the opening closely, or use a ready made cast-iron oven door.

You can either leave your cob oven as a simple, elegant dome, or sculpt over it shape you wish, as a centerpiece to your outdoor kitchen or patio. Our colleague Kiko Denzer is widely known for his spectacularly sculpted cob ovens, which have taken every imaginable form from a frog to a turkey to a huge phoenix bird with smoke coming out through its beak. His excellent and comprehensive book, *Build Your Own Earth Oven* provides full instructions for building, sculpting, and firing a cob oven, and for baking sourdough bread.

Mass Heating Stoves

Known also as the Finnish Woodstove (German *Kachelofen,* Swedish *Kakelugn*), the mass stove has normally been built of brick, tile, or soapstone. It employs a complex series of flues and chambers through which hot combustion gases flow, heating a great weight of masonry. Preheated air enters at critical points to help with clean burning. Mass stoves are usually fired only once a day, burning 20 to 40 pounds of wood in an hour or so in a fiercely hot fire, after which the whole system is sealed to prevent heat loss up the flue.

Almost certainly these stoves could be made very effectively from a monolithic clay-sand mass, though brick should probably be retained in the firebox, where damage from abrasion is most likely. At the time of writing we know of no such stove. If you are prepared to try, you could cut the cost of a mass stove from thousands to hundreds or even tens of dollars. Details and interior dimensions can be found in both David Lyle's *The Book of Masonry Stoves* and *Finnish Fireplaces* by Albert Barden and Heikki Hyytiainen.

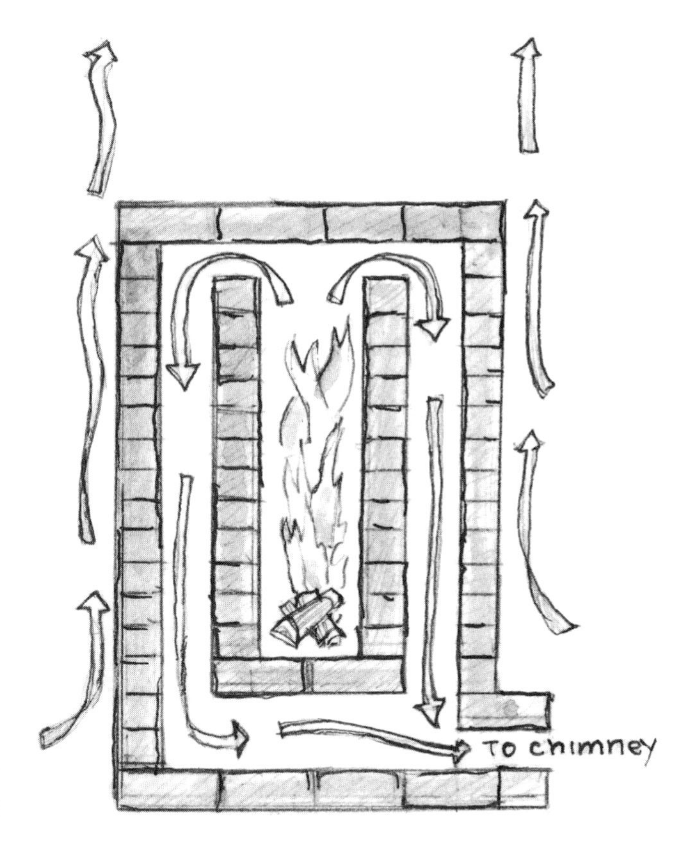

Finnish-style masonry or mass stoves are traditionally made of brick or soapstone but probably could be built of a monolithic clay-sand mix.

14 Windows and Doors

❧ WINDOWS AND DOORS . . . CONDITION MOODS, ATTITUDES, AND EXPECTATIONS IN WAYS THAT ARE PROFOUND, BOTH AS WE APPROACH THE HOUSE AND FOR ALL THE TIME WE ARE INSIDE. ❧

THE MEXICAN ARCHITECT ALEJANDRA Caballero sees houses as having characters, like humans. Not only do they have individual personalities, she says, but they have functioning physiognomies just as we do. The windows of a house are its eyes, through which it sees the world. If they are too small then the house will have no understanding of its surroundings; if they face onto squalor, the house will be essentially depressed, like a person contemplating a bad view.

Look around you, says Alejandra: consider the houses that you see. How does the door open to take you in? Everything that goes into the house enters through that doorway, the open mouth of the house. Is there a tongue that you can make your way up gradually, or do you have to pry apart the jaws and climb in quickly before the teeth close down on you?

Above all, coming to the house, does it smile or frown? Is it a clown face, a death mask, does it express a threat, a laugh, a sadness? How doors and windows are set up will affect all who visit for the whole life of the building and will condition moods and attitudes and expectations in ways that are profound both as we approach the house and for all the time we are inside.

Windows can be thought of as having four basic functions: daylighting, showing a view, collecting heat from the sun, and ventilating. In this chapter we will look at the requirements for each function by considering the special opportunities that thick and sculptural walls present. We'll also discuss different ways of making and installing doors and windows, both fixed and openable, while retaining the strength of the wall.

Industrial consumer societies have come to settle for doors and windows that fit the needs of corporate manufacturers—predictably rectangular, made to standard sizes, and available in only a few styles. Our cousins in traditional cultures have an astonishing range of doors and windows. Because of their wall thickness and flexibility of form, cob buildings give us freedom to be creative. We can find, make, recycle and adapt doors or windows to be more inspiring, and more fitting to their specific functions. Your options for sizes, shapes, and materials for the openings in your home are probably much greater than you have been led to believe, and in this chapter we will "walk you through" many of them.

KEEPING THE WALL STRONG ABOVE OPENINGS

There are two ways to support the weight of the cob above any opening. The first is to form an arch out of the cob itself. Instructions on how to build a strong corbelled arch are given in chapter 13. The second option is to install a *lintel* (also called a header), a structural member that holds up the wall above an opening. (Don't confuse this with a *lentil*; the result will be disastrous, structurally and gastronomically.) Arches are most appropriate for rounded or irregularly shaped windows that don't open and for openings without doors or windows such as a passageway between two rooms. Lintels are necessary when making large rectangular openings, especially for windows that open and for doors.

Lintels can be made of almost any material that is long, stiff, and durable. Steel, bamboo, driftwood, cast concrete—all have been used. Stones including granite, slate, and sandstone have been used as lintels, sometimes on cob buildings. In Wales, granite or slate lintels up to 10 feet long can be seen, centuries old and supporting stone upper stories. Some are long enough that a coach and horses could be driven beneath them.

For cob, heavy wood lintels seem especially appropriate—a thick milled board, a half log split down the center, a series of sticks side by side. The wood can be straight or curved. Choose beautiful pieces that can be left exposed to reveal their character. Because we move through doorways and look through windows, lintels will be seen by everyone who uses the building. The upper side will be buried in cob, so it doesn't have to be regular, flat, or handsome. The faces and lower side will be exposed, though, so choose and finish these appropriately.

Cob is monolithic and thus very different structurally from most other building systems. Wood-frame construction results in irregular loading—imposing concentrated loads wherever a post or stud stands. Likewise, to a lesser extent, brick, adobe, and stone masonry are constructed of many small, individual parts, therefore each part can exert downward force separately. Cob is more like poured concrete. The load is more evenly dispersed, so that after cob is dry, very little structure is needed above openings. However, a great deal of pressure is exerted downward while the material is still wet, both from sheer weight and through

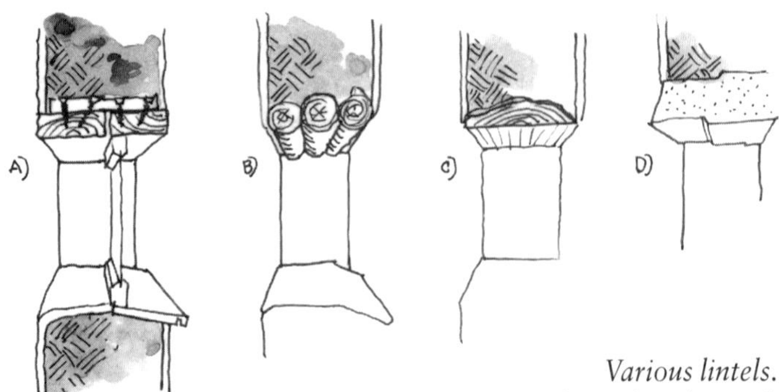

Various lintels.
A) Two thick boards cross-braced with wood, screwed in tight. B) Round poles without bark, side by side. C) "Slab wood" from a small lumber mill. D) Long, flat stone or concrete chunk.

settlement. So make provision for temporary support for all but the shortest, thickest lintels, until the cob is dry and rigid.

Lintels should extend at least a few inches into the cob on each end: the minimum overlap would be 4 inches, plus an inch per foot of length of the opening.

If a lintel is installed on fresh cob, it may settle with the wall, exerting pressure on the window or door and potentially cracking the glass or squeezing the frame. Before setting a lintel onto wet cob, raise the cob on either side of the opening to a little above the height of the opening—say, a quarter of an inch for every foot of height—to leave settlement space. Better yet, let the cob settle and dry as much as possible before installing the lintel. With unframed windows, if a gap remains between lintel and glass after settlement, you can close it with wooden molding.

CONNECTING COB TO DOOR AND WINDOW FRAMES

Doors and opening windows are normally hinged to a wooden, or sometimes metal, frame. Frames of both doors and opening windows will over their lifetime be subject to all kinds of forces, sometimes sudden and powerful: gusts of wind, slamming, kids hanging on them, perhaps even forcible entry (forget your keys again?). It is important that they stay put. Here are some ways to stabilize frames so they can't move.

Wet cob is heavy and can easily distort wooden frames. Temporarily cross-brace door or window frames *very* thoroughly before setting them in place. With doors and tall windows, also support the frame at right angles to the wall, bracing against something very solid such as the ground or another part of the building. If possible, leave the door or window right in the frame, closed, with wedges to

Temporary support for lintel.

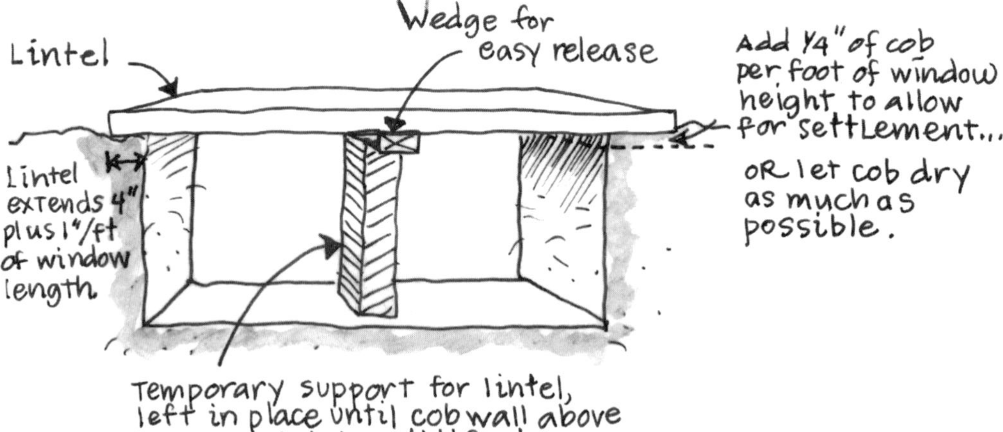

maintain a small gap between it and the frame, so that it will open and close easily.

To prevent a frame from ever coming loose, anchor it to the cob. For most small windows, before setting the frame in, it is enough to knock in a fringe of nails with their heads projecting an inch or two. You only need a few nails per foot of frame, and they can be bent and rusty. (Finally, here's an honorable use for that bucket of old nails you have kept all these years, vowing you would straighten them one day.) For bigger windows or light doors, attach a wooden stiffener to the outside of the frame, to be buried into the cob. This provides a grip and reinforces the frame against the weight of the wet cob. Scraps of 2 × 4 or 2 × 2 work well, or better, roundwood branches.

Heavier doors need more serious treatment. There are two main anchor systems: 1) A "deadman," meaning any piece of wood with irregularities to prevent it from pulling out of the cob. It could be a short piece of firewood with nails driven partway in, a short, T-shaped assembly of lumber (4 × 4, for instance), a short section cut from a thin tree trunk with branch stubs attached, or the stump and roots of a small tree. 2) A "gringo block," common in adobe construction, meaning a small, open box made of 2 × 6s or 2 × 4s, like a thick-framed drawer without a bottom. Both deadmen and gringo blocks are built into walls during construction, with one face left exposed. Door and window frames, cupboards, counters, and wooden shelves can be screwed to these exposed faces at any later stage of construction.

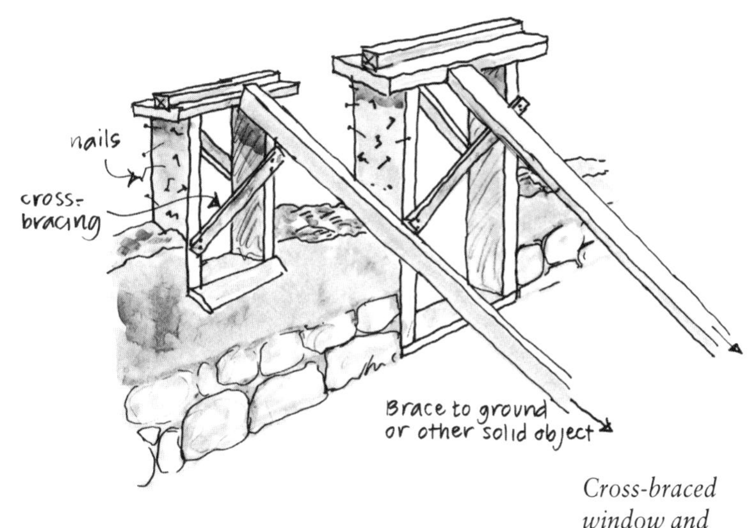

Cross-braced window and door frames.

The gringo block, simple to make, can be screwed or nailed together as shown in the drawing, using short ends of scrap wood. Gringo blocks can be built any width. Standard dimensions might be 8 inches wide by 12 inches long by 4 inches high, but the sides should be well buried in the wall. As you build the cob through the opening in the block, thumb it very securely to the wall beneath, perforating deeply. After setting the block in place, drive a small stake down

Anchor systems.

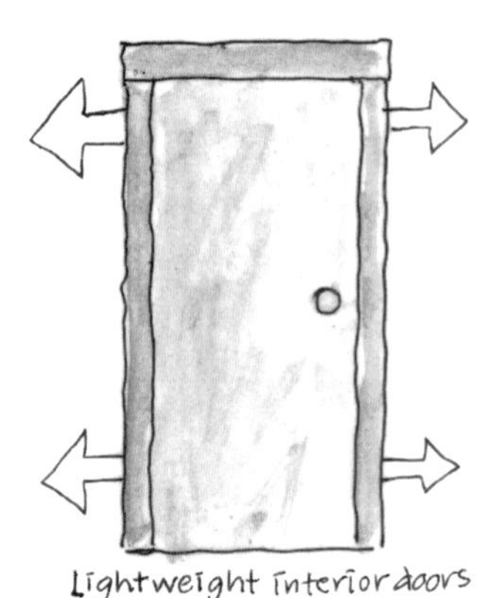

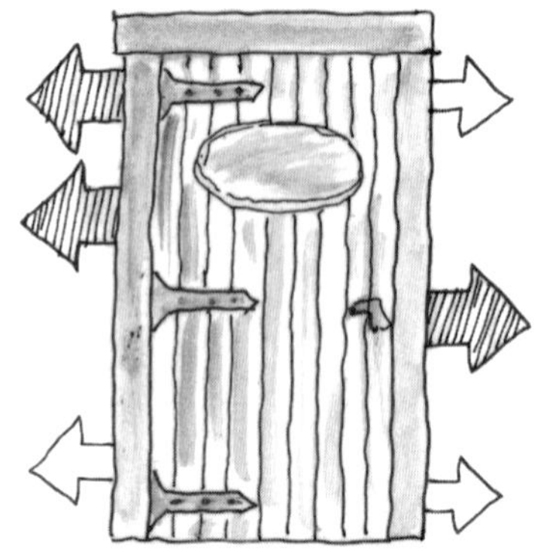

Locations of anchors to support doors.

through the cob inside, leaving a few inches of stake above the block. The stake will resist any tendency toward movement.

On the hinged side, a door tends gradually to pull away from the top of the frame, so put extra anchors there. For a heavy exterior door, put at least two deadmen above mid-height. The latching side of the frame suffers impact, which mostly affects the area of the latch itself so give this side of the frame extra attachments a little below waist height.

The most secure way to install a door frame is to set it in place before the cob walls are built. Attach the frame to the foundation, screw or bolt anchors to it tightly, cross-brace, then build cob around the frame (burying the anchors in the process) as the wall goes up. The less preferred alternative is to set the anchors into the wall as you build, then attach the frame later. In the latter case, be careful to align the anchors vertically above one another so the frame will line up with all of them. In either case, make sure the anchors will be unable to shift in the cob—if you use milled lumber, spike old nails all over it; if you use part of a tree, leave branch stubs sticking out. If the door is heavy and the wall is thin, use a long, irregularly shaped piece of wood, not a gringo block.

As with windows, when a cob wall is built up too quickly around and over a door frame, diagonal cracking in the cob will occur at the upper corners (see page 190). The cob will settle as it dries, but the frame's rigidity prevents equal settlement of the cob above it. To avoid this, build up to just above the level of the top of the frame and then wait until settlement is complete before continuing. Measure settlement until it stops. This could take a couple of days in hot, dry conditions, or a week or so in the rainy or cold weather.

It is possible to make an opening window or door that closes directly against the cob wall, without a wooden frame. Of course, the top of the opening must be either arched or linteled to carry the weight of the wall above. Gringo blocks can be set into the cob to attach hinges and latches. We don't know how these sorts of door frames hold up over time, but the cob jambs will certainly take a beating with repeated slamming of the door. We'd recommend a good hard coat of lime or gypsum plaster for a tight fit and better durability. If the door is kept closed while the wet plaster hardens, the plaster frame will conform very precisely to the shape of the door; oil the part of the door that will be against the wet plaster, so plaster doesn't stick to the door. Cushioning the door where it meets the jamb with leather, heavy felt, or the like should help protect both the jamb and the door itself, as well as ensure a better seal.

INSTALLING NONOPENING WINDOWS

Fixed windows are much easier to install than opening ones. Plate glass, wooden sashes and presealed, double-pane units can all be mounted, unframed, in cob. Use the thickest plate or tempered glass you can find, 3⁄16 of an inch or more. Thinner glass cracks easily during the building process, so should only be used in wooden frames. "Tempered" or "safety" glass is specially treated to prevent cracking and can withstand a lot of pressure—you can even deliberately bend it. Car windows are always tempered, and often have a pleasing curve. One enterprising builder in South Africa used an entire car door complete with window mechanism, leaving only the handle protruding from the wall, in order to roll down the glass.

Cob builders have enormous artistic freedom to make windows not only rectangular, but also circular, oval, heart-shaped, or any shape you want. You can use broken glass in a cob wall; it doesn't even need to be cut to the exact shape of the window. Always *tape the sharp edges of the glass* with duct tape or electrician's tape before placing it in the wall so nobody gets cut. This is also important to protect tempered glass, since its edges are particularly fragile. We usually insert glass no more than a quarter of an inch into the cob along the sides, unless we're creating an unusual shape. Once the cob dries, not much overlap is needed to hold the glass in place. If you bury it deeper, the glass is more likely to crack from stresses as the cob settles and is also more difficult to extract if it does break.

The variety of window shapes is unlimited, because cob can be sculpted around broken or unusually shaped glass.

To install a fixed window, first build up a level platform of cob at the sill height, leaving space for a pad of straw, foam or other compressible material beneath the glass and for the sill. Then build the wall at least one-third of the way up the sides of the opening to support the glass after you insert it. The outside sill should slope away from the window. In very dry climates, the sill can be formed of cob and later plastered, or in wetter places

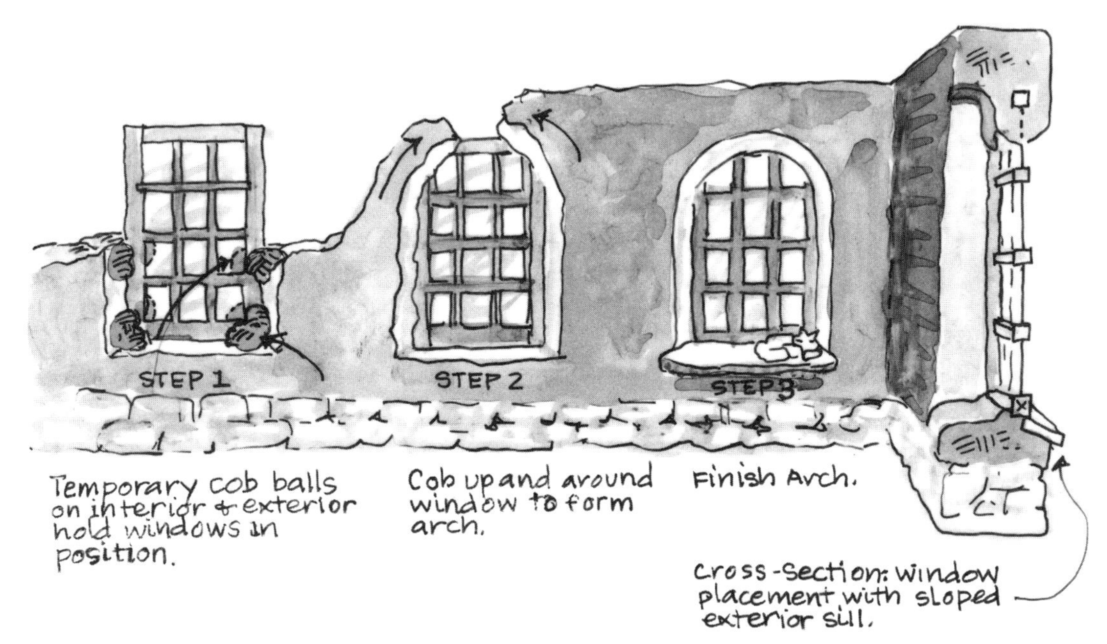

Installing a fixed window.

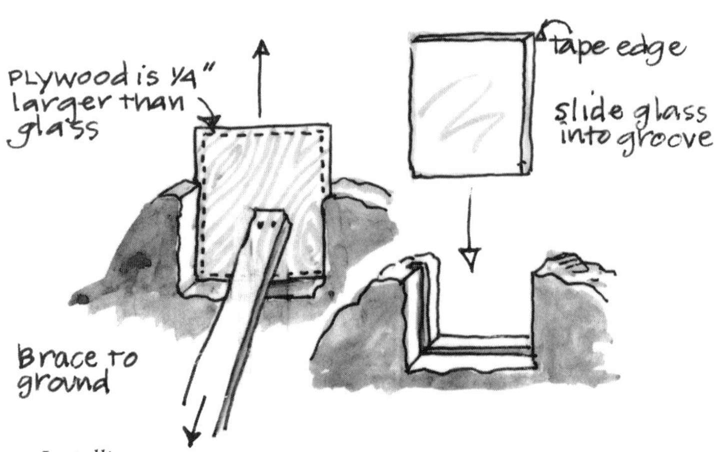

Installing unframed glass using a plywood template.

made more weatherproof with an embedded surface of flat stones, bricks, tiles, or durable wood. The sill needs to overhang the wall a couple of inches to carry water away. Set the sill material before the glass, so the glass will sit on top of it.

When building windows with large pieces of unframed glass, make a template out of plywood half an inch wider than the intended window opening. Set this template in place of the glass until the wall is up to at least half the height of the window, bracing it securely to the ground to keep it plumb. Then slide the plywood up and out, wetting the groove if necessary to lubricate it. Slide the glass down into the groove left by the plywood. If you leave the glass in place the whole time, you run a greater risk that: 1) someone will cut him- or herself on it; 2) the glass will be broken; and 3) the wet cob will not hold the glass vertical, so your window will wind up out of plumb, or fall out.

Replacing Broken Windows

For some reason, it's hard for many people to imagine how to repair a plate-glass window set directly into cob without a frame. But if you follow these steps, glass replacement is fast and easy.

1. Wet down the perimeter of the glass, inside and out, with a squirt bottle. If there is waterproof plaster, chip some of it away first. Let the water soak into the cob, then repeat two or three times over an hour. Using a wooden knife, cut the cob back a little way from the glass, all around one face.
2. Crisscross duct tape all over the broken pane, leaving a tape "handle" on one face. Then with a light hammer, carefully smash the glass.
3. Grip the tape handle and pull. Most of the glass shards will stick to it. Dispose of

Replacing a broken window. Wet the perimeter of the window opening, inside and outside, then cut back the cob on one of these sides.

Next, crisscross tape on both sides of the broken pane, leaving a loop of tape for a handle. Gently smash the glass and pull out the pane carefully by the handle.

Then cut a new pane of glass and tape the edges. Rewet the perimeter of the hole, set glass in place, and shape new cob putty around the window.

them. Pull out with leather gloves any additional glass.

4. Cut new glass and tape its edges.
5. Rewet the edges of the hole, and set the new glass in. Use cob putty, without long straw or rocks, to hold the new glass in place. Work in the fresh cob with a gloved finger, leaving the surface smooth. Trim the surrounds tidily. When it's all dried out, you can touch up with plaster, and your window will be as good as new.

Viewed from the side, A protects from high angle sun, while B brings in sky light and directs view up.

WINDOW PLACEMENT

Thick cob walls open up many previously unforeseen options for windows, including framed and unframed, sculpted into curved and irregular shapes, and "panes" of unusual materials such as mica, alabaster, bottles, or glass bricks. By careful siting and sizing of the glass and by angling of the window reveals, cob's thick walls can be used to enhance and control indoor lighting, views, and heating by sunlight.

When walls are one to three feet thick, it feels very normal to rotate a window sideways or up or down, turning it out of line of the plane of the wall. An up or downward tilt is used to direct the view, to bring in more sky light, or to reduce window cleaning if one side is in an awkward location to reach. High glass tilted inward at the top can reflect the sky to outside bypassers, a useful feature if the building is close to foot traffic and extra privacy is desirable. Rotating a window sideways can focus the gaze on specific views, and coupled with angled reveals can be used to let in extra sunlight at specific hours, or keep direct sun out. Angled panes and reveals improve privacy, and sometimes give protection from offensive streetlights or other unsavory intrusions.

Windows for Magical Light

The great wall thickness of cob buildings regulates natural light entering the building in

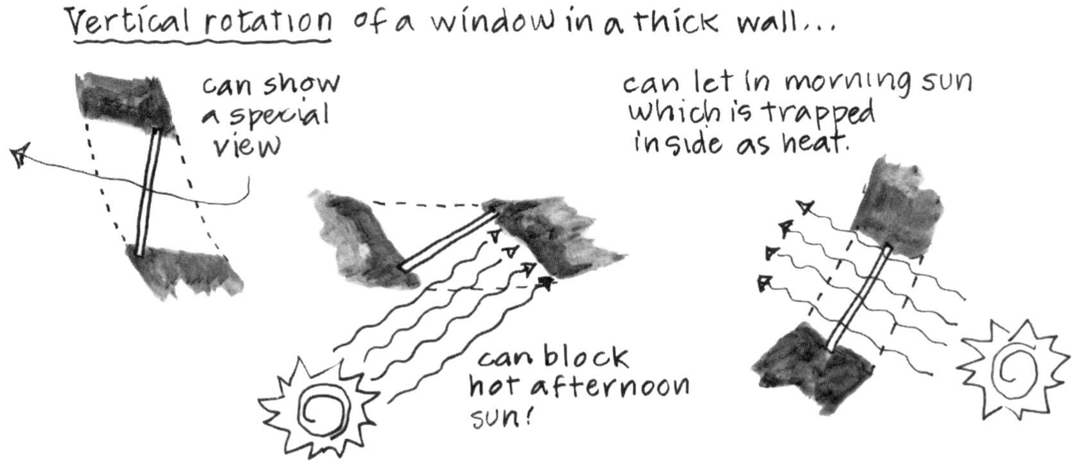

Viewed from above, various orientations of window placement in thick walls.

essential ways. Though quantity and duration of natural light profoundly affect bodily health, it is the *quality* of that light that most affects how we feel. We are a culture distinguished by how much time we spend indoors, and we mostly experience our world visually, especially in city conditions where constant acoustic fog deprives us of natural sounds, where polluted air masks natural odors, and where our sense of touch is deprived of natural stimuli. We are a product mostly of what we see, so the quality of everything we look at affects our emotional attitude. The light that comes in the windows affects us by its color, its intensity, and its direction. Sunlight sliding across the floor or the brilliance of a full moon roots us in the daily and seasonal rhythms of our terrestrial turning, intensifying our contentment and sense of security.

Similarly, the color of everything in a room is conditioned by what color of light falls upon it. The color and reflectivity of the window frame, sills, and reveals affect how much light reaches into the room, its color, and how it makes us feel.

Narrow openings in thick walls let in shafts of light, with dramatic effect. Direct sun will come in like a spotlight, highlighting everything in its path against the relatively dark walls. Indirect sunlight coming in through a deep wall will accentuate the sculptural qualities of faces, houseplants, and furniture in a way that is unimaginable with thinwall construction. Rembrandt and Vermeer understood these qualities of light: Look at their portraits, many conditioned by sunlight penetrating thick masonry walls.

Consider windows to be like task lights. Locate them in places where you will need natural light, for instance, beside the bathroom mirror, over the kitchen sink, to the left of a work desk. Windows for light should be as high as possible, oriented in any direction, though in very hot zones they are better situated to the northeast, north, and east. For maximum light with minimal heat gain, set glass toward the inside of the wall, where the depth of the reveals will shade the pane from direct sun most of the time. Be aware that in most circumstances the amount of light coming in through a vertical window will provide nowhere near the amount of sunlight from a skylight the same size (see chapter 6 for more details).

Consider windows to be like task lights, directing natural light into the locations you want noticed.

Windows for View

Since view is a function of eyes, view windows should in general be located at eye level. Near fixed seats, put view windows at sitting eye level; in kitchens or showers, put them at standing eye level; beside beds, at lying eye level. The closer viewers will be to these windows when they look out, the smaller the windows need to be and the less discomfort they will cause from heat gain or loss. Tiny "peep windows" can frame and highlight a special view. Like the landscaping in a Japanese garden, a carefully placed peep window can make an outside scene or detail more noticeable and evocative. When I was working

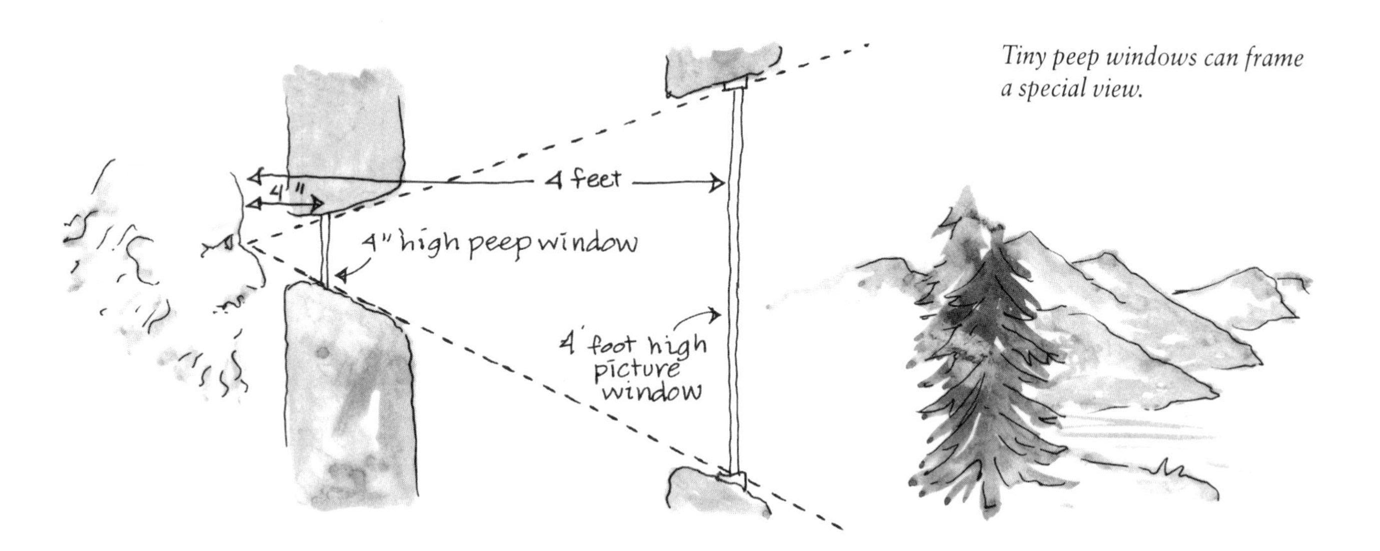

Tiny peep windows can frame a special view.

on a hybrid cob hermitage in northern California, one of the builders found a manzanita branch that had grown together and grafted onto itself. This manzanita hoop was used to frame the view of a living manzanita outside, visibly emphasizing the connection between the building and its surroundings.

Windows provide an eye onto the world but decrease thermal efficiency. Most of the heat in a well-constructed building is lost through the windows, so use small windows wherever you can. In temperate climates, this is particularly important on the north side of the building where heat loss is not compensated for by solar gain. Likewise, if your summers are hot and your best views are to the west, use small, carefully located view windows to prevent overheating on summer afternoons and evenings. If view is the only benefit, a floor-to-ceiling window is a thermal embarrassment *and* a threat to privacy.

Windows and Solar Gain

Where you position the glass relative to the thickness of the wall impacts how much solar heat that window will trap. Glass close to the exterior face of a wall will let through more direct sunlight to be absorbed as heat by the

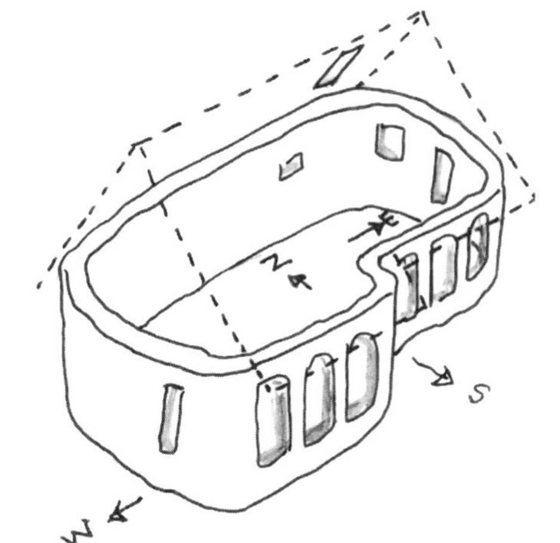

Most heat loss (or gain) in a well-built house is through glass, so reduce window area wherever possible, especially to the north and west where direct sun cannot compensate for that heat exchange.

cob walls and mass floor. If you want to reduce the amount of heat entering, for example in a west-facing wall in a hot climate, set the glass toward the inside of the wall where it will be shaded by the reveals. West-facing windows can be further shaded with small awnings, removable in the winter.

Windows for solar gain should be oriented toward the south, southeast, and perhaps east, as low in the building as possible. Use thermal shutters or curtains to keep the heat in when the sun isn't shining. (See chapter 6 for more specifics of passive solar design.)

Windows close to the outside of the wall provide deep sills, which give the impression

Placement of windows for solar gain or to provide shading.

"Cross-ventilation" is a poor aid to cooling unless there are no interior walls and winds are strong in warm weather.

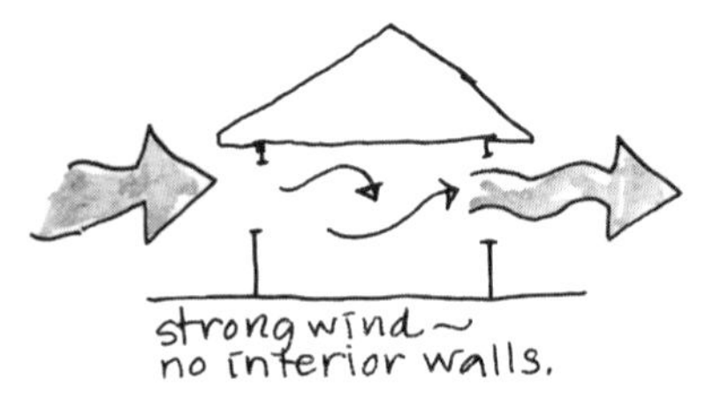

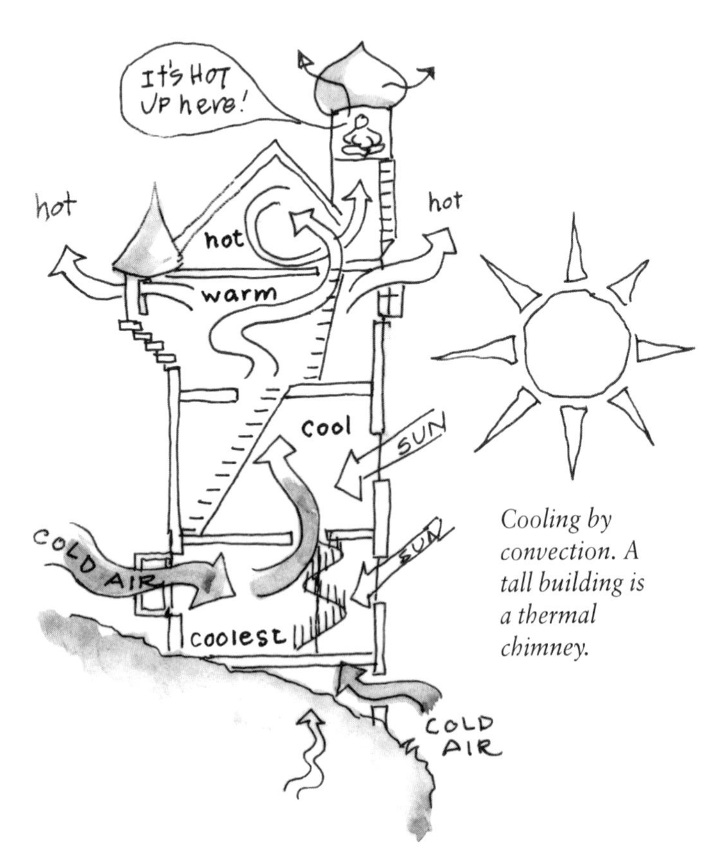

Cooling by convection. A tall building is a thermal chimney.

of a more spacious interior. This sill can be useful as a shelf, perhaps for houseplants or vegetable starts, or even as a window seat.

Conversely, windows set toward the inside of the wall create a deep exterior sill, which may be susceptible to water damage from windblown rain and snow. In windy areas, give exterior windowsills a steep slope, a thick layer of lime plaster, and/or a protective surface of brick, tile, stone, or weather-resistant wood.

Windows vs. Ventilators

Don't count on openable windows for cross-ventilation. Cross-flow, the darling of all those neat diagrams you've seen in building books, is a very poor aid to cooling a building unless you live in a house with no interior walls and experience a gale each hot afternoon. Cross-flow works well in a palm-thatched palapa on Zihuatenejo beach, but anywhere else it's largely theoretical. Even so, it's nice to put opening windows in places where you will spend a lot of time, such as behind the kitchen stove, by the desk, and above the head of your bed, so you can check the weather, hear birds singing, and enjoy fresh air.

In hot or sunny climates, the top of a room gets hotter than the bottom, because hot air rises. The building functions like a short, wide chimney, sucking cool air in at the bottom and discharging hot air at the top. To cool the building naturally, you need a big opening at the very top of the house and lots of openings at floor level, particularly on the north side or onto shaded, vegetation-covered areas. Those openings need not be windows; some can be doors, or merely shuttered openings, with or without a fan. The best ventilation is a hot attic pulling up cool air that enters the house through the basement. In Arab countries, ventilating "chimneys" in the roof pull cold air into the bottom of the house through pipes buried in the ground, and sometimes "wind scoops" are built on the roof to funnel strong prevailing winds down into the living space.

DOORS AND DOORWAYS

What does a door need to do? Closed, it isolates us from too low or too high a temperature, disturbing noises, unwanted lights, activity, visual distractions, odors. "Close the kitchen door, I can smell fried onions!" It defines territory and function. Even open, a doorway announces territory, demonstrates ownership, demands respect: "Her door was open, but it didn't feel right to go in."

Our culture is not adept at subtlety, which is necessary to heighten the magic of expectation as we move from one space to another. We crassly label identical doors to differentiate the identical boxes they open to: "Room 203—Yoga," or "Chapel, KEEP QUIET." The natural builder can handle these transitions with more sensitivity and artfulness. A door's shape, size, transparency, materials, texture, and color convey subtle messages of welcome, formality, the need for quiet, and so forth.

The recent move to standardize all house doors to 32 by 80 inches is in keeping with the constant expansion of house size that the shrinking U.S. family has to endure. If we're serious about reducing environmental impact, *everything* needs to be scaled down—smaller houses, smaller rooms, smaller spaces that are congruent with what we use them for. Large doors should be used only where necessary, in public buildings or for special access.

There is precedent for small doors, both in public facilities and in private residences: motor homes and yachts have shower doors as narrow as 18 inches. Public buses and airplanes have toilets with doors only 15 inches wide. In small buildings, exterior doors are quite functional at six feet high, or a little higher if an occupant is very tall. Interior doors can vary a great deal; for passive spaces such as meditation and bathing and for sleeping alcoves, they can be as low as five feet.

It is fine to make entry a slight challenge, or to make a kids' space almost inaccessible to adults. A small door can help even a tiny space feel spacious; a tight opening helps to expand a small room, offering subtle psychological relief after a constriction. Having to stoop slightly (bowing) to enter a sanctuary preconditions one to humility, observation, awareness, and quiet. On the other hand, there's nothing quite so annoying as banging one's head on a low door lintel, so we need a level change or visual cues to make us slow down *before* passing through a low doorway. Try a raised threshold, African-style, or steps, or an entry ramp, to emphasize a different level of privacy and seclusion. A low doorway, a narrow door, or an unusual shape of door all slow us down a little, force us to pay a little more attention to the goingthru, and therefore to be more aware of the beingin, after we arrive there.

A low doorway, high threshold, or unusually shaped door slow us down, causing us to pay more attention to the transition between spaces.

Door shape alternatives.

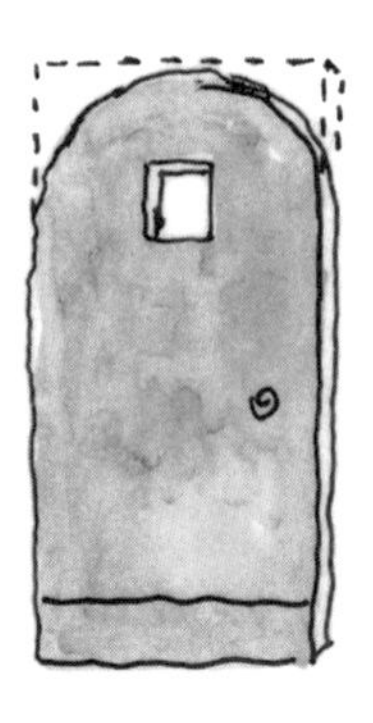

To emphasize differences from one space to the next, it may be worthwhile building a custom door, or cutting down a solid door to a custom shape. Easiest is simply to scale down a big solid door, rounding the top with a jigsaw, or slicing off a side or the ends. Another option is to build the door*way* the shape and size that is appropriate, then use a standard rectangular door to close against it. The eye will read the shape and size of the opening, particularly when the door is open. What if there's a need to allow occasional passage of wide objects? You could then put in a wide double door, split into two unequal sides. Keep one closed normally: the other can be the entrance you usually pass through. For a house entry door, a 30-inch main door with an 18-inch wing section gives you the opportunity to slide in a grand piano when both are open.

A simple homemade door, glued and screwed.

To gain complete freedom with size, shape, and material, make your own door. To make a simple, uninsulated door, use tongue and groove lumber, between ¾ and 2 inches thick. Clamp and glue laterally, screw on diagonal cross-bracing, and cut to shape after the glue sets. Insulated exterior doors can be made of two thin skins with insulation sandwiched between them, though it is hard to construct a thin, lightweight door without resorting to plywood, aluminum foil, or industrial insulation.

With each separate door, consider carefully whether to swing it in or out, and whether to hang it left or right. For privacy, swing a door inward into secluded spaces (bathing, toilet, sleeping areas), hanging the hinges on the opposite side from the room's nearest corner. In this way an occupant is warned before somebody walks in.

To conserve space in small buildings, swing exterior doors outward. Make sure there is a window in or alongside the door, and sufficient level space outside upon which to stand well back as the door opens. It's embarrassing to hospitably throw open the entry door and knock visitors off the doorstep because you didn't see them standing there.

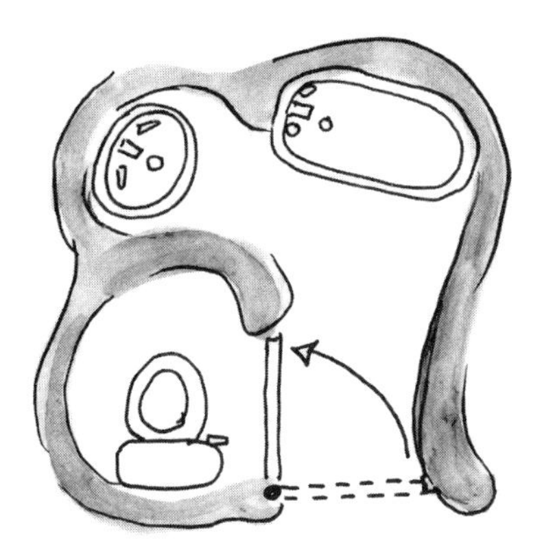

Locate doors to enhance privacy. In the example shown here, opening the bathroom door increases the privacy in the toilet area.

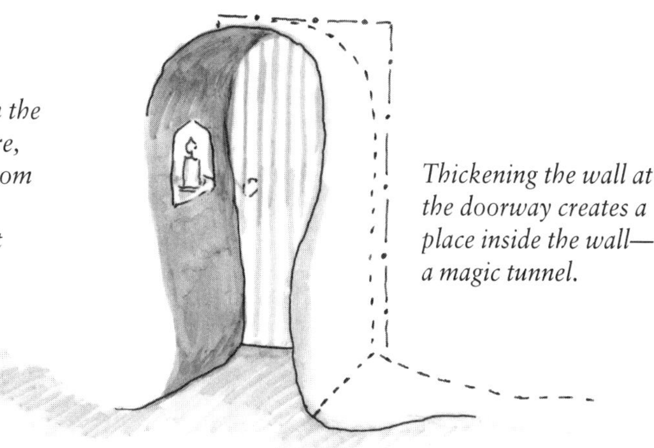

Thickening the wall at the doorway creates a place inside the wall—a magic tunnel.

Thickwall construction offers some options that make a door more than just a mechanical barrier. Traveling from one room to another through a thick wall provides a third *place*, a short tunnel between two rooms. Thickening the wall around a doorway emphasizes a passage between two places of distinct character. In a two-foot-thick wall there's room in the doorway for a candle niche or a recessed light, even for a small closet, a mirror, or a telephone niche. Moving from hangout to sleeping space becomes a journey, much more than just opening a door.

Half-Doors

Wherever you might want to spend time indoors in sunlight, a half-door is an alternative to an opening window. Half-doors, sometimes called Dutch doors, were common in Britain until about World War II and are still common in many parts of the world. They have some real advantages, thermally and socially.

The half-door offers the benefits of a door *and* a window. Working by an open window is a joy—drawing, reading, writing, or any sort of handwork. The stimulus of breezes or temperature change and the fresh air keep you awake.

The half-door has particular value in massive buildings. Because air has a small capacity to store heat (that is, a low specific heat), rapid air changes make little difference to the coziness of mass buildings. A wave of cold air rushing through a warm massive structure can steal only the heat stored in the air. Heat buried in the building's heavy structure is released slowly, so the radiant temperature of the interior stays high. Even in cool air you feel comfy.

Why not just leave the whole door open? Well, we function best with clear heads and warm feet. Driving in cold weather with only the defrost blower working can be really disturbing, as your head heats up and your feet

A half-door offers the benefits of both door and window.

cool down. In contrast, Alaskan truck drivers run at minus fifty with the heater full blast on their legs and feet, and the windows open. The stimulus keeps them alert.

With a standard door wide open in cool weather, warm air leaves through the top of the doorway and of course must be replaced by cold air, which being denser rushes in across the floor. Result? Cold feet, warm head, a good recipe for a bad mood. The half-door breaks that flow, leaving floor-level air largely undisturbed but offering refreshing coolness at head level. Half-doors are particularly useful in kitchens or workshops, where much of the work is done standing up.

If half-doors are such a good idea, how come you don't see more of them? Because in cool weather, lightweight buildings store very little heat, so major ventilation robs that heat quickly, leaving the structure cool and the radiant temperature too low for comfort. The screened porch may have developed as a uniquely American response to stud-frame buildings in breezy warm climates, replacing the half-door system that settlers from Europe knew. But a porch can only be used in warm weather.

Quite apart from the comfort and expansiveness of having a large eye-level opening, half-doors work well as dog control and toddler restraint. It is common in parts of Africa to see women leaning on their doortop, chatting with passers-by. Neighbors can casually chat through a big opening in a way that commits neither you nor they to any sort of formality.

Roofs for Cob 15

The old English saying goes, "Give a cob house a good pair of boots and a good hat and she'll last forever." The "hat" in question is of course the roof, arguably the most important part of any house. For an earthen house in a rainy climate, the roof takes on even more importance, as a leaky or inadequate roof can lead over time to severe damage in the walls. Cob doesn't dictate any particular roof style, geometry, or covering, but if your building is to last centuries, it's worth investing in a durable roof. Choose a system that fits your climate, bioregion, and building size, as well as your pocketbook, skills, and available materials.

❧ The old English saying goes, "Give a cob house a good pair of boots and a good hat and she'll last forever." ❧

In general a good roof should:

1. Keep out rain and other precipitation, shedding and directing water away from the walls.
2. Prevent unwanted heat loss and gain.
3. Protect the inhabitants and contents of the building from wind and excess light, as well as from overhead noise, dust, and other annoyances.
4. Remain in place in hurricanes, high winds, and earthquakes.
5. Resist occasional extra loading and wear from maintenance workers, snow, ice storms, fallen branches, and so on.
6. In fire zones, protect the building from fire.
7. Hold solar hot water panels, photovoltaic arrays, maybe even a rooftop deck or garden.
8. Collect water for drinking, household use, and/or irrigation.
9. Encourage, rather than deter, plant growth and wildlife.

THE PARTS OF A ROOF

If a roof is both to keep rain out and keep warm air in, it will need some sort of an airtight *membrane*. To support the membrane, *sheathing* of boards or some type of synthetic wood is normal, though natural builders try to avoid the toxins in plywood and oriented strand board (OSB). *Rafters* are the structural members that run beneath the roof, holding up the sheathing. Rafters are sometimes prevented from separating by *collar ties* or *tie-beams,* horizontal tensile members that span the whole roof, often holding up the *ceiling,* which is not necessarily part of the roof at all. A *truss* is a prebuilt assembly of two rafters and a collar tie, usually with added cross-bracing. Trusses eliminate the need for a *ridge beam,* which supports the highest point of each rafter in a standard gable or "pitched" roof. *Gables* are the triangular parts of exterior walls underneath the peak of the roof. *Eaves* are the roof's edges, which overhang beyond the walls.

Synthetic membranes such as plastic sheeting and rubber "torch-down" last longer when protected from sunlight by sod, earth, straw, sand, or gravel. If such a roof has vegetation growing on it, it is called a living roof. Composition roofing ("roll roofing") and asphalt shingles are very common, inexpensive, but short-lived and toxic roofing materials, derived from petrochemicals.

Another set of covering options includes slate, ceramic or concrete tiles, and wooden shakes and shingles. Like metal roofing, these tiles are screwed, nailed, or wired to *purlins,* wooden boards that run horizontally across the rafters. *Thatch* is a thick, tightly packed layer of reeds, straw, or palm leaves, combining roofing and insulation in one material.

Always leave a vented air space above the insulation, and a *vapor barrier* below it. This vapor barrier, usually made of plastic sheeting installed on top of the ceiling, inhibits the escape of heated air from the building and the buildup of water vapor in the roof.

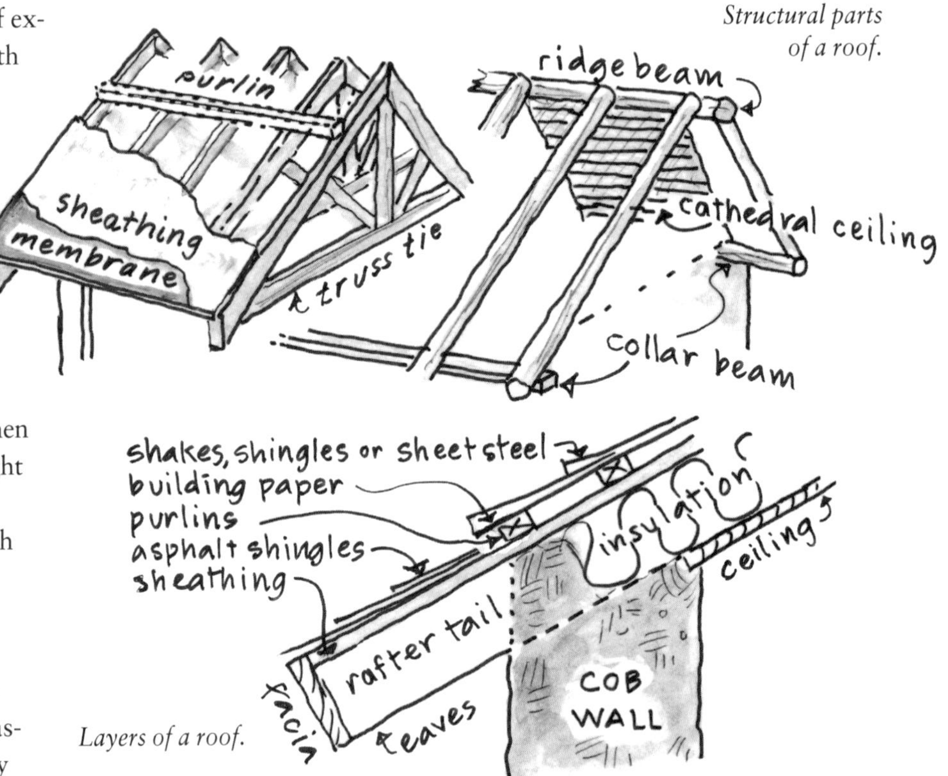

Structural parts of a roof.

Layers of a roof.

This chapter introduces the cob builder to the rudiments of roof design and construction. We'll look briefly at a few alternative roofing systems, both traditional and modern, and then at some natural options for roof insulation. Some aspects of roofing are specific to cob buildings, such as how to attach roof to walls, but most are not, so we don't cover roof building in great detail. Many fine books exist with precise instructions for home-builders who want to construct their own roof. This, then, is intended to stimulate thought, widen your sense of what is possible, and head you off from potentially disastrous directions.

ROOF DESIGN

Design of the roof is integral to designing the whole building. Leaving roofing decisions until late in the game can be exciting but hazardous. For example, you may discover you lack a ridge beam heavy enough to span the whole building, which makes you add an extra post unexpectedly in an awkward place. To avoid this sort of problem, the roof should be designed at the same time as the rest of the building. Make a detailed model of how the roof will be constructed and what will hold it up.

Cob walls are amazingly strong; they can support great vertical loading. But there are good reasons to keep your roof light, particularly in seismic areas. In earthquakes, a common cause of injury and death is falling roof members, and a heavy roof can rock the whole building back and forth (see appendix 4). *When you double the span of beams or rafters, you must quadruple their depth (vertical thickness) to carry the same load.* Roofs over narrow structures can therefore be much lighter and use much less wood. They're also cheaper and easier to build. In areas of heavy snow accumulation, try to avoid unsupported spans of more than 12 or 14 feet.

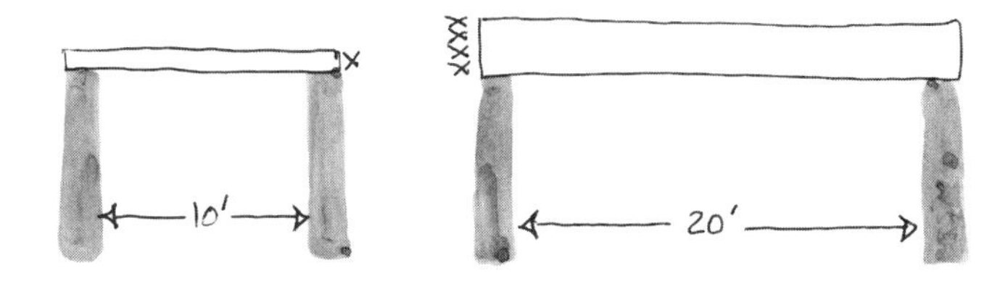

Make a detailed model of your roof design well in advance of building. When you double the span of beams or rafters, you must quadruple their depth to carry the same load.

Roof design is inextricably linked to the decision of whether to build one or two stories. The roof and foundation are generally the most expensive and time-consuming parts of a simple building. In cob buildings, they're also the parts with the largest environmental impact. For an equal amount of useful space, a two-story building therefore has a lower economic and environmental cost than a single story.

The design for a roof's structure will depend on its span and covering. The covering in turn depends on many factors, including availability, cost, weight, fire resistance, whether the roof should shed snow or collect water. Snow-shedding roofs need to be slick and fairly steep. Steel or aluminum work well; sod or shakes have too much texture. Thatch neither sheds snow nor collects water well.

Water collection is worth special thought. Rainwater is free and relatively clean, so it makes sense to utilize it. In many countries, including Australia, New Zealand, and Brazil, rainwater cisterns are everywhere; in fact, many homes have no other water source. Roofs and gutters are arranged accordingly. For irrigation water any collection surface will work, but if you plan to bathe in the water, avoid tar or bituminous finishes such as torch-down and composite shingles, and if water will be used for drinking or cooking, also avoid cedar shakes or shingles, which exude a

Protection from rain.

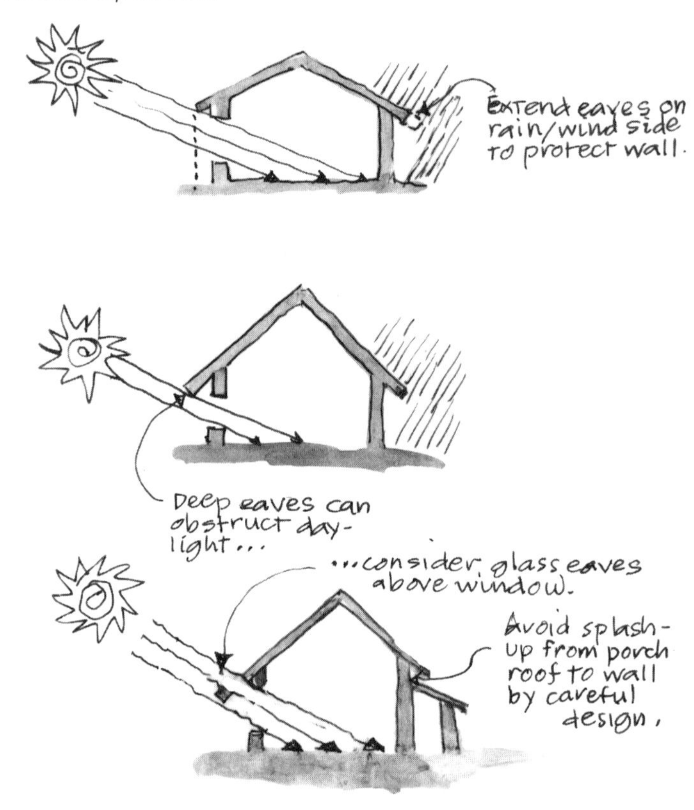

toxic oil for a year or two after attachment. Thatch gives off humic acid and discolorants, and is difficult to gutter because of its thickness. In wetter regions, consider thin sod or gravel over butyl rubber (not PVC sheeting, which is a carcinogen) or polyethylene. The quality of water from a well-established sod roof is generally very high, because the water is filtered through the roots and soil, though light rains may be completely absorbed by the grass mat. The most common roofing option for water collection is pre-enameled or anodized (*not* galvanized) steel. Slate and ceramic tile are ideal except for their weight, which necessitates extra roof structure, so for short-span roofs consider slate, tile, and sod, and for longer spans, pre-enameled steel.

A street of roof shapes: 1. pitched gable roof; 2. hip roof with dormer; 3. shed roof; 4. flat roof; 5. conical roof; 6. vault.

The most efficient way to get natural light into a building is through skylights, but leak-proof skylights are difficult to construct; get help from someone experienced. The narrower they are and the closer to the ridge of the roof, the less likely they are to leak. Several small skylights will distribute light more evenly than a single large one. Skylights facing west, south, or even east can cause overheating in the summer, so face them north or northeast, or figure out how to shade them when necessary (see diagram, page 87).

In designing the roof, try to extend eaves farthest on the side rain will blow in from. Covered porches or loggias afford good rain protection for the walls they enclose, but require careful waterproofing if the porch roof grows out from a cob wall. Deep eaves may obstruct daylight if they extend out over windows, so you might want to design in a section of glass eave directly above windows.

The Shape of a Roof

The most common basic roof shapes are: flat, shed, gabled, hip, conical, vaulted, or domed, or combinations of these. A shed roof might seem to an amateur to be the simplest to build, but we will discuss the *gable roof* in more detail because: (a) with a load-bearing ridge beam or trusses, a gable roof increases the uninterrupted span possible, or reduces the depth of rafters; (b) it is the preferred roof geometry for wall protection in rainy climates; (c) it is fairly easy to construct; (d) the load rests equally on both long walls and forces are all distributed vertically; and (e) it uses less wood. Overall, a gable roof creates extra volume inside the building, while reducing wall height.

A simple gable roof has two facing sets of rafters that are connected at the peak. Near its bottom end, each rafter rests on a wall, a post, or a beam; if this were its only support, the weight of the roof would tend to push the walls apart, so rafters also need to be attached to a beam at the ridge, or be tied together at the top of the wall by collar ties. The other option is to substitute preconstructed trusses for rafters. Their triangular geometry makes them very rigid and obviates the need for a ridge beam. Tiny buildings work best with a ridge beam; bigger ones sometimes are better with trusses.

A ridge beam usually runs the entire length of the roof and is supported by walls at each end, and if necessary by posts in between. It is simpler to construct than trusses, and it leaves the space near the ceiling more open, easily accommodating lofts, attics, and the extra space of "cathedral ceilings." For low-pitch roofs, a ridge beam is structurally easier than trusses.

There are many designs of trusses, but all use the inherent stability of the triangle to hold up the weight of the roof. They eliminate the need for posts, leaving larger, more open spaces below. Their use can reduce the amount of wood, and therefore the weight of the roof. Consider what materials you have at hand or can easily acquire. If you have plenty of long, stout round poles, a ridge beam may be the best solution. If all you have are 2 × 4s, trusses probably make more sense.

It's hard to make a rectilinear roof feel natural on an organically shaped building. One technique that we've found helpful is to use a curved ridge beam, which also increases the headroom inside without raising the wall height. If you live in the woods, find an appropriately curved tree and mill it to your needs. Such trees are often abandoned by commercial logging operations because they can't be sawn into straight lumber. If you are comfortable with carpentry, you could build a curved beam by gluing and screwing together many thin, bendable boards. A second way to soften the look of a roof is to round off the corners, although it's harder to fit gutters to rounded roof edges.

Another basic design decision is what pitch (steepness) to make the roof. The pitch of a roof drastically affects the look and feel

Options for roof-support structures. A ridge beam supports opposing rafters. Trusses can be purchased or constructed on-site.

A curved ridge beam will complement the curves of cob walls.

Attaching rafters.

Roof pitch dramatically affects the shape of a building, and is mainly dictated by the roofing material.

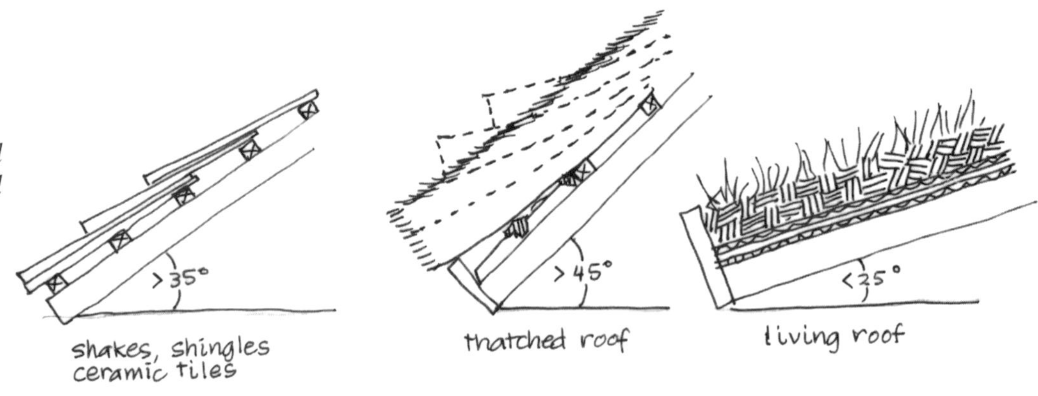

of a building, from both inside and out. A steep roof makes the roofline higher, which means you have to build the walls higher at the gable ends, and a steep roof's eaves block more sun from windows than a shallower roof. The space inside a steep roof is less useful in shape, and harder to light and heat. But roof pitch is largely dictated by what roofing material you choose. Roofs protected by small, overlapping units, such as shakes, shingles, ceramic tiles, or thatch, need to be quite steep (35° for most, 45° for thatch!) to prevent leakage of rainwater. Living roofs, on the other hand, need a shallow pitch (less than 25°), so the organic matter won't slide off.

BUILDING A ROOF STRUCTURE

If this is your first building, try to get advice on your proposed roof structure from a competent builder. Small roofs are easy—it would be difficult to make catastrophic mistakes—but if the building is more than about 200 square feet, getting a second opinion will do little harm. For rafter sizes and spacing, consult a construction manual such as *Homing Instinct* or *The Owner-Built Home*. Don't overbuild; most literature specifies structural lumber as if it were free and had no ecological costs, and building codes are even worse.

Curved ridge beam supports roundwood rafters, making a complex and sculptural roof for a cob cottage in Oregon.

The roof structure is usually made of wood, because wood is easy to work with and readily available most places. To protect forests from nonsustainable management practices, we usually build with recycled, salvaged timbers or carefully selected round poles. If you have access to forest land or plan to fell trees to open up a building site, consider having your own bigger trees milled into beams and rafters. Undermanaged forestlands are often choked by thin poles with little taper, which make excellent rafters. Select them carefully for similarity of diameter, bending strength, and straightness. Using wood in the round, with its fibers intact, gives much more strength than milled boards, so you can space round rafters farther apart than milled ones of the same depth. Sawn lumber is also more prone to decomposition, as sawing opens the cells of the wood, inviting fungal penetration. If you must buy new lumber, seek out a local, small-scale mill or buy certified sustainable wood products.

Cob buildings are almost impregnable by gales, cyclones, or hurricanes provided the roof stays on. With all but the heaviest roofs on the smallest buildings on the most sheltered sites, it's important to tie the roof down to the cob walls. Bury a deadman about 2 feet below rafter ends at 3 to 4 foot spacing, and beneath the ends of ridge beams and other beams. Make each deadman out of an 18-inch length of stout wood, such as a 4 × 4 offcut or a piece of firewood about arm thickness, with a long piece of heavy galvanized wire (#9 fencing wire is good, as is electric-fence cable) wrapped around it twice and stapled on with fencing staples. Bury each deadman in the cob with the ends of the wire sticking straight up. When you reach the top of the wall, loop the wire twice around each rafter or beam. Wait until wall shrinkage is complete before cinching down the wire and fixing it with more fencing staples.

You can either build the cob walls first, then attach beams and rafters to the waiting deadmen, or you can build your roof structure first, support it on temporary posts, and cob up around them. The latter option has many advantages. First, you can use the roof structure, with a temporary or permanent covering, to shelter you and your building from sun and rain during construction. Second, it can be difficult to calculate the varying height of a curved wall where it meets the roof. If your roof is there already, you don't have to worry about building up too high.

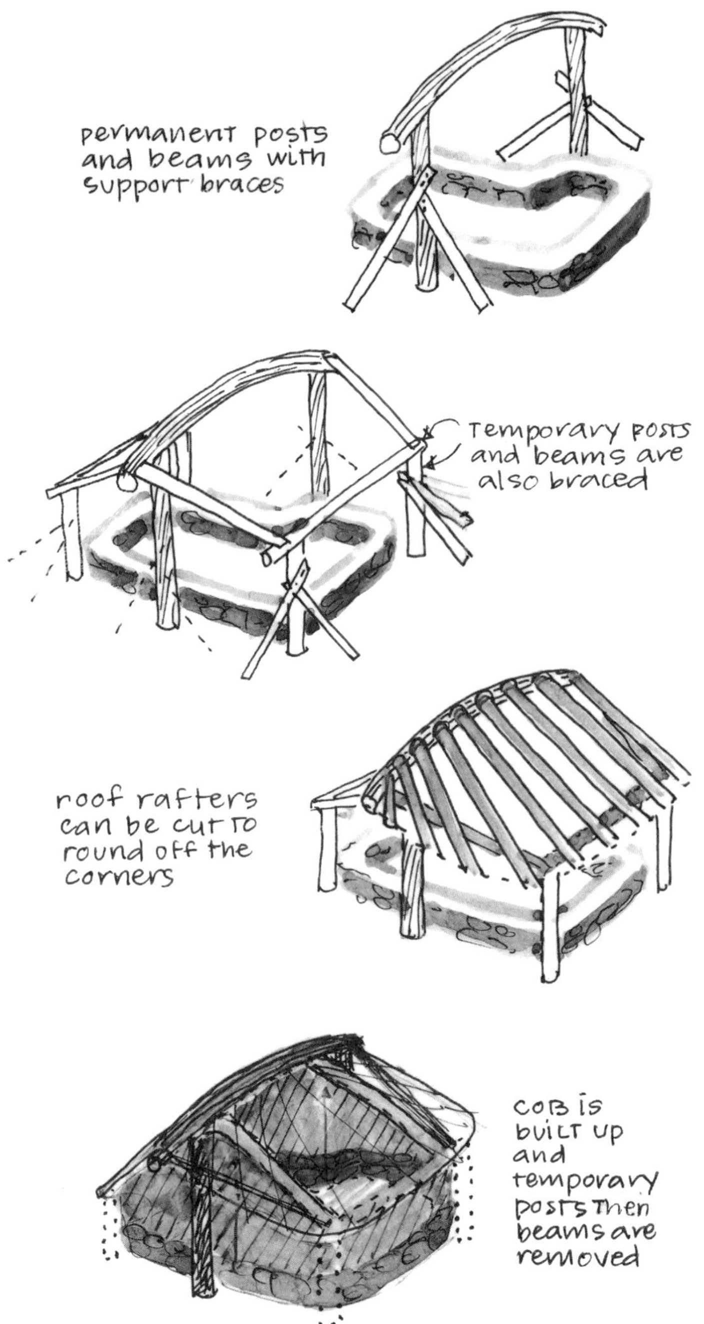

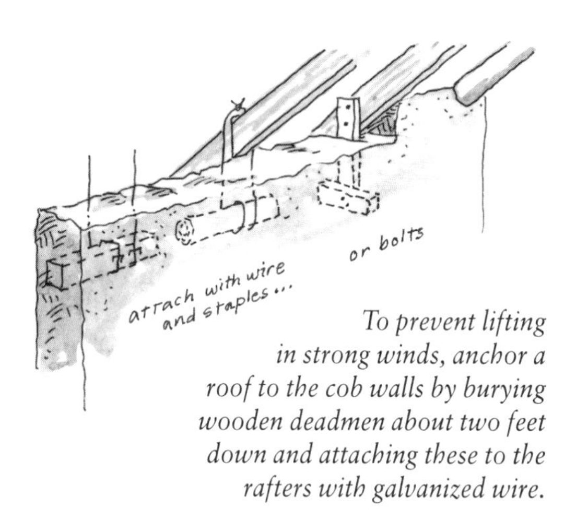

To prevent lifting in strong winds, anchor a roof to the cob walls by burying wooden deadmen about two feet down and attaching these to the rafters with galvanized wire.

But if the finished roof sheathing goes on early, that can make it difficult to finish the top of the cob walls, because you will have to build above head level in places difficult to see. My preferred system is to put up the beams and rafters early, cover them with a tarp for rain and sun protection while the cob walls go up, then remove the tarp to finish the tops of the walls, and finally install the roof sheathing and permanent covering. Ianto likes to put up beams, rafters, and ceiling, suspended about a foot higher than the wall on temporary posts, build up walls, then lower the roof and stuff any cracks with cob. Choice depends on circumstance.

A wide variety of natural materials can be attached to the rafters to create different ceiling effects.

Ceilings

The ceiling is what you see when you stand inside a building and look up. In most buildings with insulated roofs, the ceiling is separated from the roof sheathing. In addition to providing an attractive visual surface, ceilings contain and support the insulation, and inhibit the loss of warm air. Ceilings can be affixed to the undersides of roof trusses or collar ties, or placed in strips between or laid on top of rafters so that the bottoms of the rafters show. To increase the amount of space available for insulation, a separate ceiling structure can be suspended below the rafters.

Cane latillas make a ceiling in South Africa.

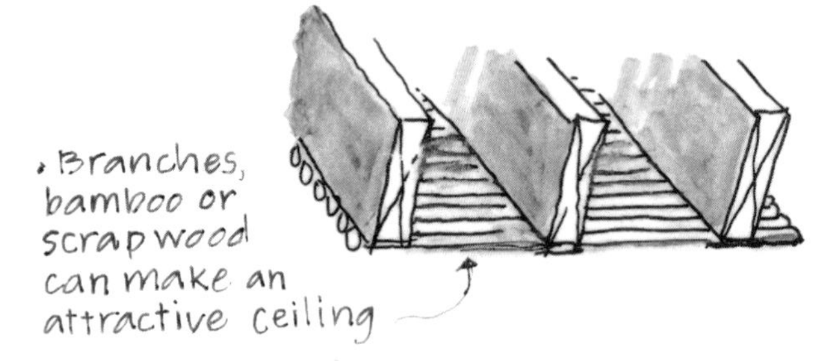

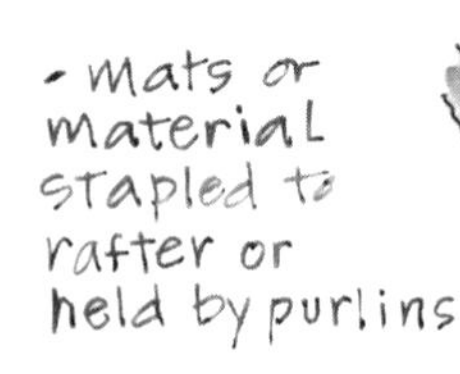

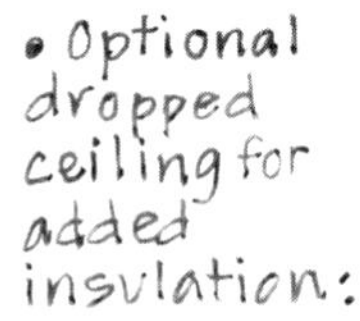

Ceilings are often made of boards or of sheet materials such as drywall and plywood. We've also made beautiful ceilings out of thin, straight branches placed close together, like Spanish-American *latillas*. The builders of the cob cottage at Heartwood Institute in northern California stapled woven bamboo mats to the underside of their rafters, creating a very elegant ceiling with little effort or expense. Another approach is to use dyed or patterned fabric such as cotton sheets or a nylon parachute, either stretched tight or hanging in loose folds with a billowing, celestial appearance. Any exposed fabric should be treated with a flame retardant to reduce fire danger. Scrap wood can make an attractive ceiling: consider offcuts and board ends, or fencing lumber—usually ¾ × 4-inch or 6-inch cedar. You may need to create a dust barrier above any ceiling with gaps. Discarded bedsheets are often available from motels, or use burlap sacks, opened out flat.

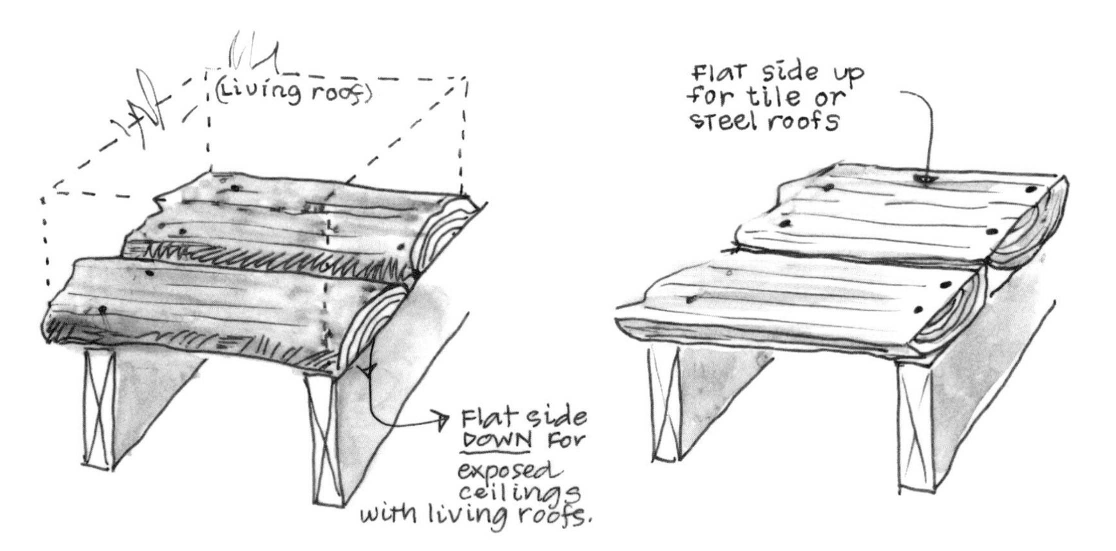

Inexpensive slab wood from milling operations can be used for sheathing, flat side down for exposed ceilings and for living roofs, or flat side up for tile or steel roofs.

Roof Sheathing

On top of the rafters, there will be either continuous roof sheathing or horizontal nailers (purlins) to support the covering that actually keeps the weather out. If you want a more natural, toxin-free roof, steer clear of anything in 4 × 8 sheets—plywood, chipboard, or OSB. We prefer to sheathe roofs with boards, so we hunt around for recycled wood, abandoned buildings to demolish, or mill offcuts. Be creative. Covering a roof can take a lot of wood, so reused or waste wood is good for your budget and reduces the demand for massive clear-cuts. In some areas, small sawmills have surplus "slab wood," the first cut off a log before milling boards. Slab is flat on only one side and can be used flat side down for an exposed ceiling with living roofs, or flat side up to provide a level surface for tiles or steel sheets. If you have access to large driftwood logs, forest thinnings, or dead standing trees, a small portable mill may for a fee slice them into ½-inch boards that can be easily molded over a curved roof structure. Pallet wood works, though it tends to be short. We have also sheathed roofs with sticks nailed side by side and have seen bamboo and carrizo (a wild cane that grows in profusion in the southwestern United States and northern Mexico) used to good effect.

As previously noted, a cob building does not necessarily dictate any special roof covering, though to keep in harmony with the natural walls you might avoid asphalt shingles or roll roofing, concrete tiles, and asbestos-cement sheets. Three roof coverings are of particular interest to natural builders, though published information is scanty, so we'll describe them here. They are sod or living roofs, thatch, and sheet metal. We will also touch on wooden shakes and shingles and a few other options for natural tiles.

LIVING ROOFS

By "living roof," we mean any roof with vegetation growing in a layer of soil or organic matter over a waterproof membrane. Traditional sod roofs sometimes had a great depth of soil to keep them green year-round. We recommend a lightweight system that requires far less wood. The soil's most important function is to protect the membrane beneath from ultraviolet light and other damage, which can be accomplished with only two to three inches of soil. Plants hold the soil in place with their roots and make the roof more attractive. A living roof will muffle traffic noises, reduce and filter runoff, and disperse the sun's heat, reducing indoor temperatures in summer, but don't count on wet earth to keep heat *in* in winter.

Living roofs are fairly easy and inexpensive to build and if well maintained can last many decades. They are very beautiful, giving the building a gentler, more natural appearance. In addition to grass, you can grow ornamental herbs and flowers or even edibles such as strawberries on your roof. Some people with sod roofs mow and water in the summer to keep them lush and green and to increase their fire resistance. We have seen sod roofs maintained by rabbits and guinea pigs, and a whole shopping center on Vancouver Island had a goat patrol mowing the roof for a while. Irrigation is not necessary; you can let the vegetation die back and resprout naturally each year. Even if the dry grasses on the roof catch fire, they will quickly burn off while the soil layer protects the wooden structure below. In 1994, a report was published in Australia on a sod-roofed, earthen house that had survived a devastating forest fire in a Sydney suburb. While every other house on the road was burned to the ground, this one suffered no more than a cracked window.

Avoid building a living roof too steep, or the sod will slide off. In general, maximum pitch is 5 in 12. If you are concerned that your roof might be too steep, affix 2 × 4 rails horizontally to the sheathing to help anchor the sod in place. Also attach a raised fascia or edge board very firmly to the lower eave. Consider carefully how and where rainwater will flow off the roof. Water running over the fascia may stain or lead to rot. Another option is to use shower pan drains, off the shelf of the hardware store, to let water safely through the membrane inside the fascia. You can attach chains or downspouts to these drains to prevent splash.

Over the roof sheathing, lay the impermeable membrane, the most critical part of the roof. You can use any material that is waterproof, flexible and sturdy, and that either comes in sheets large enough to cover your whole roof or can be sealed along the seams in a fail-safe way. We have used 10 mil black polyethylene sheeting with good results, and a sod roof we made with two layers of 4 mil lasted fourteen years. A more durable and

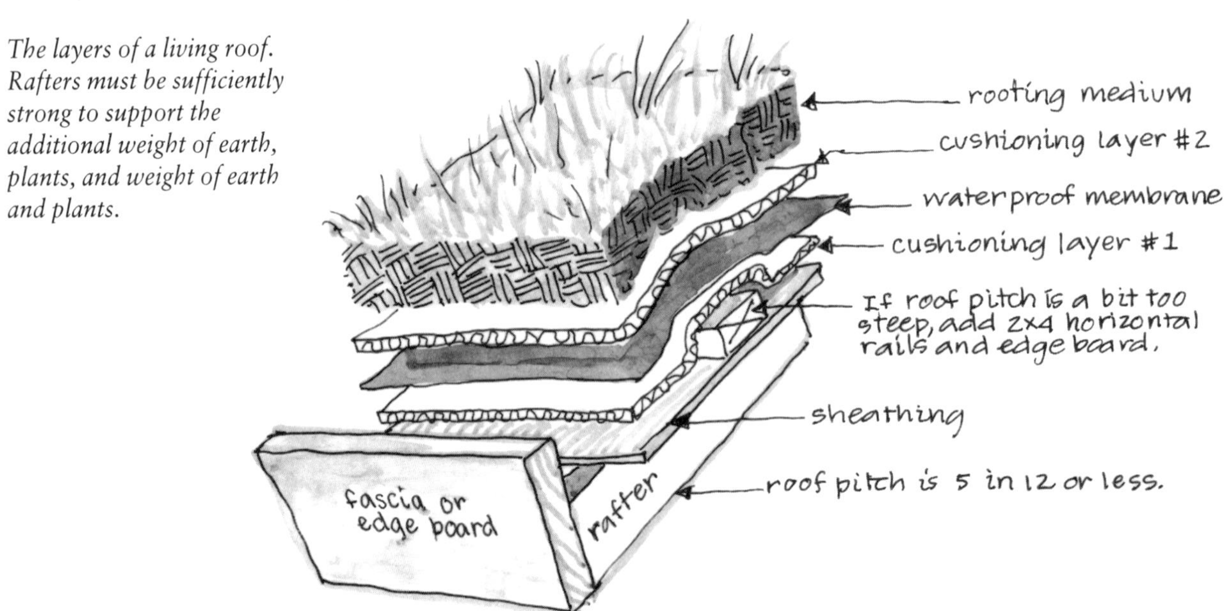

The layers of a living roof. Rafters must be sufficiently strong to support the additional weight of earth, plants, and weight of earth and plants.

LINDA SMILEY

Innovative living roof by Patrick Hennebery, Mayne Island, BC. Loose hay was seeded with wildflowers over 3-tab shingles

more expensive option is butylene rubber or EPDM, which is pieced together with special glue. EPDM is extremely difficult to pierce. It's good for about fifty years exposed, presumably longer under sod. David Easton recommends 20 mil chlorinated polyethylene, which is available in large sheets for sealing ponds. ArchiBio of Quebec reports good results with Bentonite clay, neoprene, and PVC membranes. Their favorite is a torch-down roofing called "Armorplast." There are several other commercial roofing membranes available from roofing supply stores. I hve had good results using an inexpensive self-adhesive membrane called "Vycor." Note that many of these products maybe toxic in both the manufacturing and installation stages. Chlorinated plastics are especially hazardous to the environment and to factory workers. In Wales, *natural* rubber membranes are now available.

Keep in mind that most membranes contract in cold weather, so particularly if you're roofing in warm weather, leave the membrane loose, with wrinkles and folds to accommodate movement. Use enormous care that all seams are properly sealed and that there are no puncture holes. Run the membrane up over the top of the fascia, and flash over the edge. David Easton, in "Roofing with Sod" (reprinted in *The Best of Fine Homebuilding,* with many useful construction details), writes, "I now have about a dozen sod roofs under my belt, and almost all the problems I've had with them have been due to faulty workmanship along the rakes and eaves, and around chimneys, skylights and vent stacks. The hard-learned lesson is that you can't be too careful, especially when you're burying a waterproof membrane beneath 8 inches of soil."

If you use the more fragile polyethylene sheeting, or if your sheathing is of rough boards, it's a good idea to cushion beneath the membrane with corrugated cardboard. You can find huge quantities of broken-down cardboard boxes behind supermarkets and department stores. Be extra careful to remove staples and anything else that could puncture the membrane. To prevent tears and punctures, studiously avoid walking on polyethylene once it's in place.

On top of the membrane goes another cushioning layer to protect the membrane from puncture from above. Use more cardboard, or something more flexible such as carpet scraps (check the dumpster behind carpet stores, or ask them for cutoffs), which may also give the vegetation something to root into to hold the soil layer together.

Finally, you're ready for the rooting medium. You could simply shovel soil or compost onto the roof, 2 to 4 inches thick. Thicker soil will keep green longer, but as it's heavier demands heavier rafters and sheathing. Or you can cut strips of turf and set them in place. Sod strips are particularly helpful along the ridge and edges of the roof, because they hold themselves together much better than loose soil.

Another system is merely to throw loose straw or hay upon the membrane, together with wildflower seed. The flowers' roots tie the whole thing together, the flowers look good, and more straw can be added every couple of years until a good compost has developed. I have a roof like this on my current house. The only obvious drawback so far is that decomposing straw leaches out a black, acidic liquid, making collected roofwater unsavory.

At this stage, your roof will be ready to come to life. You can seed it with cover crop, flower bulbs, or wildflowers, or simply wait for dormant seeds in the soil and straw, blown in on the wind, or transported by birds, to sprout. Over time, the ecology of your roof will come to mimic a natural meadow with shallow soil over a rock ledge. You can speed this process by using seeds, soil, or sod strips from such a place.

To maintain your living roof, make sure that the impermeable membrane remains protected from ultraviolet light. The most susceptible place is the ridge, where gravity pulls the soil away in both directions, so it's a good idea to add extra layers of sod across the ridge and for several feet on either side. After the first rainy season, go back and fill any gaps that have opened up between the sod with fine sand or soil. Do this in dry weather, when the sod has shrunk. If you check your roof every year or so, and put more soil or sod strips anywhere the covering is growing thin, you should have a functional and attractive roof for a long time.

For more information on living roofs, contact ArchiBio in Quebec; see the resources listing in the back of the book for contact information. You can read about their system in *The Straw Bale House* or in Michel Bergeron and Paul Lacinski's *Serious Straw Bale*. Also see "How to Make a Sod Roof," by Ianto Evans, in *The Smallholder* (Argenta, British Columbia; no. 82: spring, 1996).

THATCH

The word *thatch* describes any roofing system that uses flexible stems and/or leaves of plants to shed water. In Northern Europe, thatch is made from tight bundles of grain straw (usually wheat or rye) or common reed grass *(Phragmites communis)*. Most cob cottages in England, Ireland, and Wales were originally thatched. In the colonial United States, thatching was a common technique, but there are now only a handful of professional thatchers in the United States and Canada, and increasingly they use imported material. However, thatching is still widely practiced in Britain, other parts of Europe, Africa, and Latin America.

What's the big fuss about thatch? First, thatch is one of the most beautiful roofing materials imaginable. Thatch is particularly suitable to the curvilinear lines of cob buildings, because of all roof materials it conforms most easily to curved and irregular shapes. It's one of the only effective roofing systems that can be made entirely of natural, renewable materials. Thatch doesn't require any kind of impermeable membrane or even very much wood. A well-made thatch roof can last a long time: straw thatch up to forty years, and reed thatch sixty or more. The foot-thick thatch common in Britain makes not only a durable roof but

Thatch is one of the most beautiful roofing materials imaginable, yet in North America few people know the age-old techniques of thatch installation and maintenance.

also an attractive ceiling and provides good insulation at the same time.

At present there are three major barriers to the use of thatch for building in North America: a shortage of skilled thatchers; concerns about fire safety; and the low availability of suitable thatching materials. In Britain and elsewhere, thatching is a specialized craft requiring years to learn. With modern technology, simpler techniques are being developed, but thatching will always demand more skill than other aspects of building a cob cottage.

Fire is a serious concern, particularly in towns and areas subject to forest fires. However, recent research suggests that fire danger can be substantially reduced by simple additions such as ceilings that reduce airflow to the roof. Danish master thatcher Flemming Abrahamsson has demonstrated that sparks alone will not ignite a thatch roof and that a thin layer of cob beneath the thatch will dramatically retard the spread of a house fire into the roof. Other fire-preventive measures include sprinkler systems and treating the roof with flame-retardants, though these tend to leach out in rainwater.

Perhaps the biggest obstacle to using thatch in the United States is a shortage of appropriate materials. Modern high-yield hybrid grains have been bred for short stalks, which makes them unsuitable for thatching. Furthermore, combine harvesters cut the stalks high and reduce them to a crushed and jumbled mass, whereas thatching requires whole straws with their butt-ends aligned. The renaissance of thatching will require collaboration between farmers willing to grow traditional grain varieties and harvest them by hand and natural builders committed to supporting them. Another approach would be to research the appropriate cultivation and harvesting of native reeds and grasses. Hopefully, as interest in natural building increases here and elsewhere, both the technical knowledge and the materials to build thatched roofs will become more available. For more information about thatching, see *The Thatcher's Craft* (Rural Development Commission, 11 Cowley St., London SW1P 3NA England, 1988).

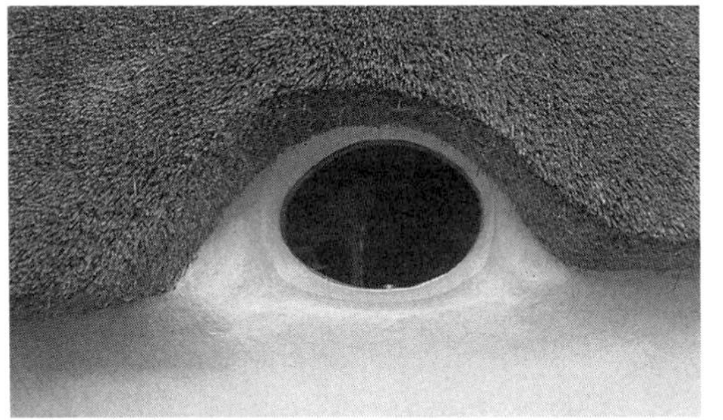

ABOVE: *Reed thatch on cob cottage in Devon.*
LEFT: *Danish style reed thatch by Flemming Abrahamsson*

Penny Livingston's cob office has a unique rain collection system on its rolling, green metal roof.

METAL ROOFING

Although it is less visually appealing than thatch or a living roof, metal roofing has some distinct advantages. It is inexpensive and, if treated right, very durable. It's also one of the fastest and easiest roofing systems to get right—even an unskilled person can install a metal roof that won't leak. Metal is easy to remove, replace, and reuse, and can be recycled when its useful lifetime is over. Because it is very lightweight, far less wood is required in the roof; rafters and purlins can be quite light, and wooden sheathing is unnecessary.

Drawbacks of metal roofing include its flat, uniform, shiny surfaces, which can appear out of place on organic natural buildings, and it is a product of the energy-intensive, polluting industries of mining, manufacturing, and transportation. There is also growing concern about the ways in which steel building components alter and concentrate natural electromagnetic fields beneath them.

A clever design to collect rainwater runoff from a metal roof with a curving edge. It is otherwise hard to gutter.

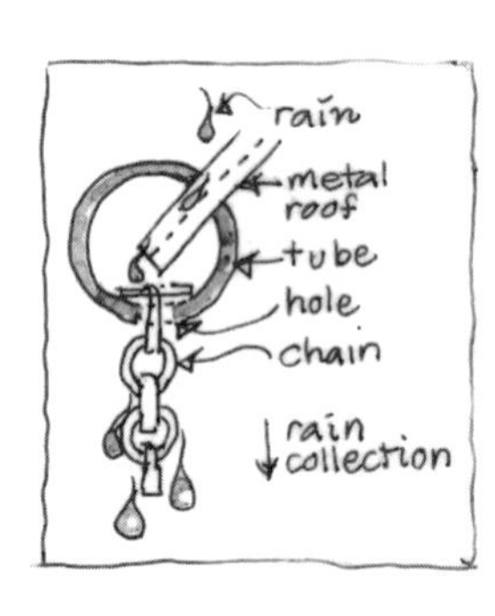

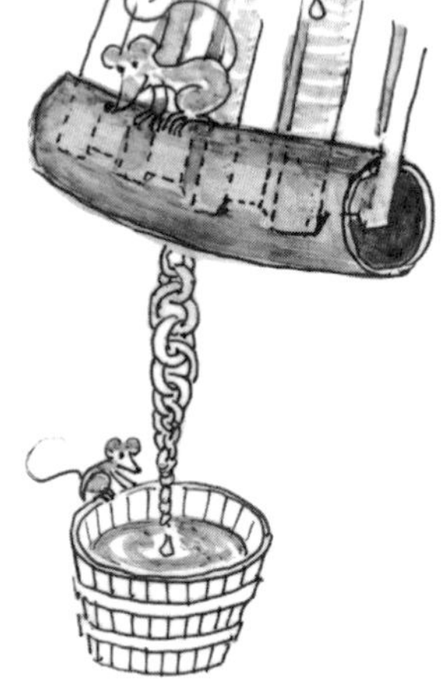

The most commonly used metal roofing in the United States is steel, usually coated on both sides with protective enamel. It comes in sheets 2 to 4 feet wide and almost any length, with various ribbing patterns to make it more rigid and to channel water downward. It is installed by screwing through the ribs into horizontal purlins below. The enemy of steel roofing is rust. Treat the roofing carefully during installation to avoid scraping the protective coating. Inspect it periodically, and if you see bare metal or rust anywhere, clean and paint these places with rust-preventive paint to extend the life of your roof to forty years or more.

There are more expensive, more durable metal roofing options such as aluminum, copper, and stainless steel. There are also cheaper, make-it-yourself options. Large metal cans like the ones that oil and other liquid foodstuffs are packaged in can be cut open, flattened, painted if desired, and installed like shingles. In more resourceful parts of the world, it's common to see roofing made from everything from 50-gallon drums to soda cans and aluminum printing plates.

TILES, SHAKES, AND SHINGLES

Many kinds of roofs, traditional all over the globe, share a common theme: small flat tiles of regular size, laid in overlapping courses to direct rain down and out. Fired terra-cotta tiles help give Mediterranean villages their characteristic cheery disposition. In the mountains of Europe and Asia, the buildings are often roofed with stone split into thin sheets: sandstone, limestone, and especially slate. Native Americans have used bark for roofing, including cedar and birch. In modern North America, wooden tiles are still fairly common. They are called "shakes" if they're split, "shingles" if they're sawn. Anything flat, rigid, and waterproof can be used to tile a roof—look around you and see what you can find. We've heard of a roof of waxed-cardboard shingles, made from broken-down produce cartons, which lasted fourteen years in the cold arid country of central Washington. A woman in southern Oregon recently patented a machine that guillotines used car tires into rubber shingles shaped like terra cotta roof tiles. Whether her idea will catch on and these tiles will become commercially available remains to be seen.

Contrast the industrial enameled steel with hand-split cedar shakes. Design and construction: Cob Cottage Company.

Shakes are split from straight-grained and rot-resistant wood. On the West Coast of the United States, shakes are usually cedar or redwood, but also sometimes Douglas fir. On the East Coast, oak was formerly used, and in Mexico and Central America pine shakes are common. Shingles are similar in appearance but are sawn rather than split. Because a saw blade slices easily through wood fibers, shingles can be cut very regular, thin, and smooth. They are generally tapered at the upper end, which means you don't lose so much steepness when they overlap as you do with shakes. But because the wood fibers in shakes are continuous, shakes are both stronger and less susceptible to fungus and rot than shingles.

Shake and shingle roofs are quite fragile. Walking on them will produce splits and leaks, so roof maintenance can be difficult. The other great bane of wooden roofs is fire. In the dry summertime, old wood turns into tinder, just waiting for a spark to ignite it. Many people who live in the forest or heat their homes with wood decide against shakes and wooden shingles for precisely this reason.

Commercial shakes are very expensive, but you can split your own with a couple of

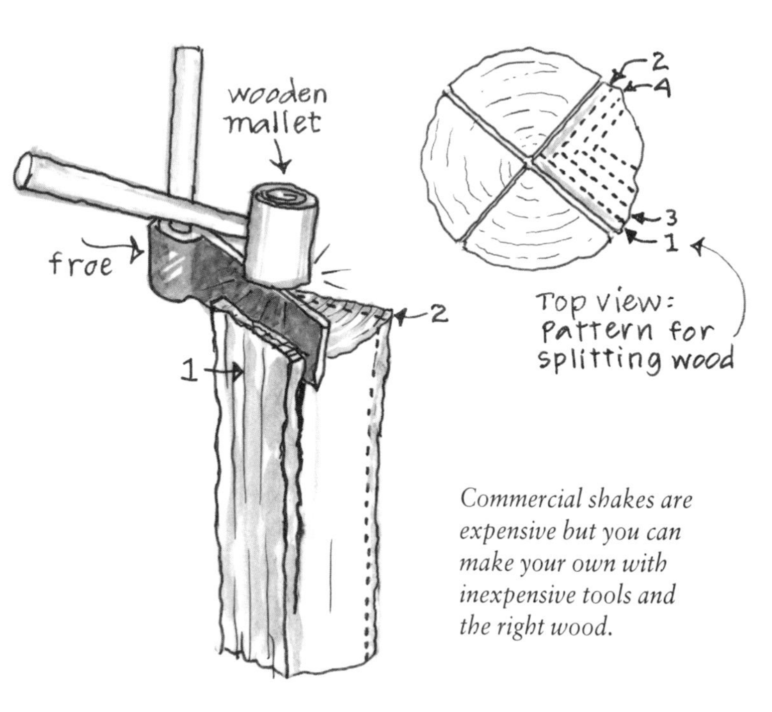

Commercial shakes are expensive but you can make your own with inexpensive tools and the right wood.

inexpensive tools. Splitting shakes is a fun and satisfying process. The tricky part is locating suitable wood. It has to be tight-grained, at least sixteen inches in diameter, and clear of knots for at least two feet. But it's not necessary to cut live trees. In many places, you can still find cedar and redwood trees that were cut down fifty to a hundred years ago; even after many decades of lying on the forest floor, or in swamps or lakes, they can be remarkably sound below a surface layer of rot. Old cedar and redwood stumps can also produce fine shakes. In Washington tate, natural builder extraordinaire Sun Ray Kelley covers entire buildings with cedar shakes up to 14 feet long and 12 inches wide, all cut from stumps left by pioneer loggers.

ROOF INSULATION

In most climates, the roof is the most important part of a building to insulate. This is because hot air rises, and also because on clear nights, horizontal surfaces rapidly lose radiant heat to outer space. In hot, sunny weather, roof insulation saves your ceiling from overheating and making the space beneath uncomfortable. Most roofing materials have very little value as insulants against heat or cold. The exceptions are thatch, which provides good insulation, and sod, which in summer helps deflect direct sun and disperse heat. Industrially made insulants, including fiberglass and rigid foam, tend to be dangerous to the people who install them and live with them, are transported great distances, and are made by exploitative polluting industries that deserve to be boycotted. In the rest of this chapter, we'll focus on non-toxic insulation materials. Unfortunately, at this point there are few natural insulants that are commercially available. You may need to make your own.

In truss-framed roofs, insulation can simply be laid on top of the horizontal ceiling. With "cathedral" ceilings, which follow the line of the roof, space needs to be allowed for a good depth of insulation either between rafters, or over a suspended ceiling below them. Sometimes, it is possible to insulate above the rafters, on top of the exposed sheathing, though a secondary structure may be necessary to hold up the roofing.

How much insulation? For once, the local building codes can be useful here. You may even want to outdo them. For most North American climates, allow for a foot of roof insulation; in coldest zones, where temperatures regularly go to minus figures Fahrenheit, think about 16 to 18 inches. Multiple layers of household aluminum foil will substitute for several inches of insulation. Rigid foam insulation is extremely effective, although it has a high embodied energy and some types are very toxic. A sheet of foam 3½ inches thick gives R-24, which is sufficient for milder climates.

Above insulation, plan on at least two inches of uninterrupted vent space so that air convection can continuously take away any condensation from the space between the insulation and the roof covering. The vent will help also to keep down temperatures at the ceiling when the sun is high. Wet insulation doesn't insulate very well and can rot the structure, damage the ceiling below, or even break through if it gets too heavy. Make sure to maintain your roof and fix any leaks as they appear.

Protect all insulation from rodents, who find it very desirable for nesting material. Organic insulants such as wool and straw are likewise susceptible to being eaten by insects and other animals. To avoid these problems, use a double layer of wire screening across roof vent inlets: one layer fine enough to keep out bugs, the other stout enough to deter mice.

One of the least toxic insulation materials now available commercially is "blown cellulose," made of recycled paper ground up and

Four ways of installing insulation above the ceiling.

treated with a flame retardant. It's a pretty good use of old newspapers, and it's easy to get and to install. Usually you rent a machine (or the building supply store loans it to you free when you buy a quantity of insulation), which breaks up the compressed blocks of cellulose and blows them through a long hose. You can install the stuff after the roof and ceiling are both in place by leaving a hole big enough to pass the hose through. The blown cellulose will fill the cavity completely but then settle, leaving a vent space on top.

Other healthy insulation materials include cotton and wool batting, which are installed like the more toxic fiberglass. Unfortunately, wool and cotton insulation are difficult to find in the United States at this time, though that should change as interest in healthy building increases.

Aluminum foil, though it is industrially produced, is lightweight, long-lasting, readily available and quite inexpensive. Aluminum foil isn't so much an insulator as a reflector; heat bounces off of it the same way light does. A single layer of foil can reflect as much as 95 percent of radiant heat. That's why fiberglass and foamboard insulation are often lined with shiny foil. Mount one or more layers of aluminum foil with the shiny side toward the room to retain heat (it will still be effective even if hidden above the ceiling, provided there is an air gap on the side facing the room), or toward the sky, below the roof cover, to reflect sun heat.

If you are prepared to make your own insulation, you have several good options. In many rural areas, both straw and wool fleece are readily available and cheap, so we'll discuss them in more detail. If none of these materials is available to you in sufficient quantity, it may be time to get creative. Insulation is basically small pockets of trapped air, so any dry material with plenty of air space in it will help. You can try things such as pine cones, dry mosses, sawdust and shavings, foam rubber from old furniture, sleeping bags, and clothes. The dust and shavings from some kinds of wood such as cedar, redwood, and pacific yew are particularly immune to vermin. When insulating with any vegetable matter, be careful of fire hazard and mildew.

Wool

Sheep's wool is one of the best natural insulators known, which is of course why we (and the sheep) wear it. Its thermal resistance (or R-value) is slightly better than fiberglass. Unlike most other materials, wool continues to insulate even when wet and is relatively flame resistant. It can be damaged by wool moths, but leaving the lanolin (a protective grease produced by sheep) on the wool, or mixing in borax, cedar shavings, or moth balls may discourage moths. Linda and Ianto live in a wool-insulated cob cottage. When they went to get the wool, Ianto asked the farmer if they would have a moth problem. The farmer looked astonished, and there was a five-second silence. Finally, he said, with great deliberation: "Weaaaal . . . I don't never seen no moths on a sheep yet!"

In New Zealand, where the sheep industry has a lot of economic and political clout, insulating buildings with wool is a standard practice. You can buy bats of woolen insulation much like the rolls of fiberglass we see here. Hopefully, woolen insulation will soon be available in the United States—but until it is, you will have to make your own. Luckily for those of us who prefer this natural, lightweight, nontoxic insulation, the world's wool market has been sluggish for about a decade. In sheep-producing areas of the United States, there are huge stockpiles of wool waiting for prices to rise. You should be able to get enough for a small building quite cheaply if you go directly to the farmer. We've had low-quality fleeces (which are still just fine for our needs) donated by sheep farmers interested in our work. You don't need a great amount; after washing and carding, a pound of wool should provide one to three cubic feet of insulation.

A traditional way of washing wool is to suspend it in sacks in a rushing stream for a long time. Avoid using soap, which removes the lanolin (that probably deters moths). Also, don't try to wash loose wool in a household washing machine. The fibers tend to get wrapped around the spindle and can cause a maintenance nightmare. Instead, wash your wool in an onion sack or other loose-knit bag. We had good results alternately soaking and stomping wool in a bathtub full of water. (Again, be careful not to clog the drain!) When clean, spread the wool out on racks to dry in the wind. Drying takes a long time and should be done during very dry weather. Finally, tease the wool with a brush to fluff it up so it will insulate better. Better yet, find someone with a carding machine.

After following this lengthy and arduous procedure on our first wool-insulated roof, the second time we used whole fleeces in more-or-less the condition that they came off the sheep, cutting off the particularly mucky parts. We worried that the manure remaining in the wool would cause an epidemic of flies and rank smells, but perhaps because the insulation was well-separated from the living space (by both a layer of plastic sheeting and a wooden ceiling) there have been no big problems. Both the sheepy smell and the fly population seemed to decrease rapidly.

Jan Sturmann came up with the brilliant innovation of stuffing unwashed wool into plastic garbage bags, which are then layered into the ceiling as insulation. The plastic bags not only contain odors and discourage moths, but also serve as a built-in vapor barrier. Bags of fleece can be stapled between rafters overhead, instead of first having to build the ceiling to contain them.

Straw and Straw-Clay

Because it is cheap and widely available, and because you need some on-site anyway for making cob, straw is an obvious candidate for roof insulation. Tightly baled straw is quite flame resistant and unpalatable to most life-forms as long as it stays dry (although it does make a cozy home for rodents). Two-string bales laid flat offer about R-30 for their 14 inches of depth, exactly what building codes demand in colder North American climates. Bales are fairly large, semirigid units, which can be supported on top of or between roof trusses or robust rafters. Keep in mind that bales will add about 10 pounds per square foot to your roof weight, equivalent to a foot of wettish snow. The roof will require extra strength, so keep roof spans short if you plan to use baled straw. This heftier structure will use more wood, but balanced against the long-term savings in energy use, the extra insulation may be a good ecological investment.

In earthquake zones, having so much weight overhead might not be a good idea. Also, if the roof should for any reason leak, bales could become soaked and very heavy, with potentially disastrous results. In very mild areas with reliable sunshine, you could settle for less insulation by packing in flakes of straw obtained by breaking up bales. Flakes certainly weigh far less, but are more of a fire hazard.

If you are concerned about insect or rodent pests, or about fire, consider treating the straw with a clay slip. This is a traditional system called *leichtlehm* in German, "slip-straw," "straw-clay" or "light clay" in English. It is currently being popularized in the United States thanks largely to Robert Laporte, who is based in Santa Fe. Although the clay helps retard fire and discourages insects and rodents, straw-clay is heavier than pure straw, so is not such a good insulant. The extra weight also intensifies the structural concerns mentioned above.

Tossing light clay or slip-straw, like salad.

To make straw-clay, clay is agitated in water to make a creamy suspension called slip. Pour a little slip onto a pile of loose straw and toss with pitchforks, like tossing a salad. It doesn't take much slip—just enough to get it all evenly coated. The mix can be pressed into place on top of a ceiling and left to dry before covering, or made into blocks on the ground, using a form that fits them tightly between rafters. For more details, read Robert Laporte's booklet, "MoosePrints: A Holistic Home Building Guide" (Natural House Building Center, Santa Fe).

16 Natural Floors

❧ An earthen mass floor . . .

is an object of daily beauty. ❧

In North American houses, floors are most often made of wood—wood boards on a wooden subfloor on wooden joists, sometimes on wooden beams. The 2,200 square feet of the average new house eats up a lot of trees, about 11,000 board feet, or an area the size of a square a hundred feet on each side, over one inch thick. Most of this wood will be hidden from view until the day of demolition; the beauty and wisdom expressed in its knots and grain will never be known. Perhaps in fifty years' time, some bulldozer driver will briefly glimpse the joists and subfloor as he shoves the whole house into a pile for burning. What a waste!

A good floor needs to ensure smooth passage—you don't want to stumble on edges, irregularities, three-inch steps, or steep little ramps. It should be easy to clean and maintain, not be dusty, sticky, or porous. In most parts of the place where you live, the floor should be inviting enough to sit on, play on with little kids, do yoga on, or snooze on. To bare feet it should feel warm in winter, cool in summer. It should be solid and quiet when you walk. Floors, particularly those used with fitted carpeting, are the worst causes of allergies—dust, bacteria, molds, spores, pollen, and a gradual release of toxic gases. A healthy floor surface needs to be washable, with any rugs loose enough to be regularly taken out and shaken. And of course it should be pleasing to the touch and sight and sound and smell.

This chapter is about how to build the floor of your cottage yourself, using whatever is most available and most appropriate. Building cob means building heavy means passive solar means a heavy floor; a search

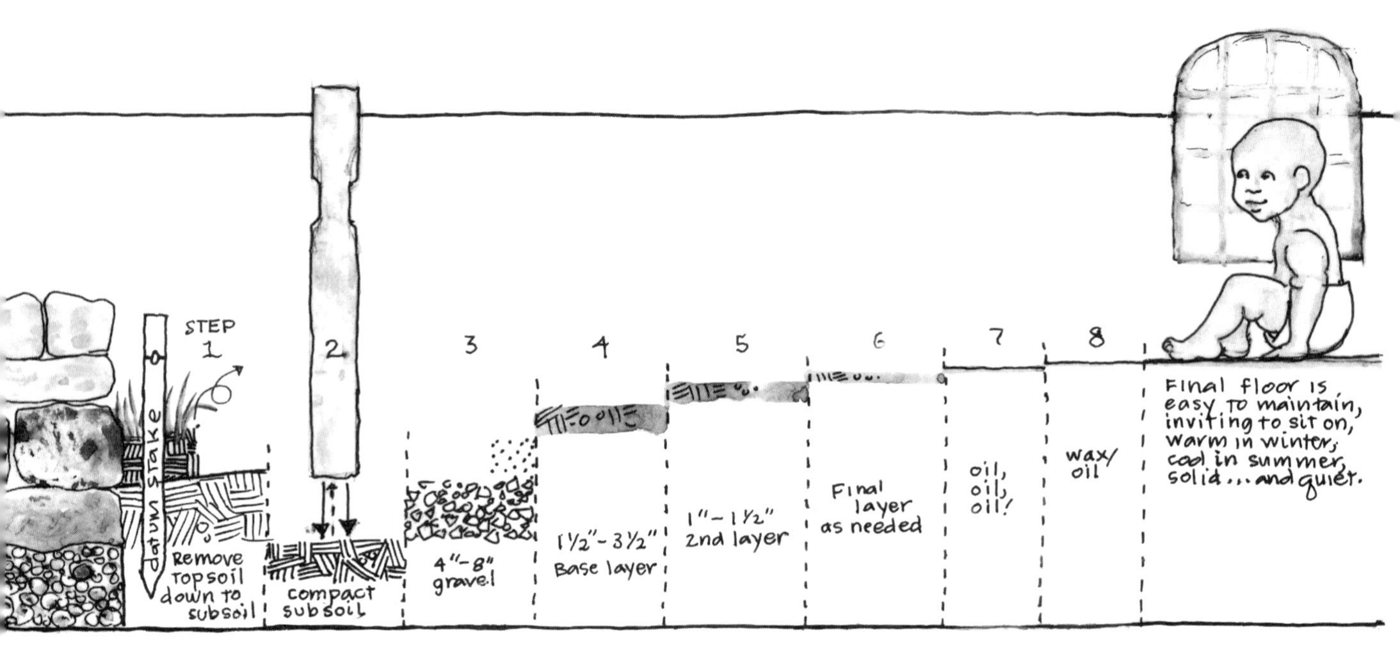

for available material usually leads to the ground beneath your feet. So this chapter is mostly about how to build earthen mass floors that are beautiful, cheap, durable, waterproof, quiet, nontoxic, and easy to clean.

Dirt floor—sounds unclean, squalid! But we're not talking dirt floors here. A well-laid adobe floor draws appreciation, even disbelief from visitors: "You're kidding, that's an earthen floor?" It resists staining and scratching, is easier to keep clean than a wooden floor, should need almost no maintenance for the rest of your life, and is an object of daily beauty. It will feel comfortable and be solid and stable, quiet to walk on, level enough that you can play marbles, and a constant reminder of the colors of the clay you put into it. Very quickly the phrase *dirt floor* will go out of your vocabulary.

MASS FLOORS

In choosing to build with cob, you are also choosing to build a passive solar building; heavy buildings and storing sunshine go together. In climates where winter heating is necessary, as much mass as possible should be placed where it can absorb energy from the sun and re-radiate it later. In very hot climates, mass should be in cool, shady places. In either situation, floors make an ideal location for mass storage. You can put as much mass in a floor as you want, by making the floor as thick as you want, without losing usable space. The floor often receives more direct sunlight than anything else inside a building. And because feet are highly sensitive to temperature, your comfort level is increased more by a warm floor than by warm walls. You may also build a pipe system into your floor, so that it can be heated by hot water from a solar collector, furnace, or stove ("hydronics").

You can make mass floors from concrete, earth, flagstone, brick, or tile. Each of these materials has its own advantages and disadvantages. In conventional construction, it's common to see exposed concrete floors in basements and outbuildings, and most floors of brick, tile, and flagstone (and even many wood floors) have concrete slabs underneath. If you want to avoid the cost, toxicity, and inflexibility of concrete, or if its health and environmental side effects give you second thoughts, there are alternatives. As this book is about earthen construction, we will give you details of three proven systems for earth floors, all of which conform to the needs of a good floor: poured adobe, tamped cob, and adobe blocks.

Flagstone floor creates continuity through glass door from kitchen into patio. To right, wooden step down to earthen floor. Design and construction: The Cob Cottage Company.

Preparation for a Mass Floor

In a cob building, there will be plenty of mud during construction. Don't attempt to build a finished floor until the cob work is complete, including the rough plastering—everything underfoot will get dirty. However, having the subfloor leveled in advance will give a good surface to use while you're building the walls and roof, and the foot traffic of construction will beat it flat and compact. But bear in mind as you build, standing on the subfloor, all the levels will change several inches when you finish your floor. Don't let the temporarily low floor level fool you into setting windows or ceiling heights too low. When building poured adobe floors, we often install the drainage

An earthen floor needs a base of compacted subsoil 8 to 12 inches below the level of the finished floor.

layer and base coat *before* the walls. This creates a flat, level surface to work on, closer to the finished floor elevation. The thick base coat will also dry much faster due to increased air movement before the walls go up.

Earthen floors use no vapor barrier beneath them. Impermeable vapor barriers inevitably lead to condensation and the build-up of liquid water on one side or the other, which can reduce the life span of your building. Just as in walls, we believe that it is essential to allow moisture in floors to travel freely. Instead of an impermeable vapor barrier, a layer of a few inches of gravel prevents "rising damp" from saturating the earth layer.

These are the steps we follow to prepare for mass floors:

Dig Down to Solid Subsoil: A long-lasting earthen floor needs a solid subfloor that will not subside. Because humus is spongy by nature and hard to compress, it is essential to remove all topsoil and organic matter (including roots) from under the floor. Dig down to undisturbed subsoil and make a flat platform 8 to 12 inches below the level of your finished floor. This subfloor can slope somewhat toward the outsides, so that any water which gets onto it will run off into the drainage trenches.

Compact the Subfloor: If the subsoil platform you have created is not absolutely rock-hard, it needs to be tamped. You are creating a rammed earth base of highly compacted soil that extends deep below the surface. You can use a pneumatic tamper, but we prefer a homemade tamper made from a log with hand-holds carved into it. The height and weight of these simple tools can be custom-selected for the people using them. A subfloor tamper should have a base about 3 inches in diameter. Don't use the sort that looks like a metal square on the bottom of a shovel handle; it has too much surface area to provide good compaction.

Hand-tamping is tiring and takes a long time. Be patient and don't wear yourself out. Plan to warm up every morning with not more than ten minutes of tamping. When the subfloor is fully tamped, you won't be able to sink the tamper any deeper into it. This condition is impossible to achieve if the ground is too wet, as mud does not compact. If the subsoil is too dry, it will turn to powder, so before you can tamp it effectively, you may need to soak it down and leave it to absorb water until it is just slightly damp. Before building your floor, be sure to let the subfloor dry thoroughly or you may later have settlement that could crack your floor.

Install a Layer of Gravel or Drain Rock: To create a capillary break so that dampness won't rise from the subfloor into the floor proper, bring in clean gravel or drain rock, ¾ to 1½ inches. If you have to use crushed rock, screen out the fine particles. This layer should be at least 4 inches thick—more is better. Spread the gravel out level and tamp well. If you have trouble with the tamper kicking gravel around, try a lighter, broader tamper such as the kind available in hardware stores. Take time, using a level taped to a long straight board, to get the top surface of the

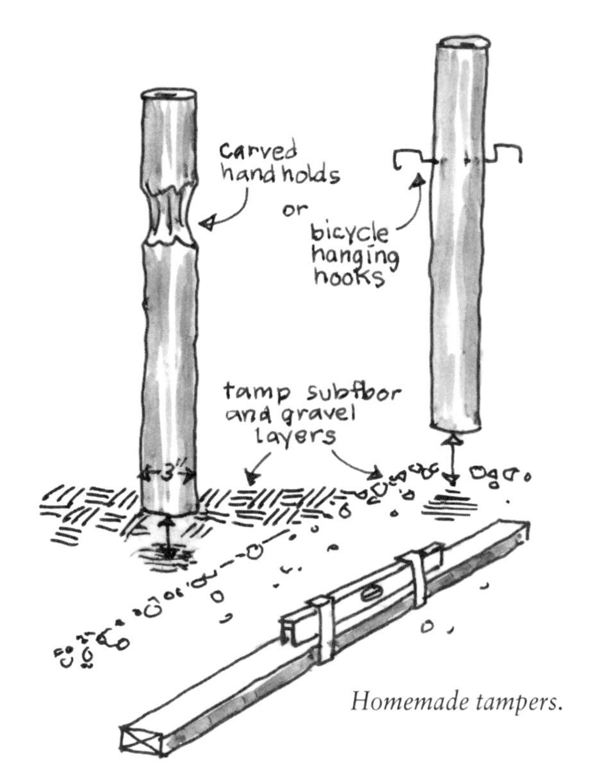

Homemade tampers.

gravel precisely level. This will save a lot of work later.

Under-Floor Insulation: Insulation in passive solar buildings is generally more effective around the perimeter of the building, within the foundation trench, rather than under the floor itself, except with tiny buildings in extremely cold zones where it may be of benefit to use both (see chapter 10). If you choose to insulate beneath the floor, lay that insulation on top of the gravel. Foamboard insulation is rather spongy and makes laying a rigid floor difficult; some types are toxic. An alternative would be 2 to 4 inches of vermiculite or perlite, with just enough clay to stick it together.

Poured Adobe Floors

Poured adobe floors are traditional in the southwestern United States. Poured and tamped floors are built up of several strata of progressively finer material in progressively thinner layers, to give a floor depth of 3 to 8 inches, depending upon the size of the room and amount of impact the floor may suffer. Here are instructions for a three-coat floor, 5 inches thick, for heavy use in a medium-sized room. A thinner floor can be poured in only two layers.

Poured adobe floors dry very slowly. They are in the coolest part of the building, where there is the least airflow, and they have only one side exposed to the air. Don't lay an earthen floor in damp conditions, or if damp conditions are expected within a month or so. Unless you live in the desert, plan to put in at least the base layer as early in the dry season as you can, though the finish layer can go in much later.

The mix is similar to cob, except wetter, with more sand and with, if possible, some gravel. A handful of mix should sound really crunchy when you squeeze it and contain 50 to 80 percent sand and gravel. Always make a test sample, at least 3 feet square, to check your proposed mix for hardness and cracking before you pour an entire floor. The trickiest part is getting the right proportion of clay. Too much and your floor will crack; too little will leave it soft and crumbly.

It is important that each layer is laid totally level, or the layers above may crack through uneven shrinkage. Make sure, too, that each layer is completely dry before adding the next.

The Base Layer: Make a mix about the consistency of stiff cake batter in a mixing boat, a concrete mixer, a large tarp, or a wheelbarrow. This is the structural base to your floor, ideally made of coarse gravel (up to a couple of inches in diameter), some sand, and a little clay, though almost any sticky soil will suffice. Add lots of full-length straw, straight from the bale. If you work with crushed rock, get "1½-inch minus." Separate it through a ½- or ¾-inch metal screen. The coarser portion will go into the base coat mixture; that which falls through the screen is for the

Two-by-four boards are used to hold the floor mixture in place. Boards for the base layer go in one direction with the second layer at right angles.

second layer, which needs finer material. Or you could use "¾-inch minus" in both.

If the stones you used in the drainage layer are large, your mix might run down between them and clog the spaces. To prevent this, lay some breathable material such as old sheets, newspapers, or feed sacks over the gravel. Don't worry if it seems flimsy; it only needs to last a day or two, until the mud begins to dry.

Find two straight lengths of 2 × 4 for levelling boards; set them on edge on the floor about 2 feet apart, parallel with the wall where you're going to start pouring. For a thinner floor in a small space, lay the boards flat, giving 1½ to 2 inches of base layer, instead of 3½ to 4 inches. Level both along their length, and relative to each other. Shovel your mix between the 2 × 4s, and run a straight screed board across the top of them, scraping off any excess mud. Make sure the space between the 2 × 4s is full to the top and evenly screened, but leave the surface rough.

After packing the mix, level across both levelling boards using your screed board.

Remove the levelling board that is farthest from you. The mix should be firm enough that it doesn't slump when you pull the board out. Fill in the gap with more mixture after roughing the edges so it doesn't leave a straight seam that may crack. Reposition the board you pulled out a couple of feet in front of the one you left in place. Check it again for level, then fill with mixture, tamp it down, and screed off as before. Continue leapfrogging the two boards until you reach the door. Be careful not to mud yourself into a corner!

When you are finished, you should have a continuous surface, flat and level but somewhat rough. If you used coarse gravel in your mix, there may be surface voids. Leave them, they help the next layer to bond. Avoid walking on the base layer until it becomes rigid. This will take a couple of days to weeks, depending on the weather. If necessary use heaters and fans to dry it out. Heating alone is ineffective to dry a new floor; air movement is needed.

The Second Layer: The mix for the second layer should be smoother than the first, without gravel larger than ¾-inch. If your base layer cracked at all, raise the proportion of aggregate and fiber. Use chopped straw and/or fresh cow or horse manure, for finer fiber. You can chop straw with a leaf-mulching machine, a weed-whacker in a plastic barrel, a lawn mower, a chain saw, or a machete. To avoid a bottleneck during installation, prepare the materials in advance.

When the first earth layer is thoroughly dry, check to make sure that it is still level. Apply the second coat the same way as the first, using levelling boards only 1- to 1½-inches thick, and thoroughly wetting the dry surface for good adhesion, immediately before each batch is added. If the base coat isn't dead level, level the new guide boards by sprinkling sand to raise them or by chipping away at the floor to lower them. Flatten the surface with a trowel or a board, but leave it rough. To eliminate cracking, try not to lay the 1-inch boards parallel with the 2 × 4s but at right angles to the first ones. Always roughen the edges of the mud when you pull each board out, before filling in the hole.

There is another way to get a level surface without cracks. Drive nails into the dried base layer every couple of feet over the whole floor, particularly along the walls. Checking with an accurate level, drive each nail in until its head is level with those around it. You probably want to do this as you go, so you don't trip over the nails or move them. Use long nails, which are less likely to be accidentally nudged. A quick dip in white paint will leave them more visible in even the darkest corners. Apply the mud mix with a trowel to a depth where the trowel barely skims the top of each nail. You can either leave the nails in place or pull them out as you go.

The Finish Layer: The final adobe layer will bring your floor up to wherever you want the final level to be. If you planned for 5 inches of earth floor with the first coat 3½ inches thick and the second an inch, that leaves you half an inch for the final layer. To avoid cracking, the finish layer should be between a ½ and ¾ of an inch thick. Remember, wait until all the heavy construction and plastering have been finished before installing the finish layer, then keep completely off it until it dries.

For the finish layer, you can select a special clay soil of a color you like. When you oil it, the color will revert to that of the wet clay. The addition of a small quantity of powdered psyllium husks to the finish layer helps give a bouncy elasticity, but slows down drying. Psyllium husks are available at yard and garden centers and at natural foods stores. The finish layer mix should look good enough to eat, as smooth and creamy as cake batter. The smoother the mix is, the smoother your finished floor will be. To achieve this, put your soil through an ⅛-inch screen. Sand should be sieved through a very fine window screen, or use beach or dune sand. Chopped straw should be sifted through an ⅛-inch screen until it is about the size of shredded coconut, or use cow or horse manure, soaked overnight and grated through a screen to break up lumps.

If you use screed boards when applying the finish layer, be very careful to roughen up the seams they leave behind before filling with mix, or cracks can form. Alternatively, use the nail trick described above, pulling them out as you go or punching them down so they don't show through the floor, or use a long, accurate level. As always, wet the floor surface to ensure good bonding immediately before pouring more mud.

Smooth the mix carefully as you apply it. The best tool for this is a pool float, an oblong, rounded, flexible steel trowel. The proper use of a pool float or other trowel takes practice. You might want to install your first poured adobe floor where a little unevenness won't bother you too much, or practice on the second layer.

Once your poured adobe floor is dry all the way through, apply oil and wax finishes (described later in this chapter). Until those finishes are dry, keep everyone, including animals, off! Unsealed earthen floors can be very easily damaged by impact and abrasion.

Tamped Cob Floors

Perhaps not quite as satisfying to lay, and requiring a little more labor, tamped cob dries faster than poured adobe. It is more suitable for sites where the subfloor is not completely dust-dry; for soils lacking much clay; for laying at a time of year when drying will be slower; or for places where humidity is always high.

The technique is very similar to poured adobe, with these differences.

❧As with poured adobe, three layers are usual. Layers 1 and 2 are both tamped into place with a wooden hand tamper or a mechanized compactor, which can be rented. Tamp layer 1 strongly; be a little more gentle with layer 2. The finish layer will be poured adobe, floated exactly as described above.

❧The base layer needs a high proportion of crushed rock, or shillet, graded from about 1½ inches to sand size. The best base will contain 50 to 80 percent crushed rock and not more than 20 percent clay. In some areas a ready-made mixture is sold as "road base." Call paving contractors or quarries, and check a sample of exactly what you will receive before you buy it. Road base is cheap, and delivery can usually be ordered right to your site. If you prepare your subfloor and drainage layer before building the walls, you might be able to back the truck right up and dump the load inside the foundation, to save double handling.

❧Tamped cob needs very little water in the mix. Layers 1 and 2 should stand up in a rigid heap without slumping under gravity, and without dirtying the leather uppers of your shoes if you jump on the heap and be hard enough to walk on as you lay it. Layer 2 should be of finer material than the base layer.

❧Tamped cob contains less straw, though a proportion of straw is important to give it tensile and shear strength. If, over time, the subfloor should subside or move in any way, straw in the mix will make cracking less likely.

❧We suggest that you make a test mix and lay about a square yard first, in a place that is not too critical, to get a mix that compacts well and creates a rigid floor.

An Adobe Block Floor

Adobe bricks can be used to make an attractive and durable floor. It will be slightly irregular and not quite flat, but have visual interest and texture that are lacking in a poured adobe finish. Blocks have the advantage that if for some reason the floor needs to be dug up (a leak in an under-floor pipe for instance), the excavation can be repaired almost invisibly. Blocks can also be precast and dried, for installation in cool, damp weather. If you cast your own blocks, you can make them whatever shapes will be most pleasing to you: square, rectangular, triangular, and hexagonal are most versatile, and they could be any size. It is sometimes worthwhile to slightly vary the color of adobes by using two different clay sources in different proportions. The effect when laid can be very pleasing and give life to the floor. Blocks should be at least 4 inches thick and contain a lot of straw.

It is essential to lay adobes on a stable, solid base floor. Use poured adobe or tamped cob for your base floor (unless you *really* need to use concrete). Set the adobes in ½ to 1 inch of cob mortar that has been through a ¼-inch screen. Use generous mortar joints of very sandy clay-sand mortar, without fiber. Select blocks for their flattest or most interesting faces—the tracks where a cat ran over the wet surface or a print of a big leaf have enduring interest.

Oil and Wax Finish

When the earth floor is dry all the way through, it is time to apply the oil and wax that will make the floor durable and waterproof. Until then, avoid walking on it and scuffing it up. The more coats of oil you apply, the harder the floor will become. New Mexican earth builder Anita Rodriguez uses seven coats of oil on her floors, and she guarantees them for life. She says you can walk on them in stiletto heels or soccer cleats and ride bicycles or play ball on them, although she won't guarantee them against chopping wood on the floor with an ax.

With a rag, brush, roller, or soft sponge, apply at least four coats of boiled linseed oil. It is important to use boiled, not raw, linseed oil. (In some countries, Mexico, for instance, earthen floors are stabilized using old motor oil. While this might be okay for a well-ventilated space such as a carport, it will give off toxic fumes, so we can't recommend it.) Apply each coat until it begins to puddle, then wipe off any excess. The first coat should go on full strength. Wait for it to dry before applying the next coat. The second coat should be thinned with 25 percent turpentine, citrus thinner, alcohol, or mineral spirits. The third should be 50/50 oil and thinner, and the fourth 25 percent oil, 75 percent thinner. The reason for this is to improve penetration so you don't end up with an eggshell-thin hard layer on top of your floor, which can chip off to expose soft earth beneath. As each layer of oil is applied, it clogs up the open pores in the earth, which the water left behind as it dried, making the surface harder for the next layer to soak in. Warming the oil or the floor before application makes the oil penetrate better.

When your final coat of oil is dry, which may take many days, you will have a hard, durable, water-resistant floor. Pour a little water on it and try scrubbing with your fingers—you will find that the water stays clean, but soaks into the floor slowly. Your floor is now functionally finished, durable, and attractive, but if you want to be able to clean it with a wet mop or spill colored liquids without staining it, you will have to wax the floor.

Make a paste by melting together 1 part beeswax with 2 parts linseed oil. It doesn't take much wax to cover a large area, so mix only a small quantity at first. While the paste is still warm, rub it into the floor with a clean, lint-free rag. After it has dried, you will be able to pour water onto your earth floor, and it will bead up like magic. The wax will wear off over time, so you may want to reapply it periodically, perhaps every few months or every year.

Other Materials

In areas where there will be heavy traffic or lots of water, such as mudrooms, bathrooms, and workshops, you may prefer a more durable floor than you can make with earth. Bricks, tiles, slate, wood blocks, flagstones, glass blocks—all can be used in mass floors, separately or together. All require a solid base. They can be bedded on pure sand, lime mortar, earth mortar, or cement mortar; tiles thinner than 1½ inches are more likely to stay down if you use cement-sand mortar and grout. All have similar thermal characteristics, except wood, which stores less heat and is difficult to heat up, and glass blocks, which are hollow and therefore don't transmit heat well, either.

Brick, tile, and flagstones are sometimes available from demolition sites, free or quite cheap. In hilly areas flat stone can sometimes be found in old quarries, riverbanks, and roadcuts. Mining riverbeds is not recommended, as this can harm the stream ecology.

Woodblock floors can be made from blocks of dimensionally stable lumber. Your blocks need to be quite large to make the project worthwhile—a floor 20 feet square would take 1,600 6 × 6-inch blocks. Another option is to use slices cut from dead trees, carefully cutting 4- to 6-inch thick rounds off a log, preferably of hardwood that has recently died on its feet. Using interlocking rounds of two completely different diameters closes up the gaps better, and each round can be trimmed in place to give a tight fit. Another variation is to use offcuts from boards, which sometimes can be found at firewood price from lumber mills or planing mills. Make sure wood is thoroughly seasoned and very dry throughout.

HYDRONICS AND HYPOCAUSTS

If your cob cottage requires supplemental heating, consider heating the floor. This is not only thermally efficient (a warm mass at the bottom of the building can easily heat the entire space by radiation and convection), but is also a sensual delight. Our feet are one of the most sensitive parts of our bodies to temperature, and they love the feel of warm earth beneath them. It's reminiscent of a sunny day at the beach.

There are two ways to heat a mass floor: hydronics and hypocausts. Hydronics systems rely on hot liquid (usually water) flowing through pipes buried in the floor. Hypocausts use hot gases from combustion for the same purpose.

Hydronics are becoming quite commonplace in conventional buildings. There are good systems commercially available that are fairly inexpensive, easy to install, and fail-safe. They rely on long loops of narrow, flexible plastic pipe, without any joints or junctions, which are buried in the floor. The standard usage in conventional construction is to pour a concrete slab over and around these pipes, and then often to lay tile or even wood flooring on top of that. But exactly the same hardware can be installed in an identical manner in a poured adobe floor, as we have seen done several times with excellent results.

The source of hot water can be either a standard (electrical, propane, or oil-fired) water heater, or a more sustainable option such as a solar water heater or a thermocouple on a woodstove. The biggest drawback of hydronics systems is that they generally require pumping to make the hot water flow through them in a continuous loop. It may be

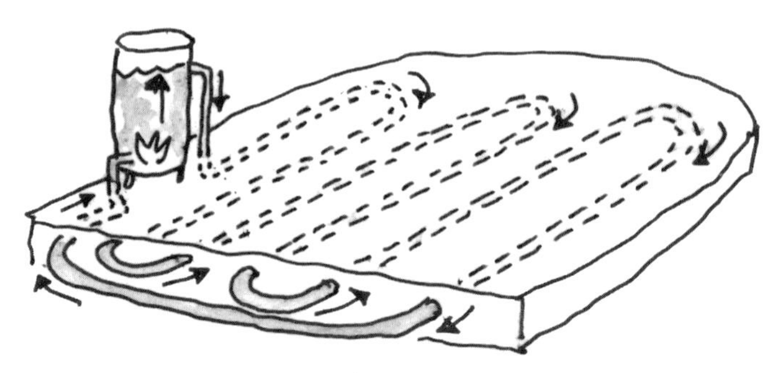

Hydronics is the use of buried pipes to circulate hot water through the floor, which results in radiant heat from the mass below.

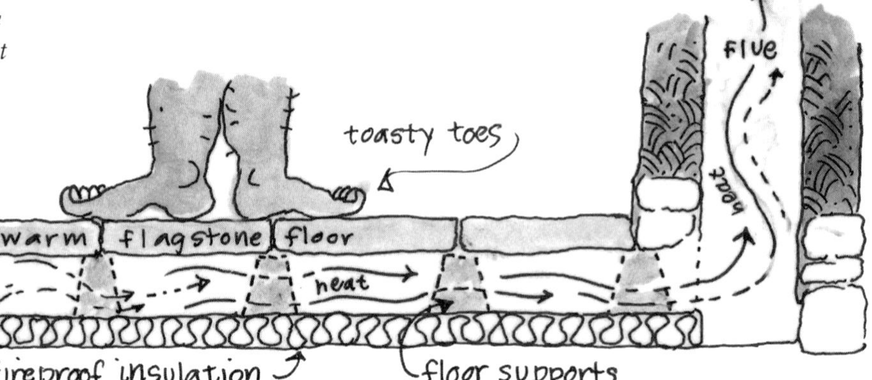

Hypocaust systems heat the floor by circulating hot air. Although unusual today, they were used successfully by the ancient Romans.

possible to sidestep this need with a thermosiphon, if you have a sloped site where you can place your water heater downhill. Consult a plumber or someone who has experience with hydronics before installing your own.

Hypocausts (the word derives from the Latin for "burning beneath") are largely unknown in North America, but were used extensively, historically. Hot gases from burning wood (or conceivably gas or oil) flow through a duct system beneath the floor. The Romans used hypocausts to heat colonial villas in the miserable, cloudy dampness of their British outposts.

Consider whether you want under-floor heating before constructing the foundation; supply pipes may need to be sunk beneath the subfloor. Heating pipes normally would go above any insulation, buried in the structural part of the earthen floor. Pipes could also be set in a layer of loose sand to allow for expansion or contraction with a couple inches of adobe poured on top. Before covering pipes, test them under pressure and full temperature to make sure there are no leaks.

SUSPENDED FLOORS

A floor that has air on both sides is called "suspended," and in North American houses is usually made of wood. Suspended floors have advantages, though they use a lot of wood. They are lightweight, they can feel warmer to the touch, and in hot tropical climates they don't hold the heat of the day into the evening. From a purely practical viewpoint, an advantage is that building contractors know how to make them. Normally a suspended wooden floor is constructed of boards or composite wood (plywood, chipboard, oriented strand board), laid across horizontal joists (thick boards on edge). Great rigidity is achieved by using tongue-and-groove (T and G) boards.

A suspended floor has an insulated cavity beneath it. These floors use a great deal of wood.

Groundfloor joists sometimes rest on the foundation, sometimes on beams; upstairs joists are attached to the house's walls or overhead beams. Beams or joists for lofts and floors can be supported directly by the cob walls. Load-bearing beams should extend as far into the cob as is practical—a minimum of 8 inches. With training in wooden construction and a concern about earthquakes, our initial inclination was to anchor the ends of these beams as strongly as possible into the cob, using various kinds of deadmen. When we mentioned this to Alfred Howard, one of the people most knowledgeable about traditional English cob, he was aghast. He told us that the tried-and-true method was simply to insert beams into cob with no anchor whatsoever. This allows the cob walls to move somewhat over the centuries without pulling the building apart as it moves and settles.

Because the joists need to be deep for strength, the whole floor is likely to be deep, adding extra height to the building's walls. Clearly there are advantages to reducing the depth; in some circumstances this is easy.

❧If spans can be kept short, joist depth can be reduced. Joists, like any beam, need depth proportional to the square of their length.

❧It's better to use more joists with less depth than to use fewer joists which are deeper, though in total that can use more wood. Closer

Short spans of flooring, for instance, in lofts, need no joists if the flooring is robust.

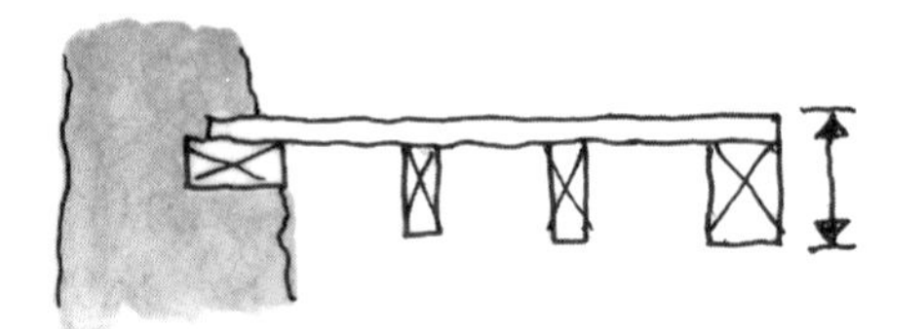

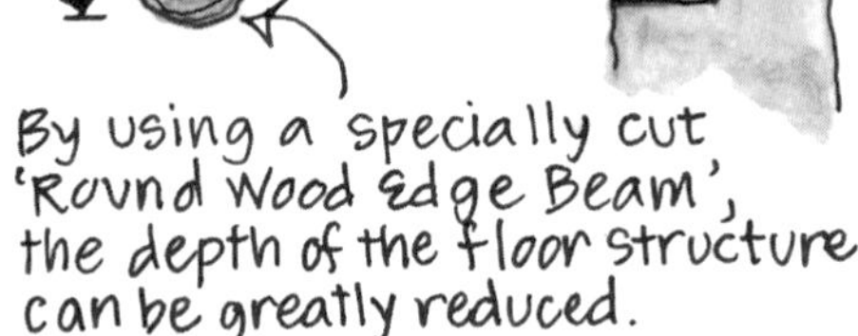

spacing also offers more flexibility in dimensions of flooring and prevents individual floor boards from bowing under load.

❧For lofts, roundwood joists can look attractive, and are stronger than their milled equivalents. A roundwood joist at 5 inches in diameter is much stronger than a nominal 2 × 6 of the same depth, though it is also heavier.

❧Short spans of flooring, for instance, for lofts, need no joists if the flooring is robust. Tongue-and-groove 2 × 6s or 2 × 8s can easily span eight feet, sometimes ten feet, with acceptable deflection under load, without intermediate support. Test them in advance.

❧Cob walls can accept a floor built directly into the wall or set upon a thin ledger board built into the cob. A standard detail for loft floors developed by The Cob Cottage Company is shown in the diagram above. One end of the boards sits on the ledger, the other on a roundwood edge beam with a V-shaped notch chainsawed out of it the whole length. Total floor depth can be as little as 3 inches, compared with 9 inches if a standard 2 × 8 joist is used. The 6 inches in savings means one less step in the staircase, and 6 inches less of wall height. On a 1½-story building 12 feet high, this reduces wall height by 4 percent, and likewise reduces heating, maintenance, cleaning, and roofline height.

Plasters and Finishes 17

> ❧ Natural plaster is a key to living in earthen houses. In the Heart House the brightness bounces around in the building. Even cloudy days can feel sunny. ❧

Now the walls are up, the roof is on, windows and doors are at least framed, and the base of the floor is laid. You are ready to plaster. Earthen buildings survive some climates fine without exterior plaster; in others the surface will erode or very slowly flake away. The interiors, without some kind of plaster, would be dark and dusty. Even well-trimmed cob walls are crumbly enough for parts to rub off, hard to keep clean, and too rough for comfort in any place where you might be touching them.

In North America, and increasingly all over the industrial consumer world, most builders submit to giant destructive industries by using their products to cover walls. Exterior stucco is generally Portland cement and sand; interiors are "drywalled" with gypsum board or plastered with gypsum plaster then painted with paints of variable and often unknown toxicity. The problems associated with Portland cement in stucco are discussed in appendix 3 and in this chapter. Making and transporting gypsum and paints are insensitive industrial processes involving many toxins and high energy use; they lead to air and water pollution, and human illnesses. There are now small companies addressing these problems and producing nontoxic paints and building products. See Resources.

Many wall finishing systems are possible with more natural materials. In this chapter we will explain a few of the most common natural plasters that we ourselves have experienced. Our culture has largely forgotten the immense store of vernacular information on

plasters; only recently has there been a resurgence of interest in the craft. We learned traditional limework from European builders, adobe plasters and *alis* paints from artisans in the American Southwest, and clay-dung "litema" plasters from craftspeople in southern Africa. The field is ripe for experiment and research. Good luck!

Exterior plasters are generally called *stucco.* In Britain, stucco is likely to be called *render* or *rendering,* and in Australia and New Zealand, *pargeting.* We've also heard interior plasters called pargeting. Canada seems to use all the terms interchangeably. In this chapter we will call all interior surface finishes *plaster* and all exterior plasters *stucco.* We'll use the verb *to plaster* to mean applying a thick paste either inside or outside a building.

EXTERIOR STUCCOS

In mild climates with little wind, generous eaves will protect most parts of a cottage quite adequately. Our own homes and most of the buildings we've done in western Oregon have no stucco. Many people have a preference for leaving exterior cob well trimmed exposed, as its patterns and colors can be aesthetically rewarding. In southern England, in a fiercely windy climate, cob farm buildings are often unstuccoed on the north and east sides, away from driving rain. Some have stood for centuries, completely unprotected.

On windy, rainy sites, even well-made cob gradually spalls off and erodes. The loss is roughly proportional to how much rain is driven against it, how forcefully, and how often. The English research that suggests "loss of face" at one inch per century seems optimistic. Walls exposed to driving rain should probably be protected, particularly if frost and rain are likely to occur in the same day. Even without erosion, a wet wall is a cold one, because the heat needed to dry it out is likely to be robbed from the building. Unprotected walls may raise your heating costs.

While stucco may keep rain from soaking into or eroding a wall, it needs to allow the escape of water vapor from inside the building. Cob walls are porous; water vapor generated inside the building by activities such as breathing, cooking, and bathing gradually works its way out through the wall. To avoid trapping this moisture in the wall, exterior stucco needs to be at least as porous as the interior plaster.

Exterior plasters, all over the world, have traditionally been of five kinds: *earth, dung-clay, lime-sand, lime-clay,* and, in very arid zones, sometimes *gypsum.* Simple *earth* stucco alone is not durable; without stabilizers such

Weatherproofing is essential with earthen walls. Long eaves help protect from rain; a stucco finish on the windward side will shield walls from blown rain.

as casein or flour paste it needs regular repair or replacement, or incorporate a high proportion of fiber, usually straw, or the dung of cattle or horses. Traditional New Mexico stucco is a *clay-straw* mixture, with very little sand, and is replaced every 1 to 5 years. In sub-Saharan Africa *clay-dung* is ubiquitous, even today. We can vouch for its durability and will describe how to make it. Permanently damp conditions rot out fiber, so people in the windy wet climates of the British Isles have used lime for centuries, possibly millennia, both as a stucco (mixed with sand) and as *limewash or whitewash* paint, sometimes emulsified with tallow. *Lime*-based stucco is very durable and can last centuries. *Gypsum* allegedly does not tolerate either frost or repeated soaking.

Don't Use Cement Stucco

Don't put cement on cob structures (or on straw bale, wood, or even brick). Plasters must permit the cob underneath to "breathe." *Avoid using cement stucco, impermeable vapor barriers, tar, or oil- or latex-based paints on the outside of cob buildings.*

Larry Keefe, the secretary of the Devon Earth Builder's Association, tells a story about cement stucco in Devon. An elderly couple sleeping in an upstairs bedroom of their English cob farmhouse awoke to a dull thud. It was dark out, and they could see nothing outside. With daylight they awoke again to find the entire gable wall had fallen out, leaving only a skin of interior lime plaster between their bed and the sky. The lime, being porous, had allowed water vapor to pass through to the outside, where its escape was blocked by a cement stucco covering. Condensation had built up and run down the inside of the stucco until the base of the wall was soaked, and the wet cob had lost its strength and collapsed in the night (see appendix 3 for more explanation).

Sixteenth century house near Tavistock, Devon. Lime stucco over cob. Five hundred years of ferocious North Atlantic gales and driving rain have left it undamaged.

Our own field research in England, the United States, and New Zealand indicates that cement stucco is a major cause of the collapse of cob buildings. Devon Historic Buildings Trust agrees: "The most usual way in which these problems manifest themselves is when the cement render parts company with the cob and at the same time develops hairline cracks. Rainwater penetrating the cracked surface is unable to evaporate and will, over a period of time, run down and soak the foot of the wall. It is at this point that the wall carries its maximum load, and is therefore most vulnerable." Throughout the American Southwest there are many similar examples of the tragic loss of beautiful old adobe buildings due to cement stuccos.

INTERIOR PLASTER

Interior plasters can brighten a house by reflecting natural light. Plasters near to light sources such as window reveals and skylights, behind electric lights, and in candle niches or lamp shelves should be as pale and reflective as possible. Yet sometimes a darker, richer color is called for. In recesses, on ceilings, around fireplaces, carefully chosen pigments can enrich the mood and enhance snugness. Here is the place for those bright orange clays or a touch of red iron oxide in the otherwise stark white limewash. Paler washes of transparent pigments can change the mood or define a special place within a room, perhaps layered one over another to create finely graded color changes across a wall or ceiling.

How plaster feels conditions your tactile environment. You probably don't care how rough the ceiling is, but near projecting corners, at the head of the bed, inside the shower—these places need to feel friendly, sensuous, interesting. The parts of buildings we touch with our hands and naked skin should make us feel good.

All plasters have odors, so pick interior finishes you will like to smell. While still new, their aroma may dominate the house. You may want to pay special care in places close to the nose such as alongside a bed, where the plaster will get damp or hot, or where food is stored.

Freestanding cob wall badly damaged by concrete stucco, Devon. Note rock foundation, brick stemwall.

Unexpectedly, one of the most serene house smells is that of dung-clay plaster, a dreamy hint of summer hayfields that pervades the sleeper's senses on the floors of Africa.

Not all finishes are washable. Splash zones around sinks or baths need extra protection and a surface that can be wiped down. Gypsum works well in these places, as does a fine lime-sand plaster, a casein wash over kaolin plaster, or tiles, wood, or flagstone. On surfaces that get less wear and less moisture, we prefer an earthen finish plaster, painted with *alis,* or lime whitewash, sometimes with added pigments.

Earth makes the least expensive plaster, as well as the easiest and most pleasant to work with. It has very little embodied energy or toxicity. It produces no waste for future generations, and is the easiest kind of plaster to change, remodel, rework, and reshape later on. We always recommend a base coat of mud plaster to fill in and smooth the wall, even if you plan to use gypsum or lime plaster for your finish surface, because mud adheres to cob better than any other material does.

MATERIALS FOR STUCCOS AND PLASTERS

Earth—the basic ingredient. Just as for cob, earth for plastering needs enough clay content that the plaster will really stick to the wall. For a base coat, you may need to screen the soil through a ¼-inch sieve. For finer finish coats, use an ⅛-inch or even 1⁄16-inch screen. There are two ways to sieve clay soil—either powder dry or as a liquid slip that you can pour through the screen.

Clay. For most earthen plasters, found clays are quite adequate (see chapter 8 for where to prospect). If you want a light-colored plaster, both white *kaolin* and gray *ball clay* are available as dry powder in fifty-pound bags from pottery supply stores. Using a kaolin base,

INNER LIGHT

BY IANTO

LIVING IN A rain forest may sound romantic, and it is, but it certainly has its drawbacks. Where Linda and I live, on the dripping wet Pacific slope of the Cascade Mountains, Douglas fir dominates the landscape. Douglas fir must be one of the darkest-foliaged trees in the world, so it is hard to get natural light into dwellings, particularly through the eight months of wet season. As a consequence, most western Oregon houses feel like caves. A combination of fir up to three hundred feet high, constant cloud cover, big single-story houses, and the long eaves so characteristic of this bioregion makes for electric lights on all day. Natural light is so important to keeping people cheerful that every chance should be taken not to dominate its subtle moods with artificial lighting.

Visitors to our cottage come with expectations. They know that we live in a "mud hut." Sure enough, at first viewing the exterior may be elegant in shape and texture, but without stucco the exposed cob confirms some of their prejudices. One step inside and their whole attitude suddenly brightens. They're entranced, stunned, in awe.

"Take a seat," says Linda, and they sit, gingerly, on the cushions of the white plastered bench. Sometimes we are six, eight, even a dozen, packed into this tiny room. We serve tea; they marvel, discuss, question. Nobody comments on how small the space is, in fact the opposite, how spacious, and

White-painted window reveals and pale walls reflect light in Linda and Ianto's kitchen

those knowledgeable in the building trade make ridiculously big guesses about the square footage.

Often it's a gray day, wisps of cloud wandering across the broken ridgetops, big trees poking momentarily through a cloud, as in old Chinese paintings. Then, finally, somebody will notice. The space inside is bigger than the building was outside—magic! Then: "How d'you make it so *light* in here?" "What d'you put on the walls?" Or, "My house is really dark, what d'you do to get all the light?"

Natural plasters are a key to living in earthen houses. Our walls are painted with *alis,* all natural, clean, a creamy white with sparkles of mica. The ceilings are white, as are the window reveals, even the built-in bench and part of the woodstove. Brightness bounces around in the building. Even cloudy days can feel sunny.

lovely tints can be achieved by slowly adding bright-colored clays you have found. Within a mile of The Cob Cottage Company office we have found gray, cream, red, yellow, orange, pink, and even green clay, all in roadcuts and foundation trenches. For really bright colors, try masonry pigments that are made from powdered mineral oxides. For *alis* and fine finish plasters, soak clay in water for several days, completely homogenize the mix, then pour it through a ⅛- or 1⁄16-inch screen. Use this thick slip as a base for mixes.

Sand. Sand quality for base coats is not critical. The same coarse sand you used in your cob will work well. In finish plasters, however, you need a much finer texture; look for fine, rounded sand such as that from dunes or beaches. For specialized uses, quartz sand can be purchased. It is pale, and can help with luster in finish coats. "Mason's sand" is cheaper, less refined, often dark colored, but fine-screened.

Straw. Straw for plaster usually needs to be short and flexible. The simplest method is to cut it with a machete or even with scissors. You can take a sharp handsaw or a chain saw to an entire bale, though the fiber will tend to dull the teeth and clog the engine. One of the best tools for chopping straw is a manual or motorized leaf mulcher or wood chipper. Other options include a lawn mower driven over a broken-open bale, or a weed-whacker in a (preferably plastic) garbage can. Straw for finish plasters should be chopped fine and rubbed through an ⅛-inch screen.

Dung. Animal manure is an excellent source of short, pliable fibers. Fresh dung contains digestive enzymes that help to plasticize clay. In West Africa, cow dung is gathered fresh and set to soak in the clay pit. After three days of fermenting it stinks, yet when dry has very little odor. What smell remains is quite pleasant. Both horse and cow dung are used, almost worldwide; it is possible that the manure of other ruminants would work—goats, llamas, or sheep, for instance—though they tend to masticate finer, so their manure has shorter fibers. If fresh dung is not available, dry will work, but the fibers need to be intact, not decomposed. Dry dung should be rubbed through a screen to break it apart. If you have trouble, try soaking it first.

"Clean out your refrigerator when making a batch of mud," says Carole Crews.

Flour Paste. You can buy wallpaper paste, or you can easily make your own from cheap white flour. Add a cup of flour to two cups of cold water, while another six cups of water comes to a boil. Whisk up the flour and water to a paste like pancake batter, then slowly stir it into the now boiling water, and cook until thick and translucent, stirring constantly. Don't make more than you can use in a day, as it spoils quickly and smells bad.

Other Materials. To make plasters stickier, a number of materials can be added. *White glue* in the water helps bind a mix, as does a 10 percent proportion of *milk*. The milk can be powdered, fresh, or sour. Carole Crews of Gourmet Adobe in Taos, New Mexico says, "Clean your refrigerator when you're making a batch of mud. Leftover noodles or oatmeal put through a blender, old bear mush, polenta, mashed potatoes, spoiling milk or yogurt will all strengthen the mud."

TOOLS FOR PLASTERING AND MIXING

Plastering has its own set of special tools (see drawing on page 257). Most important are *trowels*. Over time anyone doing plasters accumulates a tool kit containing a wide range of trowels, but initially you can get away with a rigid steel rectangular plasterer's trowel and a pool float, a flexible, steel trowel with rounded ends. Flexible stainless steel or plastic plastering trowels imported from Japan give the smoothest finish to earthen plasters. A cheap but effective substitute can be made by cutting a plastic disk from the flat lid of a yogurt containers. Scraps of burlap or other coarse, absorbent cloth are also good for smoothing earth finish plasters, as are large tile sponges.

To carry a load of wet plaster right to the wall, use a plasterer's hawk, a square, flat tray 12 to 16 inches on a side with a handle beneath. Make your own from a square of rigid plywood with a 1½-inch dowel handle, or you can buy a lightweight professional model new for $15 to $20.

To sift the ingredients for adobe plasters, you will need a series of metal screens mounted on wooden frames. Have on hand ¼- and ⅛-inch mesh and a finer window screen (preferably of stainless steel, not plastic), each piece about 2 by 3 feet. If you work alone, mount the framed screens with legs on one end so that the screen stands up at a diagonal. Otherwise, set them up with handles so that two people can shake the screen back and forth over a wheelbarrow or tarp. A freestanding screen works well for big volumes, a two-person screen for smaller quantities or for fine finish plaster materials.

What will you mix plaster in? A wheelbarrow works fairly well, or for larger quantities a plywood mixing boat. In either case, a couple of swan-necked hoes are useful for mixing ingredients. Very large batches of plaster can be made in a tarp-lined straw bale pit like the ones described in chapter 11, or you can rent a plaster mixer.

To wet down a dry surface prior to plastering, a spray bottle is useful, as is a big soft brush. Thick, rubber dishwashing gloves are essential for handling lime and also good for applying earth and clay-dung plasters.

Various sponges will be useful for applying and burnishing finishes, and a piece of scrap foam carpet underlay from which you can cut 3 × 6-inch rectangles, is ideal for applying lime or gypsum plaster in tight spaces, niches, and holes. Finally, for an *alis* clay-slip finish, you will need a stove and heavy cookpot to make flour paste, a wire whisk, and a large, clean bowl or 5-gallon bucket. For applying *alis* or limewash, you'll need brushes in a range of sizes, up to about 8 inches in width.

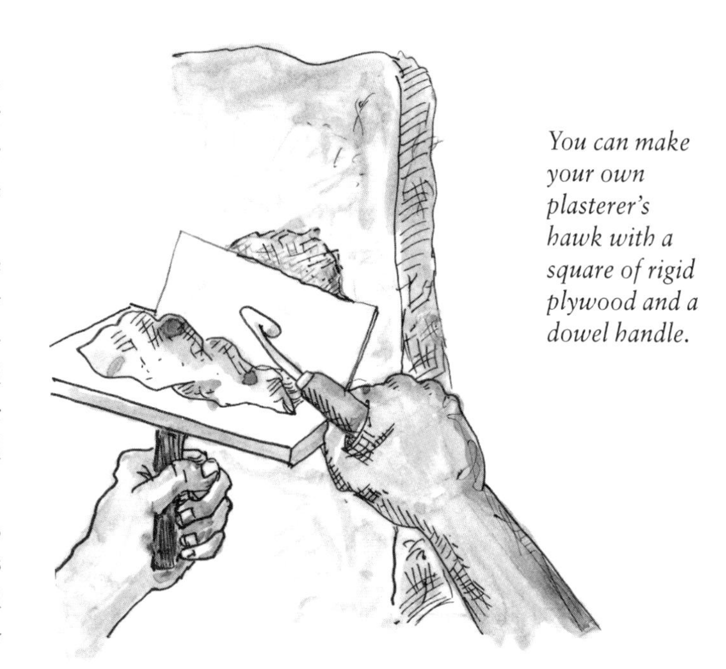

You can make your own plasterer's hawk with a square of rigid plywood and a dowel handle.

PREPARING A WALL FOR STUCCO OR PLASTER

We can't emphasize enough: Taking care with shaping and trimming the walls as you build saves time in preparation for plaster. Train yourself and your helpers to trim and shape walls while they are still wet, after any danger of splüging or movement is over. It's a plasterer's worst nightmare to have to spend days of hard work avoiding breathing dust to prepare a rock-solid wall that is completely dry. It's hard on both body and tools. Rob Pollacek, and innovative cobber, has found a power saw to be a great time saver in shaving his beautiful red clay walls in Nevada City, California.

In order not to waste plaster, the wall should be carefully shaped. An irregular depth of plaster tends to cause cracking that, though not so important in a base coat, could

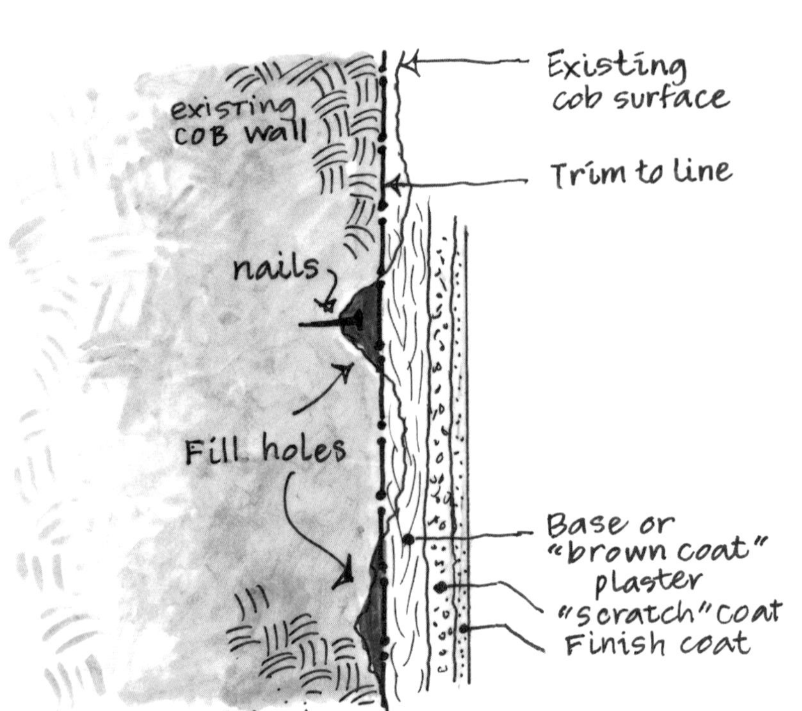

To conserve plaster and facilitate application, the wall should be level and smooth before finishing.

prevent the use of single-coat plasters. To make plastering easy, try to level the surfaces. Spend time just looking at your wall, changing viewpoint, getting an idea of where you would like it projecting, hollowed, sculpted, gently curved.

First trim off major high points with spud, saw, or machete, wetting the wall if necessary. Trimming is easier than rebuilding, but if there are deep holes, you'll need to fill them. Premoisten the surface, then slap small sticky cobs, dipped in clay slip, into the holes, pressing hard and smearing the edges carefully. If a section of the wall is very smooth, roughen it with a hatchet, or a claw hammer.

EARTH PLASTERS

Earth plasters are traditional everywhere earth walls are found, notably in the American Southwest and Africa, but also in Europe, Asia, and Latin America. Basic earth plaster (also called mud plaster, "adobe plaster" in the Southwest, and *dagga* in the Mideast and Africa) is a mixture of clay soil, sand, water, and straw. It is much smoother than a normal cob mix, being made of screened soil and sand and chopped straw or manure. It is also much wetter than cob, about the consistency of cake batter. Earthen base coats are generally high in clay to make them stick, while finish coats contain more sand or straw to prevent cracking. Because clay soils vary so much from place to place, we can't give you a fail-safe recipe. Always test your plaster mix by applying some to a small area of wall (a few square feet) or maybe a sample on a board to dry in the sun or an oven. Be aware that fast drying may cause some cracks. When dry, check for cracking, adhesion, and hardness (and, with finish coats, color and texture).

To make a base coat or "scratch coat" of earth plaster, first screen your clay soil through a ¼-inch wire mesh. Unless it is terribly coarse, your sand won't need to be screened. As you mix, add the clay soil to the water, and slowly stir in the sand. Then add lots of chopped straw.

In the early years of our work, we practiced and taught the use of several thin coats of earthen plasters. We find now, with improved techniques, that a one-coat earthen plaster is all you need. For one-coat interior earthen plaster, add a huge amount of chopped straw. Expect to use 2 to 3 buckets of straw, pressed down by hand, for a big wheelbarrow full of plaster. When you grab a handful of earthen plaster, it should stick more to itself than to your hands, but when you throw it against a cob wall from a distance of several yards, most of it should stick. If it doesn't behave this way, adjust the mix with more sand or straw (if it sticks to your hands), more clay (if it doesn't stick to the wall), or more chopped straw (if it's too loose).

Exterior earthen stucco that will be exposed to weather requires an additional level of care. One method contains little or no sand, as sand behaves like scouring powder, grinding the clay off the wall when rain runs down it. In New Mexico, we have seen exterior mud-stucco plasters without any sand, made with huge quantities of long straw. That way, when the clay washes off the surface, the wall will be left with a shaggy texture not unlike thatch. The exposed straw sheds water and protects the mud underneath. Carollee Pelos and Jean-Louis Bourgeois's adobe home in Taos, finished in this fashion, went seven years before needing another coat of stucco, despite the fact that their roof has no overhang and Taos has heavy rains in summer.

As always, trim and thoroughly wet the wall before you start plastering. To apply a scratch coat, smear the plaster across the wall, applying pressure with the heels of your hands. Thick rubber gloves will prevent your hands from getting scraped up in the process. Keep your hands or gloves wet to prevent the plaster from sticking to them, but avoid a shiny clay surface, which can cause cracking.

Plaster can also be applied with a trowel, the use of which takes practice. Put a load of plaster on a hawk and push the hawk tight against the wall. Use the trowel to shove some plaster off the back of the hawk and smear it vertically up the wall, pushing hard with the trowel. For best results, use long strokes. Wet the trowel frequently so the plaster doesn't stick to it.

Regardless of which technique you select, apply your plaster in a fairly thin coat (no more than ½ an inch thick, except for one-coat plasters) to achieve good bonding and minimize cracking. Wait for the base coat to dry thoroughly, then wet the surface again before applying the finish coat. If the base coat cracks, add more sand to your finish plaster.

The smoothness you can achieve with finish plaster depends on how finely you screen ingredients. Usually, we run the clay and straw through a ⅛-inch screen and the sand through a 1/16-inch window screen. To get the smoothest possible finish, put the clay through a fly screen, too. Finish plaster should go on quite thin—no more than ¼ of an inch. Depending on how smooth a finish you want and how uneven your wall started out, you may need a second coat of finish plaster.

GENERAL TIPS FOR PLASTERING

- However well you think you know a plaster recipe, try a small test patch and leave it to dry before committing to an entire wall.
- Similar materials have similar drying patterns, and are therefore less likely to crack or separate. Earthen plasters for instance can be applied immediately onto cob that is not totally dry, whereas gypsum would set rapidly into a hard inflexible plaster that would separate as the cob dried and shrank. After a wall is completely dry, there will be very little movement, so most kinds of plasters should adhere well.
- Some plasters need to dry slowly, lime, for instance, so shade them from direct sun and spray them periodically with water. Some will mold if they are not dried fast, including litema, dung-clay and single-coat straw-clay plasters. The addition of borax or hydrogen peroxide to the plaster mix retards mold growth.
- All plaster coats except the finish should be left rough-surfaced, so the next coat will stick.
- Always dampen surfaces you are about to plaster, be they a cob wall or a plaster undercoat except for an *alis* finish.
- Plasters are sticky and can dry hard. Cover floors and furniture carefully. Use tape along edges of glass and woodwork. Clean off any spilled plaster thoroughly while it's still wet. Check again after your clean-up job dries; pale plasters show up more after they are dry.
- Keep plaster coats thin and even in depth, to avoid cracking. Traditional plastering advice is "many thin coats," usually less than ½ an inch (1 cm), or ¼ of an inch for finish coats, though in very dry conditions a single thicker coat of clay-sand-straw plaster may work.

Techniques and tools for finishing plaster work.

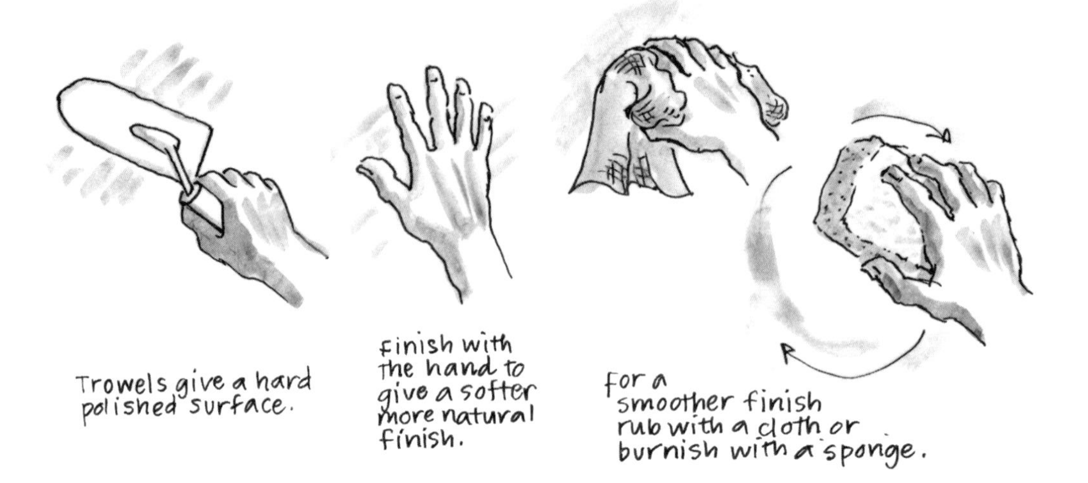

Trowels give a hard, polished finish. For a softer, more natural feel, apply finish plaster with the hand. To make it smoother, as it begins to set rub it down with damp burlap, leather, or other coarse cloth, using broad, circular motions. This evens out the surface and removes larger grains of sand that might otherwise fall off later. You can even burnish the surface with a sponge as it begins to set hard.

The addition of small quantities of non-water-soluble substances to earth stucco can increase its water resistance without seriously reducing its breathability. Robert Laporte of the Natural House Building Center adds 1 per-

ONE-COAT KAOLIN PLASTER

BY LINDA

IT WAS EARLY MAY when my friend Carole Crews and I flew to a small remote island in British Columbia to do the finishing plaster on an exquisitely crafted building of cob and natural materials. It was the third cob building I had played a part in building on this island. We were honored to be part of a community team effort of craftspeople and builders in the creation of this remarkable work of art. To build the structure, clay, sand, rock, and wood had been collected on-site or on the island. Driftwood posts came from the beach. The wood for the cabinets and windowsills and countertops was from an old-growth cedar tree blown down in a storm. Granite for the foundation, patio, and steps leading up to the building had been selected by hand and transported by boat to the site, each piece shaped and carefully placed by professional masons.

To make decisions on what kind of finish plaster to recommend, I assess the resources and the time available. In this situation we recommended that instead of hiring a team of people who had never plastered before to do an earthen base coat and then to fly me in to do the finishing touches, it would be time- and cost-effective to eliminate the base coat entirely and for us to do a one-coat kaolin plaster over an already prepared, shaped, and trimmed cob wall. We chose white kaolin clay because the plaster needed to be light and soft. Because kaolin is very nonexpansive, it can be applied in thicker layers without cracking, making one-coat plasters possible. All the "dirty" work was done, windows and doors and earthen floor and cabinets were in. Everything must be clean and free from dust for a white finish coat.

The plaster contained very fine local beach sand, homemade flour paste, kaolin clay bought in

cent linseed oil by volume to his earthen finish plaster. Another favorite plaster additive among those who have access to it is the mucilaginous juice of prickly pear cactus (*nopal* in Spanish), which has been used traditionally in Latin America for waterproofing. Lime is also sometimes added to earth stuccos to increase their weather resistance, as are gypsum, rye flour, animal manure, alum, and gum arabic (see *The Earth Builders' Encyclopedia* for recipes). White glue is said to help, at about 1 percent by volume.

LITEMA CLAY-DUNG PLASTERS

Litema (pronounced dee-TAY-ma) is a Sesotho word for clay-dung plaster. In Africa, dung plasters are ubiquitous. They appear as highlights around doors and windows, and are used to create colored geometric patterns on the outsides of houses and mosques. Entrance steps are often finished with clay-dung, as are inside floors where people sleep, fireplace surrounds, and even shelves. We use *litema* as a decorative and protective coating, particularly on areas that are likely to get scraped or knocked, such as around windows and doors. It has a pliable, fibrous texture and is moderately resistant to rain. It dries with little odor, so it can be used either inside or out. This is an excellent use for brightly colored clays that you may have discovered on your travels.

If your reaction to the word *dung* is "Oh gawd, what now? They're telling me to work with mud, now they want me to get shit on my hands!"—you can relax. Very few pathogens can transfer from either cow or horse manure to humans, and clay-dung is very enjoyable to work with. It *will* leave your hands smelling of manure for a day or

fifty-pound bags, wheat straw finely shredded in a leaf mulcher, a splash of oil so the plaster would slide off the trowel nicely, and a little bleach or borax to prevent color bleeding out of the straw into the mix and the water.

We made small mixes in a wheelbarrow using a giant whisk and a hoe, with 70 percent sand and 30 percent clay. We started by whisking together the water, flour paste, oil, and borax, then added the kaolin clay. Next we hoed in the sand, bit by bit. Last we mixed in plenty of chopped straw, essential in a one-coat plaster, adding more water—until it reached the consistency most satisfying to apply, not too wet and not too stiff. It was a pure joy to use small, very fine Japanese finish trowels. Some people may prefer larger trowels so they can apply greater amounts of plaster more quickly. That's okay for a coarser plaster, but we were creating a very elegant finish, and the small trowel was a perfect tool for the job.

The application went on beautifully. It instantly transformed the room from its cavelike roughness to give a very fine light and soft feeling. Even though we had all been satisfied with our small test samples, dark minerals in the sand left the room with a gray hue. When the plaster was dry, we decided to give it a warmer feel by applying a tinted casein wash. We mixed ocher pigment with water and stirred it into a casein mixture. Painted on with a brush, the casein wash dried quickly to leave a lightly polished look. It was faster to apply than an *alis* paint, and gave extra water resistance to areas like the kitchen that needed splash protection. The light ocher color brought out the blond straw in the plaster and complemented the natural woodwork.

Sometimes people can't imagine themselves living in a cob building until they see it plastered. The plaster creates a feeling of lightness and euphoria, allowing everyone to relax and breathe deeply in such an exquisitely finished space.

This Litema window surround lasted five years on a very exposed wall; 50% horse dung, 50% local red clay.

two, so either oil them first or wear gloves. Dry *litema* has almost no odor, and what it has is pleasant.

Traditionally, fresh cow manure is used. For the squeamish, or in case fresh dung is not available, we have found that dry cow or horse manure works well enough, as long as the fiber is still intact and strong. Combine the dung thoroughly with an equal volume of wet clay, then leave the mixture to "ferment" for several days. Allegedly, enzymatic reaction between the manure and the clay creates better bonding for a more durable plaster. As always, try a small test patch before plastering large areas. If it cracks, add more manure.

To apply *litema* over cob or earth plaster, first wet the surface. Then simply smear a thin layer of the mixture on the wall with your hand, pressing hard for good adhesion. Fresh manure gives *litema* a smooth, almost greasy consistency, making it easy and pleasant to apply. *Litema* is slow drying and can't be applied in damp weather or it rots.

Manure also makes excellent fiber for use in earthen finish plasters and the final layer of earthen floors. If you don't like to work with fresh dung, you can get washed manure from dairies or wash it yourself. Simply spread manure out in the rain for a few days. You will end up with the finest, cleanest fiber imaginable, which can save you a whole lot of time chopping and screening straw.

GYPSUM

Gypsum (calcium sulfate dihydrate) occurs naturally as a soft crystalline rock or sand. It is also produced as a byproduct of evaporating seawater for salt. It is heated to create builders' gypsum, which is used for plasters.

Several kinds of commercial gypsum plasters are available as dry powder in bags. "Structolite"("base plaster" or "underplaster") has perlite or vermiculite mixed in to make it lighter. Because it sets more slowly, and is somewhat rough and soft, it is generally used as a base or scratch coat. In contrast, gypsum "finish plaster" sets much faster and harder. Gypsum plaster can be mixed with sand in proportions as high as 3 parts sand to 1 part gypsum to save expense and give the plaster more texture.

The advantages of gypsum plaster are that it sets very hard and quickly doesn't shrink or crack much, and has a low embodied energy compared to lime or Portland cement. It is bright white, so needs no further painting. It is very hard and stands up well to scrubbing, making it ideal for kitchens and bathrooms. The main disadvantages are that it is relatively expensive (especially finish plaster), it performs poorly as an outdoor material, and it sets *so* quickly that it can be difficult to work with.

Working with gypsum plaster is the ultimate exercise in organization. Everything needs to be totally prepared in advance—walls smooth and damp, tools at hand, no distractions. To slow down the setting, mix the gypsum with really cold water and/or a little lime putty. Make a paste by adding water to the powder and stirring to break up lumps. (If you like, you can add pigment to the dry powder.) Because it sets so quickly, mix only as much as you will use in 5 to 10 minutes. Keep your tools and mixing surfaces very clean; exposure to hardened plaster will make the mix set faster, as will overstirring.

With a wet trowel, you can get a very smooth, polished surface. Or you can apply a wetter mixture with a brush or a sponge for a more textured finish.

LIME-SAND PLASTERS

Burned lime has been used for more than seven thousand years as the basis for mortars, plasters, and whitewash. The process of making it is somewhat unecological and energy-intensive, but it is much more flexible and breathable than cement, making it very suitable for plasters and paints.

Lime is made from calcareous rock such as limestone or chalk that is heated in a kiln to about 900°C, turning the calcium carbonate ($CaCO_3$) into calcium oxide (CaO) or "quicklime." Quicklime is then "slaked" by mixing it with water, a hazardous procedure that releases instant heat, boiling the water and splashing out caustic lime. Be careful if you slake your own lime; start with a very small amount.

The resultant *lime putty* is calcium dihydroxide, $Ca(OH)_2$. After soaking in water (preferably for a minimum of six weeks), this putty can be used for making plasters, lime washes, or mortars, or can be dried out and sold as bagged, hydrated lime. The latter must be soaked in water and turned back into lime putty before it can be used. Dried and exposed to the carbon dioxide in air, lime putty very slowly (over a period of up to three years) turns back into calcium carbonate. Plastering your house with a lime-based stucco effectively sheathes it in a thin limestone skin.

Lime mortars, plasters, stuccos, and washes are common on British cob buildings. For many centuries they have protected cob walls from unbelievably harsh weather along the North Atlantic coast. Unfortunately, one can't expect quite such good results using bagged lime. In their informative leaflet, "Appropriate Plasters, Renders and Finishes for Cob and Random Stone Walls in Devon," the Devon Earth Building Association (DEBA) explain, "Bagged, hydrated lime powder is a very poor substitute for either [ready-made lime putty or quicklime that you slake yourself]. Even when soaked in water to produce a putty, performance is noticeably less good. If the bagged material has stood at the builders' merchant's or their suppliers for some time it loses its strength." Bagged lime (sold as "type S" or "type N" builder's lime) is generally all that is available commercially in the United States. But until the natural building movement creates sufficient demand to make ready-made lime putty available, at least bagged hydrated lime is better than cement.

To make lime putty from hydrated lime, stir the dry powder into a bucket or barrel not more than two-thirds full of water, until the mix is smooth, the consistency of a thick milkshake. Wear a mask to prevent breathing lime dust. Cover the putty with water, put a tight lid on the container, and leave it for several weeks before using it. You can store lime putty indefinitely, and it will only get better as a building material. Top it off with water occasionally to keep the surface from drying out, or it will begin to set.

Lime-sand plaster is a mixture of lime putty and sand with little or no additional water. Proportions of lime to sand vary from 1:5 for a rough scratch coat to about 1:1 for a finish coat. Novices should start with a mix of 1:3. The strongest lime-sand plasters use "well-graded" sand with particles of different sizes. For a fine finish, use very fine rounded sand. Mix the sand and putty with a mason's trowel or a hoe on a flat, clean surface or in a mechanical mixer. DEBA recommends storing lime-sand plaster mix in plastic bags or airtight containers for at least two weeks before use. Remove only what you need for that day, then remix it on a clean board. Chopped animal hair, hemp fiber, and chopped straw are sometimes added at this point to prevent cracking, though there are reports that lime's causticity destroys cellulose.

The wall must be dampened with a brush or a spray bottle before plastering. To improve adhesion between the earthen wall and the lime plaster, wet down the wall with "lime water," made by mixing 2 or 3 percent lime putty in water.

Traditionally, lime-sand plasters are applied in two coats, each no more than 3/8 of an inch thick. The first coat is generally "rough cast" for best adhesion. DEBA explains, "Stand about 2 feet away and side-on to the wall and flick the mix [using a small coal shovel or trowel] with a backhand stroke. When the first coat is 'green hard' (i.e., firm to the touch but not bone dry), rub over it with a wood block. Use a circular motion. This process knocks off high grits and pushes others back into the render. When the backing coat is still green hard, wet it and apply the final coat. It may be necessary to add a little water to the mix of this final coat to ensure that it spreads evenly." The final coat can either be rough cast or troweled. DEBA recommends using a wooden rather than a metal float.

Lime plaster needs to dry slowly. Protect the surfaces you are working on from direct sun and drying wind, both during and after plastering. One strategy is to hang wet fabric or sacking a couple of inches from the newly plastered surface. Also protect new lime plaster from rain and frost.

As interest in lime grows, good written information on materials and technique is be-

The lime cycle.

coming more available, both in England and the United States. The bible of the lime revival is *Building with Lime: A Practical Introduction* by Stafford Holmes and Michael Wingate. If you don't thirst for such a depth of knowledge, most of the practical information you will need can be found in Charmaine Taylor's booklet *All About Lime*, *Lime in Building* by Jane Schofield, Tom Brown's paper on line from The Cob Cottage Company, or Bruce and Liz Induni's *Using Lime*. Another good source for natural builders is Issue #29 of *The Last Straw Journal,* compiled in 2000 by Athena and Bill Steen. See Resources.

WASHES AND PAINTS

Some plasters, including gypsum, lime-sand, and kaolin clay plaster, provide a durable, light-colored finish by themselves. Most earthen plasters, however, are dark in color, difficult to clean, and susceptible to wear. To increase their brightness and durability, coat them with a natural paint.

There are many types of natural paints. We will describe in detail our favorites: *alis* and lime wash. Both are inexpensive, easy to make, and give consistently excellent results. By adding pigments, you can achieve any color you want—a home decorator's delight.

Alis (also spelled *aliz,* rhymes with "a fleece") is a traditional, all-natural interior house paint that can be applied with a brush or sponge, then polished to an amazing luster. The word derives froth Spanish *alisar,* to make smooth, to polish. We learned about *alis* from Carole Crews, who has been a professional *enjarradora* (adobe plasteress) in Taos, New Mexico, for twenty years. Carole has developed her own variation on traditional *alis,* which she generously describes on the next page.

Alis and other natural paints can be applied over earthen plaster and most other wall surfaces, including drywall, wood, even wallpaper and synthetic paints. If the wall surface is very smooth and hard (gypsum plaster, for example, or high-gloss paint), first paint on an adhesion coat to prevent flaking and blistering. This adhesion coat can be made of flour paste and sand or white glue and sand, thinned to the consistency of paint. For people who rent or own conventional houses, natural paints can reduce the toxicity of their homes and provide an easy and rewarding introduction to natural building.

Limewash or Whitewash

Those picture-perfect white cottages on the Irish coast and in the hillside towns of Crete, and the white sections of half-timbered Tudor mansions, all owe their stunning brightness to whitewash. Until the invasion of synthetic paints, mostly in the 20th century, this cheap, beautiful white paint was homemade. In many countries it still is. Here's how to make your own. Paint a whole house for $5.67!

In its most basic form, limewash or whitewash is simply lime putty watered down to the consistency of paint. (See "Lime-Sand Plasters" on page 270 for more on lime putty.)

Apply limewash with a paintbrush. To get an opaque white finish over a dark mud plaster usually requires several coats. You can tint your limewash by mixing in finely sieved clay slip the color of your choice, although in our experience this reduces both its durability and water resistance. An alternative is to add a commercial pigment.

Traditionally, oil or sheep's tallow have sometimes been added to limewashes to make them more water resistant. Often the fat was added at slaking time; the heat of slaking emulsifies the oil. In "Appropriate Plasters, Renders and Finishes for Cob and Random Stone Walls in Devon," DEBA stresses that this should be done on external walls only. Their directions are: "Warm the limewash in a metal container, add the oil and continue

ALIS

By Carole Crews

IMAGINE LIVING INSIDE a polished clay pot. The same lustrous surface can grace your cob walls when you use *alis.*

Clay slips are as old as the most ancient civilizations; examples of their use have been published in books such as *Spectacular Vernacular* and *An African Canvas.* Smooth clay taken from riverbanks or found dry in Mother Earth's special pockets has long been used to enhance the beauty and cleanliness of walls not only through colorful and decorative designs but by filling in pores and roughness.

A clay slip is like skin to a building. It should breathe, not hold in moisture, and it should be smooth and fine in order not to collect dust and grime. Slips are used in much the same way as modern paints. It's a simple matter to redecorate and have a fresh new color to enjoy from time to time, although some people might prefer to put it off a decade or two.

To mix the *alis,* use a large whisk or a paint or plaster mixing tool attached to the end of a drill. Large plastic tubs are available that hold twenty to twenty-five gallons, but a 5-gallon bucket will work for small batches. Five gallons will cover a lot of wall area.

While fine clay is the main ingredient of clay slips, an important addition is mica, though an inferior version of *alis* can be made without it. Mica is like a molecularly flat sand that is smooth instead of gritty. It has lubricating properties and is used for that purpose in drilling oil wells. *Alis* is smoother to apply, easier to polish, and more lustrous and beautiful with the addition of mica.

Some clay comes with mica already in it. In the old days in Taos, there was a micaceous clay called Tierra Blanca that was found in a special cave; the women would apply a thin coat of this on the walls every spring with a sheepskin, to clean and freshen the surface after a winter of heating with wood. If a little clay is shaken up in water and it takes on the appearance of metallic clouds, this clay very likely has mica in it. Larger flakes, when available, may also be added to clay slips to add further interest and beauty, as can bits of chopped straw if you're careful to remove those pieces that contain the hard, rounded joints.

If you have no mica, add a proportion of 30 to 50 percent very fine sand or silt to the clay to keep it from cracking. You can create surface texture with bits of chopped straw or various hulls, seeds, or flower petals you might find in the wild. If you're adding such materials, the first coat could be sponged smooth and subsequent coats applied more thinly, with less sand or silt added, needing only slight polishing with the damp sponge. Or a final wash of clay with a little flour paste could be applied, but not polished at all.

There are some beautiful clays to be found by roadsides in places such as the Painted Desert, but if you have no ready source, buy white kaolin from the pottery supplier and tint it with naturally col-

heating until it has blended with the mixture. Use no more than 1 tablespoon of raw linseed oil to every 2 gallons of limewash."

Other Natural Paints

In addition to *alis* and lime whitewash, there are many other traditional recipes for natural, durable paints. Most natural paints are made up of a *binder* that sticks the paint together, a *pigment* that colors it, and an *extender* that gives it the proper texture and consistency. Paints can be either water-based or oil-based. Boiled linseed oil is a cheap and readily available oil that works well for paints, though it does yellow somewhat when dry. Oil-based paints are highly water resistant, so you may wish to use them in wet areas such as bathrooms and kitchens. In general, however,

ored clays, powdered pigments, or liquid additives sold for coloring cement and grout. Purchase more exotic hues through art supply dealers.

An average recipe for *alis* to go over a troweled mud plaster would be: 1 gallon water, 4 to 6 cups cooked flour paste, enough predissolved pigment to reach desired shade, 3 quarts white kaolin powder, 2 to 3 quarts mixed sizes of mica, ½ to 1 quart fine sand, and ½ a cup finely chopped straw (optional). Buttermilk is a good addition and can be substituted for double the amount of wheat paste, making an even tougher finish. A little sodium silicate keeps the clay in suspension longer. If mold is a problem in your home, add a little borax. If you want to keep this mixture more than a couple of days, add a couple of tablespoons of household bleach to keep it from souring. This recipe may be multiplied to match the capacity of your container, but avoid huge batches because they are difficult to mix and will spoil if you don't use them.

Alis can be brushed over a smooth, dry browncoat mud plaster, floated structolite or gypsum finish plaster, drywall, or an older clay finish that has become soiled. The natural-bristle Chinese brushes used for staining work well, and 1-inch-wide ones are good for details. Do not wet the wall before applying the *alis* or it will go on thinly and unevenly. If you put *alis* on a wall that isn't yet dry, it will stain.

Before starting a major *alis* project, tape any edges of woodwork and cover the floor. Cracks along the edges of wood or glass can be filled with a little slip into which sand has been added, turning it into a plaster. Use a palette knife, small trowel, or plastic disk to apply this "smoosh," then brush on the *alis* from a small bucket with a 3- to 5-inch brush. Always do a test patch, especially if you are using clay you found, as shrinkage rates vary and some clays are entirely unsuitable for finish work.

Let the first coat dry thoroughly, then apply a second coat. When the second coat has firmed up but is not yet dry, rub it lightly across the grain of the brushmarks with circular strokes of a damp tile sponge. If the sponge drags, let the wall dry more. Keep dipping the sponge in warm water then squeezing it out, to clean excess clay off the mica and straw. This process will smooth out roughness and create a glossy clean surface. Sometimes you can get by with only one coat of slip, and in that case, it would usually be polished, but try this only if your slip coat is almost the same color as the plaster beneath.

Let leftovers dry into "cookies" that can later be used for repairs and touch-ups. Reconstitute these by soaking the broken-up chunks in water and then pouring off the excess water.

Very fine sand such as silica or beach sand should be added to *alis* if the surface to be covered is rough or very porous. On the first coat over floated structolite (gypsum base plaster), one would use 20 to 40 percent sand in the mix, in addition to the mica. Second coats require less sand, often none if mica is used. Later coats used to freshen the wall over the years could be purely of clay, mica, and wheat paste or buttermilk.

water-based paints are preferred in cob building because they allow the walls to breathe.

One of the most durable and easy to make water-based options is casein, or milk paint. You can either buy casein powder at an art supply store or make your own from curdled milk. To make your own casein, add 2 tablespoons of low-fat sour cream to a quart of nonfat milk and allow to thicken in a warm place for a day or two. Then curdle the milk either by heating it or by adding vinegar or lemon juice. Separate the solid curds from the liquid whey by straining through a cheesecloth. Then blend the curds with a tablespoon of borax dissolved in ¾ of a cup of warm water.

This curd-borax mixture (or powdered casein dissolved in water) becomes the binder for your casein paint. It should be refrigerated and used as quickly as possible. Mix your pigment into a paste with an extender such as lime putty, kaolin clay, or diatomaceous earth, then add water to the desired consistency. Then combine 3 parts of this pigment/ extender mixture with 1 part casein. Note that mineral pigments available at ceramic and masonry supply stores are the most persistent; plant-derived pigments often fade with age.

Casein paints are very durable and highly water resistant. They may be applied onto almost anything, though porous surfaces including most plasters may need to be primed first with a mixture of casein and water. To make casein paint even more waterproof, you can whip oil into the curd mixture in place of an extender, then add pigment and water. Oil will reduce the opacity of the paint, creating a glaze-like effect.

Water-Resistant Finishes

In situations where cob will be exposed to large amounts of water, none of the finishes discussed above will hold up for long, except lime-sand plaster. Although lime plaster is very weather resistant, it may crack and let liquid water seep through, so without inspection and maintenance it may not always protect cob from eventual saturation and collapse.

The linseed oil and beeswax finish described in chapter 16, for sealing earthen floors, could possibly be used on walls as well. Herta Sturmann, a cob builder in South Africa, reports good results from finishing her cob sinks this way. Another option is to set ceramic tiles into a bed of cement-sand or possibly lime-sand mortar. Tile mosaics make a beautiful finish for exterior cob benches.

However, truly waterproof finishes are likely to lead to the same problems that cement stucco creates by trapping water inside. Until experience proves otherwise, we should caution you not to rely on plasters or finishes to protect cob in very wet conditions. Instead, use careful siting and design, high foundations, and wide roof eaves. If you come up with something else that seems to work, please let us know.

ONWORD: BRIDGING THE INNER AND OUTER WORLDS

BY LINDA SMILEY

"Smiley Sun Face" sculpted by Deanne Bednar and Michele Brooks. Material? Cob, of course.

WE HAVE COME FULL CIRCLE WITH our invitation to you: While building your cob cottage, consider your own life and your unique process for change.

This section addresses the "builder as artist" and how hand-sculpting your house of cob can enhance the health and well-being of you and your community. I have shared several specific processes and exercises I call Intuitive Design to help you gain the confidence to create your own sacred space and healing architecture. I have gathered stories to show how cob building has improved the quality of people's lives.

LINDA'S STORY

We all create our own personal myths and stories. I was a Smiley Valentine's baby, born 4 pounds, 11 ounces, could fit in a shoebox they said, but the hospital put me in an incubator instead. Maybe that explains why to this day I prefer snug-fitting houses. My life journey also includes learning the deeper meaning behind the smile in my last name, Smiley.

I am grateful to my parents for my childhood home on Melody Lane among the apricot orchards in the rolling hills of northern California. This place set the foundation for my world view and work, which weaves together relationships between human health, well-being, personal growth, psychology, ecology, and architecture.

As a California country girl, I played close to the earth and was an adventurous explorer. It was in this early environment that I developed love and appreciation for landscapes, natural beauty, adventure, water, and sun.

One of my most memorable school experiences was a project that allowed me to get my hands into the earth and build an adobe village. If my fifth-grade teacher had been able to look into my future and tell me I would grow up to be a professional mud-pie builder, I probably would have believed her, laughed, and said, "Sounds like fun to me!" I still smile and laugh about my chosen profession.

❧ COB IS NATURE'S HEALING ART FORM FOR EXPRESSING ONE'S WHOLE SELF AND SPIRIT . . . WHILE HOUSING ONE'S DREAMS. ❧

Later my natural but hidden interests were in dance, art and, I realize now, even architecture. I found myself habitually drawing very simple, gnomish-style houses. I was also fascinated by my father's brilliance in earth sciences. However, the path I followed in college was in the people-related fields. I started off in child

COURTNEY ROGMANS

psychology but found myself very impatient to work with children and bored with many of the texts and tests. So I changed to therapeutic recreation, where I could immediately work with people in creating therapeutic programs.

I was drawn to the spirit and definition of "recreation": to re-create; to form anew in the imagination; to restore to health; to refresh the strength and spirit after work.

Although attracted to the first several of these definitions, I never understood the last. Why is there a division between leisure and work? Why does one have to have recreation "after work" rather than the process of work itself being creative and renewing? Little did I know that I would choose a lifestyle where I would be able to integrate this spirit of restoration into my work with The Cob Cottage Company and sculpting houses.

In my mind, restoration also has to do with Nature, and therefore all the teaching programs I have designed have integrated a common theme of connecting with Nature. The university didn't teach me this; I discovered that using Nature as a tool to grow and learn about myself and others came to me intuitively.

Linda looking through cob cottage window to the beautiful cob landscape beyond. Cottage Grove, Oregon.

I went through strong growing pains learning how to make this simple principle work best. My first job after graduation was working in a girls' group home. One evening I took a group of teenagers up to the mountains to enjoy a sunset, but when we returned home they were more interested in holding me at knifepoint and stealing my car. I will never forget that year of working with troubled teenagers. As I grew into my work, I found that being a "naturalist" was a simple way to connect with almost everyone. Nature gave us a common ground. Later I designed, directed, and taught my own environmental education program, the Four Seasons Children's Camp, for twelve years in the forested coastal range of Oregon.

Fascinated by life stories and what makes people tick, I informally studied Dr. Arnold Mindell's work in process-oriented psychology with Gary Reiss (author of *Changing Ourselves, Changing the World*). I integrated one of Arny's basic concepts into my work: *By following people's intuitive processes you can help to find the gifts in their unique, natural way of healing.* While working in a counseling clinic, I combined this process approach with my use of Nature as a healing tool and created programs for kids and teens.

I have over time integrated my own natural way of being and woven a building process for change with cob sculpting. By working with my own personal process with this clay-like medium, I discovered that the building process itself becomes a "healing achitecture." Cob is Nature's healing art form for expressing one's whole self and spirit while housing one's dreams with sacred shapes.

I now live in a heart-shaped cob house, but years before I ever heard of cob, I had a dream that I was standing in my own heart-shaped cottage looking out a window to the beautiful landscape beyond. These days I enjoy creating our cob garden sanctuary courtyard outside the Heart House door—truly a healing space for whatever ails us.

Our cob cottage and garden walls form a container holding the beauty and grace of a natural shape, a curve sculpted of rich, golden-colored clay soil. Within this space is an abundant sweetness of fruit: peaches, grapes, oranges, lemons, apples, and figs, all thriving in the microclimate generated by the cob's thermal mass. Mason bees, butterflies, and other wildlife congregate among the flowers, fruits, and herbs. The courtyard creates a place of sanctuary where we experience the inner warmth of spiritual renewal and well-being. We are surrounded by walls that contain the embodied energy of one hundred people's hands and feet. One of my greatest pleasures has been to watch how cob workshops and the building process have generated a community of people. This community we call the Cob Web has created partnerships, lifelong friendships, mutual support systems for sustainable lifestyles, inspiration, and empowerment.

SCULPTING SACRED SPACES FOR WELL-BEING

The cob builder is an artist, sculpting a sacred space with earth as an expression of his or her individual journey toward wholeness and health. From the first thought of a cob house, through the completion of the building process, creating with cob is making a symbolic and personal work of art.

When Ianto and I began our first cob addition to a small wooden cabin we found that making our own building materials became a sacred process. We enjoyed digging clay soils in the early morning and building into the late afternoon. The satisfaction and enjoyment I felt in my body was similar to the feeling I had as a child when I hand-sculpted my first clay pot. We were, in fact, hand-sculpting a living, life-sized container, a cottage home. Our tools were simple: our hands and feet. We had no idea that working with clay in this way would so resemble the actions of a sculptor, and that we would immediately and continuously feel like sculptors ourselves.

One of our first experiences came out of the question: "What would be the easiest way to make a round window?" An amazing thing happened when we realized that we could put a piece of glass directly into the wall and sculpt around it to create any shape we wanted. This is a very different process from constructing a wood frame, and indeed, it was very easy.

I found that sculpting a house with clay became so simple and natural that I did not need to have a background in carpentry or building. I could approach it as an artist and as a human with innate ability to create a

> SACRED SPACE IS . . . transparent to transcendence and everything within such a space furnishes a base for meditation. . . . When you enter through the door, everything within that space is symbolic, the whole world is mythologized.
>
> To live in a sacred space is to live in a symbolic environment where spiritual life is possible, where everything around you speaks of the exaltation of the spirit.
>
> This is a place where you can simply experience and bring forth what you are and what you might be. This is the place of creative incubation. At first you might find that nothing happens there, but, if you have a sacred place and use it, something eventually will happen. Your sacred space is where you find yourself again and again.
>
> —JOSEPH CAMPBELL, from *The Temple in the House* by Anthony Lawlor, AIA

Linda and Ianto built the first cob cottage in Oregon in 1989. They lived here for four years.

dwelling. The accuracy demanded for wood construction was not needed within this flowing style of building. Like humans themselves, each sculpted cob building is "unique" and "one of a kind," containing the memories, the hands and feet and laughter, of those who helped shape this building. Building with cob gives a great sense of empowerment that connects us with those first experiences we had with clay in our childhood and leads us onward to becoming the sculptors of our own responsive and natural living environments.

INTUITIVE DESIGN AND MAGIC SPOTS

Our second cottage was not intentionally designed to be a cutesy heart shape. Rather, it was intuitively designed to create spaces around our activities, the way a glove fits around a hand. We needed a small space that could be built quickly to house the activities of cooking, sleeping, eating and socializing, and "desking." We had an emotional need to feel nurtured and embraced. So what symbol appeared on the back of the envelope, as Ianto sketched a design, but a heart — a shape that embraced separate activities and symbolized being embraced in a warm hug.

THIS TINY HOUSE

To every one of you
entering this sacred place,
may you and peace of mind
meet face to face.
Away from worldly stress,
if but for a minute
may you leave your cares behind,
as you venture inside
feeling only stillness
wrapped around your soul,
something almost magical
making you feel whole.
Where else can you find it?
Perhaps on a mountaintop,
that spiritual uplifting
that makes you look and stop.

We found this anonymous poem (which I have adapted slightly) on a gnomish Roman-style entrance to a tiny church on the windswept Welsh bluffs overlooking the sea.

Another example of sculpting intuitively is our yin-yang window. We placed a jagged piece of glass in our living room wall and sculpted a round window between two corbelled bookshelves. A visitor walked down the lane to the cottage, looked through the round window, and spotted the symbol. He greeted us and said, "I love your yin-yang window." I said, "Please show me." Sure enough, from one place along the path, this symbol could be seen, a daily reminder to integrate balance into our relationship and into our day.

In designing and building your home, follow your intuition, your dreams, and their symbols. Fantasies about home are often processes that are trying to happen. As you explore your inner landscape, you will gain an awareness of your personal psychology as it relates to architecture, ecology, and the environment.

I teach a therapeutic design process that uses meditation and sculpting cob models. Cobbers are guided in their personal process work towards self discovery and growth by gaining awareness of the importance of ecstatic qualities found in the natural environment. Well-being is enhanced when these discovered qualities are consciously integrated into the design of a hand-sculpted cob house. I call this process Intuitive Design.

INTUITION IS THE ABILITY TO USE ENERGY DATA TO MAKE DECISIONS IN THE IMMEDIATE MOMENT. ENERGY DATA ARE THE EMOTIONAL, PSYCHOLOGICAL AND SPIRITUAL COMPONENTS OF A GIVEN SITUATION. THEY ARE THE HERE AND NOW OF LIFE.

—Caroline Myss, *Anatomy of the Spirit*

Intuitive Design evolved from my background of more than a decade of experience observing children and their relationships to natural building in "Magic Spots" and from my training in Process Orientation Psychology (known as process work). My Intuitive Design process was inspired by Arnold Mindells's book on personal discovery, *Working on Yourself Alone*.

A favorite activity I learned in an environmental education conference was called Magic Spots. This activity evolved into a major part of the Four Seasons program. We would go to a special outside place in the natural environment. I would define the boundaries and ask the children to find their own Magic Spot where they could see and hear me. They were

LEFT TO RIGHT: MIGUEL SALMOIRAGHI, KIKO DENZER, LINDA SMILEY

LEFT: *Bas relief on his cob studio, by Miguel Salmoiraghi.* CENTER: *Little girl's bas-relief sculpture on the Bake House. Onenta, New York.* RIGHT: *Yin-yang window, an intuitive design. Heart House.*

instructed to sit still in one spot and observe what happens there. This meditation exercise led into developing their Magic Spots into homes modeling after Nature Architects, such as beavers, birds, woodrats. As the children built these magical places through play and discovery, I noticed that they emphasized the qualities that made them feel good and they followed their intuition.

We each have a lifetime's experience as intuitive designers. In building and creating Magical Spots as children, youth, and adults, we can explore our boundaries in relationship to ourselves, friends, and family while solving problems, healing emotional wounds, and experiencing joy through this playful and creative art form.

Do you remember the special places you designed as a child? Remember the fort in the living room when you used anything you could find to create a sanctuary—a space just for you? Remember the spaces you made in the closets, basements, attics, garages, and in backyard trees? Maybe you turned old outbuildings or chicken coops into your own secret abodes: GROWN-UPS KEEP OUT.

As a teenager your sacred space might have been in your room; on your bike or skate board; or in your car. As adults your sacred space may be on the computer or watching a movie, or the space created through music, art, meditation, prayer, yoga, reading, communing with nature, running, dancing, swimming, or perhaps the space you share with family and loved ones.

We have all woven our unique blanket of place memories. Every thread, rich in remembered shapes and textures, symbolizes places that have had a powerful role in shaping our lives. We can metaphorically open this memory blanket at any time to rediscover and integrate its sacred qualities into our designs.

As a recreational therapist I used Magic Spots as a therapeutic tool to provide children and youth inner sacredness through sculpting their own important places in the natural environment.

I once worked with a program designed to guide teenagers through difficult transitions from traumatic childhoods to adulthood. The project was a team effort between the high school, counseling clinic, and a youth employment program. Participating kids received science credit and a paid job as well as counseling services for building a nature trail that featured their Magic Spots, interpretive nature activities, and artwork.

I asked the kids to find their own Magic Spot in either forest or meadow and to spend time there until the magic of the surroundings came alive for them. Linnea, a young woman, chose a cool cluster of young firs, a little island of shade in the middle of an open meadow filled with wildflowers. This sacred space backed itself close to the trail, yet had qualities that felt embracing, secret, cozy, and peaceful, reflecting this young woman's need to feel a part of the group and also her need for a personal retreat in which to feel her individual strength. She connected deeply to it, and shared it with two new friends, Rose and Molly.

Linnea named her space Shalude and gave the name a meaning: a secret language between girls. Together the girls enjoyed the meadows, wrote poetry, played flute, talked about their problems in privacy, and carved a bench with their names and the name Shalude on it. Linnea drew the wildflowers in the meadow and created activities to focus in on the magic of this place. She left a journal in a carved-out cavity in a tree stump so visitors could write down their experiences and she could read what they wrote. The three girls completed the program and went on to be nature guides, taking younger children on discovery hikes on the trail. They produced a fun video of their work. Magic Spots helped to

LINDA SMILEY

COB COTTAGE COMPANY

LEFT: *Autumn LeBank and friend find their magic spot in a cave.* RIGHT: *Adam Large communes with a great blue Heron in his magic spot!*

empower them and create for them their support group and potentially lifelong friendships. These three friends, now adults, frequently walk the trail to this day, sharing testaments to the power of place, love, magic, and friendship.

INTUITIVE DESIGN EXERCISE

Often I find that we can discover qualities for well-being in places that call to us. I developed a design exercise I have been using in my workshops since 1993 to help strengthen intuitive skills. Here is another way that everyone can design, by making models using a cob sculpting mixture. The experience can help to bring to the surface hidden needs, dreams, and essential qualities tucked away into memories of one's scared and magical places.

This exercise can help you use the tools of meditation and intuition to design spaces that integrate qualities essential to development of your true nature, health, spirituality, and relationships. I have integrated Arny Mindell's theories of dream body work, childhood dreams, inner work, and sensory channel perceptions into this model-making exercise. First, find an outdoor spot where you can sit or lie, where you feel drawn to be. Let intuition be your guide. Take this opportunity to recall one of your childhood magic spots in nature, woven into your memory of the places you have known. If a childhood memory doesn't come easily, remembering any special place in your life will do. Begin to experience the place's important qualities and its meaning to you once again—like a dream still waiting to recur. Allow yourself to notice the sensations of being there.

Feel it. What feels right about this remembered place? What's the magic that makes you feel good? Where do you sense this feeling in your body?

See it. Allow yourself to see the shapes, the quality of light, and the colors. Notice the floor or ground, the walls, roofs, and eyes (windows and doors). What do you see here that you wish to have more of in your life?

Hear it. Remember the soundscape. Are the sounds melodious, flowing, staccato, chaotic, lyrical, or is this place silent? Turn up the volume and listen to what happens. Are you hearing internal or external sounds? What are you not hearing?

Move with it. How are you moving in this space? Are you still? Are you moving inward or outward? Pay attention, and watch for any movement that feels right at this moment. What part of you wants to move? Let yourself explore this movement and exaggerate it. What if you were to live out your day like this movement? What could happen if you were to build this kind of movement into the design of your cottage? What would it look and feel like?

Relate to it. How would integrating these qualities you remember into your design of a sacred space be a right direction for you to take in relationships to yourself, loved ones, family, or community, environment, and with the world at large? Are there any similar qualities between the spot you have chosen to imagine in this exercise and a childhood magical spot?

Expand it. What is your connection to the world while being in this space? How is it just right for you?

Write about it. If you like, write about your experience of this exercise as well. Make a list of the qualities you have imagined, as though you are selecting seeds that you wish to grow to become your own sacred space.

WRITING IN A MAGIC SPOT

I SIT IN A well-protected cove at Face Rock at Bandon Beach on the southern Oregon coast. My back rests against a solid support of basalt rock. My body is warmed by the morning sun positioned in the Southeast. The cove embraces me on three sides with basalt rock. My feet soothe my soul by contact with silky sensual sands. I hear the sounds of power in the ocean waves. I smell the salt of the sea and taste the blowing gritty sands. I feel the movement of the wind blowing sand across my body. I see to the west the spectacular view of Face Rock in the ocean. The Indian legend of Face Rock is a tale about Chief Siskiyou's beautiful daughter Ewauna, who kept her face toward the friendly moon. The image of her face lies still solid and peaceful among the chaotic swirling waters around her. I feel nourished, uplifted, and inspired.

My unconscious prescribed to me, through my choice of place and my relationship to my spirit, exactly what I needed for this personal healing retreat weekend.

The next step is *integration.* Take the key qualities in your writing and ask yourself what's right about these qualities for you. How do you plan to incorporate them into each day and into the design of your sacred space?

MAKE A MODEL

Magic Spots express dreams trying to come true. Sculpting a model will allow your mind to rest—untrained hands know just what to do. Following your intuition and observing what your hands can't resist sculpting, bring forth your own natural way of healing yourself and the inner landscapes of your being. When you consciously integrate these symbols and their meaning into the structure of your home, you create a place that can lift your spirit and soothe your soul.

To make an intuitive model of the sacred space that will be your home, first create for yourself a carefully chosen environment in which to work. Then, prepare and collect modeling materials that appeal to you, for instance:

- Cob sculpting mixture, a finely screened mixture of clay soils, sand, straw, and water.
- Miniature natural building materials: geological materials such as small pebbles, stones, sand, and mica, and biological materials such as driftwood, sticks, bark, sod, bamboo, and straw for thatched roofs or straw bale walls.
- Mood setters—tiny scale plants, flowers, and water. Don't forget color!

Prepare a platform for your model for ease of transportation. Make or find a base using a piece of flat wood or some other durable material. A good size is between two to four feet, more or less square. You may also want to have ready a writing journal, pens, and colors.

When you are ready, take a ball of cob sculpting mix, hold it in your hands, and sit still. Remembering the qualities of place that felt ecstatic and good to you in the previous exercise, observe your body's feelings, sensations, and reactions. Trust your hands to sculpt your model. Observe what your hands want to do. See what wants to happen, and let it happen.

Use other natural materials that you have collected. Notice the materials you choose and how they are right for you. Remember to place an object or cob figurine in your model to represent yourself and give scale. Allow your model to feel like it is growing out of the landscape.

Remember to build in any foundation, floors, walls, windows, doors, and roofs that you imagine. How are these parts of a house symbolic of yourself? You may want to sculpt a single quality, instead of making an entire model of your cottage. Don't worry, don't rush. You can come back to this process again and again.

Share your model in your cob workshop or with a friend, telling stories about your sacred space or Magic Spot. Notice how you and others approach this model. Notice what qualities are trying to be expressed through this cob sculpture.

COB BUILDING AND HEALTH—THEY BELONG TOGETHER

From 1993 through 2002, during our first nine years as teachers of cob building and cob living, we taught more than one thousand students. We hear from our alumni again and again, "This workshop has changed my life." The changes come in the form of personal growth in relationships with family and community, lifestyle, deconsumerization, health, attitudes, liberation, empowerment, and fulfillment.

Building with cob is a process that gives the builder direct physical contact with Earth. Building becomes a meditation. Primal and tribal work begins to happen. The mud dancing, body reflexology, and community building that occur during a workshop open, strengthen, relax, and adjust the cobbers mentally, physically, and spiritually in mysterious and unexpected ways. The work tends to generate the healing power of love, as even chronic illness responds to healing, and needed changes in career and life attitudes begin to be made.

Cob model: note fir cone for scale.

Once the cob seed is planted inside your being, it keeps on growing and you keep on transforming from the inside out. With every breath you take, a life-changing force brings you closer to the true spirit of home. Once you have opened this sacred door, you can enter in to an internal place where spiritual life thrives, a place where relationships with self, soul, spirit, place, Nature, family, your tribe, and the world at large can exist in harmony. A place where you can always return home.

Not long ago, I was working in the office in our Big House (a passive solar stick frame). Sun in wintertime Oregon is rare, and I just knew, as the rays shone through those Big House windows, that I had to be in the cob cottage instead. I put a message on the phone machine: "It's a beautiful sunny day, gone home to the cob, catch me there." I ran over to

LEFT: *I love cob.* CENTER: *Doreen Hynd, cobbing at age 75.* RIGHT: *Kids cobbing during workshop in at Kuruna, Australia.*

MICHAEL G. SMITH

the Heart House and immediately gave a sigh of relief, feeling sun and hearing water and bird sounds. I felt joyful, relaxed, grounded, and connected to the Nature that surrounded me. I heard myself saying, "Now I am really enjoying my work." I was aware how cob helps me to feel deeply happy, which leads me to healthy attitudes and living a healthy lifestyle—all ingredients for well-being.

At the cob cottage, I soon got a call from David, a workshop alumnus, who said, "We want to come and take your pyromania workshop. We need to get fired up again. Our house is hurting us. Our house is designed to keep us away from being outside." His partner, Bonnie, had taken her first cob workshop three years ago, and then David took a thatching workshop. They rose to Ianto's challenge—If you want to live simply, halve your income, quit your job, sell your house, and grow a garden.

David says, "I made the decision that I want to live to be vigorous and a hundred years old. I have been going through major healing. I decided if I'm going to live to be a hundred we'd better move. So we are selling our house and are going to buy land in the country in your area and build a cob and natural home." (They have taken the first step, and they moved to the country.)

We don't have time *not* to establish a sacred relationship with Nature and her natural materials of clay, wood, stone, water, air, light, and life.

Mud Dancing

The expression "mud dancing" was named by my nephew McKenzie James Wilson at the age of five, when he asked me, "Can I go mud dancing with you?" Now we often use that term to describe our technique of mixing and foot treading cob ingredients. Using a partner, a tarp, a mix, and body movement, students experience the joy of making their own building material. It's fun for both partners to jump on the mix and dance together, but often one will turn the ingredients on the tarp while the other does a solo jig, changing positions when a dancer tires.

Mud dancing is not only fun, good, and free exercise, but also may have positive health benefits including help in losing weight and strengthening bones.

Ianto improved his back through mud dancing. For years Ianto sat on "the donut," as we called it, a small wheelbarrow innertube, to relieve painful sciatica from an old injury. After a building season of mud dancing as a daily ritualized activity, Ianto one day threw away his innertube and no longer needed it.

Mud dancing is a bit like jogging; the dancer/builder breathes fresh air in an outdoor setting, while increasing the heart rate, oxygen, and blood flow. The chemical phenylalanine is released, producing pleasure and euphoria, creating a state of natural high. One day a friend, a natural builder, brilliant inventor, and chocolatier joined us for lunch. While eating his homemade chocolate we discussed how cobbing and eating chocolate have similar euphoric qualities. He commented, "If you can become as addicted to cobbing as you primally are to chocolate, then you are inspired."

The mud dancer benefits therapeutically through movement, sound, meditation, body reflexology, sense of community, and direct contact of bare feet and hands with the earth. Bonding the sand, clay soil, straw, and water under the dancer's treading feet creates a rhythm, marked by the sounds of sticky clay and crunchy sand. In the relative silence of a nonmechanical worksite, a meditational dance forms, fostering an awareness of the union of breath and body movements with the tactile feeling of familiar natural materials.

Dancers can feel with their feet how each material plays its important role of providing either compression, tensile strength, or "stickum" to hold the composite together. The dancer feels and hears the clay particles bonding with the sand and thereby knows when the clay has been worked well enough to add the straw. The use of one's sense of sound and body feeling (proprioception) in this process awakens the body into a kinesthetic manner of learning. A new relationship is born between the building materials, the spirit of the mud dancer, and the place.

There is no substitute for the therapeutic and health-building effects of direct body contact with clay soil, sand, and straw. Treading cob with bare feet and applying cob to the walls with bare hands lets the clay draw out toxins from the body. Barefoot treading is a massage to the feet. By applying pressure to the feet and hands and working reflexological and acupressure points, the entire body can be stimulated, toned, and released from tension.

IF YOU FEEL STUCK IN LIFE,

START MOVING YOUR FEET.

— Anonymous dance ethnologist

We organize times in our workshops for drumming music to be played for the mud dancers. In New Zealand, we had a fiddler. On Lopez Island, Washington, it was a women's marimba band, and more than fifty islanders mud danced, a nice way to channel all that up and down movement into the building of cob cottages. At our Breitenbush Hot Springs workshop "Building the Meditation Cottage," the mud dancers were honored to dance to special guest musicians Tute and

With commitment and determination, mud dancing is possible for almost anyone.

There is some of the same fitness in a man's building his own house that there is in a bird's building it's own nest. Who knows but if men constructed their dwellings with their own hands, and provided food for themselves and families simply and honestly enough, the poetic faculty would be universally developed, as birds universally sing when they are so engaged?

—Henry David Thoreau, "Economy," in *Walden*

Irene Chigamba, Zimbabwe's world-famous mbira players.

The elation of mud dancing is often linked to building relationships and to the community spirit that is created by a group of people having fun together, making new friendships, learning new skills, expressing creativity, and sculpting free-flowing forms. A kind of tribal and primal—and healing—work begins to happen.

COBBERS' STORIES ABOUT DELIGHT AND QUALITY OF LIFE

Many cobbers have stories that tell about the impact cob building has had on improving their health and relationships and on increasing the level of delight in their lives. I would like to share some of these stories with you.

Joan

Joan Murphy was given by a friend the gift of a two-week Cob Cottage Company workshop on Mayne Island in British Columbia. She called us beforehand to make sure that there was electricity on the site so that she could hook up her life-support oxygen machine, and asked if she could bring a chair and just watch. Joan, in her sixties, has a rare terminal lung disease called primary pulmonary hypertension (PPH).

Joan came to the workshop accompanied by her oxygen machine and her caregiver. She was not planning on physically participating in the workshop, but as she watched, her excitement grew and she unhooked from her oxygen machine for a few minutes each day, jumping in and joining the mud dancers. In her ecstatic state, she didn't feel the need to turn on her oxygen one night.

Joan's new love for cob gave her a dream of building herself a cob cottage with her children and grandchildren helping. A goal to live for. Joan said, "The reason I keep myself going on this planet is because I am so excited about cob. I don't know what more I want to do with the rest of my life than promote cob." Her teenaged granddaughter said, "If Grandma dies before her cob cottage is finished, I'll finish it myself for her."

Joan spent "a thousand hours" at her drafting board, designing her cob home and its placement on her son's land. Her children, now grown, married, and with their own children, became excited about her wild dream of building a "mud hut" after seeing a cob cottage. As the family's enthusiasm and their loving support for Joan's project continued to increase, creative and beautiful materials began to appear on-site. Using clay and natural materials improved Joan's health and emotional and spiritual well-being.

I could almost see Joan glowing during a phone conversation when she told me about going out to look for a Mother Mary sculpture for her cottage. The secondhand store attendant said they didn't have one. But above the door, she saw a dusty, dingy Mother Mary figure. She took it home, gave it a cleansing bath, and restored its spirit of purity and beauty. She plans to sculpt a very special altar for the statue in her new home.

Mike

Mike Carter and his wife, Carol, began building their cob and thatch cottage after Mike was diagnosed with melanoma in 1995. They had come to the conclusion that it was time for Mike to start to do some of the things he had always wanted to do with his life, one of which was building his own house. During the cob-building process, while also holding a position as a corporate manager in a semiconductor company, Mike began to suffer many physical effects from lack of sleep and the stress of a high-pressure job in a "vicious corporate environment." Carol and Mike had discussed at length how to simplify their lives so that they could finish their building, which was going very slowly at the time, trying to figure out how to get out of their cycle of high income, expenses, and fundamental insecurity.

Mike decided to quit his job and to start full-time work with his newly formed "Cob Crew." He noticed that he never left the site depressed or stressed out. He said, "A bad day cobbing was better than a good day in a cube under fluorescent lights."

"Building with cob has been one of the most significant experiences of my life," Mike says. "It has conquered depression, making me physically fit and mentally engaged. We have made many new friends and learned much by undertaking this activity. There is a great sense of empowerment in knowing how to build your own shelter—the knowledge that people had so recently had taken from them is returned. There is a long journey still to make, and I want cob to be a big part of it."

Elisheva

Elisheva Rauchwerger, who suffers badly from environmental sensitivities and chronic fatigue, has found cob building to have great therapeutic benefits for her. She says: "Cobbing gives me energy and restores my sense of well-being. Spending my entire day outdoors, in the natural world, breathing fresh air and doing simple, rhythmical workwork that induces a meditative state relieves depression and stress and calms my mind. Performing physical labor promotes deep breathing and sweating, which detoxifies my body and provides me with the opportunity to build my strength and endurance. The exercise and stretching inherent in the building process brings warmth to my body, relieving soreness in my muscles and decreasing my sensitivity to cold.

"My body can recuperate working in a natural environment far from automobiles as well as other sources of pollution—most especially poor-quality indoor air caused by outgasing from toxic building materials. Free from everyday common chemical exposures, I am relieved of headaches, nausea, chronic neck and shoulder pain, and the resultant mental/emotional spillover of irritability, anxiety, and an inability to think clearly.

"I always listen to my body, for there is a delicate balance to be sought. Ever mindful of the ebb and flow of my energy reserves, I must weigh the benefits of exercise and detoxification against the health consequences of overdoing it and adjust accordingly."

GAIL BORST

Mike Carter's cob cottage with African thatch, Austin Texas.

Tricia

Tricia McDowell discovered sculpting a cob grotto in her garden sanctuary to be therapeutic and healing. I call this Cob Art Therapy. She writes,

"Looking back, I see clearly now that, for me, it was the hours I worked alone out there that brought about the deepest healing. Whether I cobbed in the cool of early morning, in the searing heat of the afternoon, by sunset's glow, or by the gentle light of the moon, a clear purpose was emerging. To continually envision this structure as the perfect, harmonious blend of male and female energy, at a time when my marriage seemed to be in grave danger, demanded that I dig deep into my soul for courage and strength. You can't sculpt or articulate what you don't feel in your heart.

"For many years Forrest and I have been evolving the concept of the sanctuary garden by gradually creating our own two-acre ridgetop garden (within a larger twenty-two-acre nature sanctuary) to be a sacred and peaceful place for all creatures. As beautiful as it is, though, we had known that something was missing. We needed a larger, covered sitting area that was protected from the weather. We began to envision a roundish structure with a cedar shake roof, opening wide on one side, with a broad, flowing view down across the garden and out to the towering firs and distant hills beyond. If it hadn't been for serendipitiously meeting Linda Smiley and getting so excited about cob construction, we probably would have gone a more traditional route and built a wooden gazebo.

COB COTTAGE COMPANY

Elisheva and dance partner, mud dancing.

"As it is, we knew next to nothing about the actual cob process, and thus got ourselves into what the poet Rumi would call 'a large foolish project' that far surpassed the intended limit of our time and energy. Lucky for us that the Universe conspired to fool us into such a noble and transforming endeavor with such a skillful and enthusiastic guide as Linda.

"By the time we were all mixing cob and forming the round wall and curving bench, I began to feel better than I had in many months. Over a five-week period, we worked long, hard days—often eight to twelve hours—with very little time off. Sometimes we had lots of help; most of the time we did not.

"I added small, oval, stained-glass windows near the top of the undulating wall: cobalt for the male energy, rose to symbolize the female energy, with marbled, cobalt/rose, heart-shaped windows in between. Below the four heart windows are niched altars to hold various sacred and natural objects. On the outside of the wall are two bas-relief grapevines that intertwine and wind all the way around the structure, each rising up to one large cluster of grapes on either side—the symbolic fruit of our efforts. Within the sculpted grapes, Forrest and I bravely decided to embed the wedding rings we have worn for over fifteen years. We were challenging ourselves and each other to bury the old conflicts and negative attitudes and re-envision our marriage in a new light. The belief that we could in fact attain this higher ground demanded a total leap of faith. Since then, not only has our faith turned out to be justified, it has been richly rewarded beyond all expectations. It is as if, through the sanctuary structure (which we now call "The Grotto"), we have lived into our dream of harmony the way Rilke suggests we "live into our questions." The questions, fears, and doubts have been symbolically laid to rest in the wall.

"In the later stages of construction, I began to mix cob and sculpt wearing a beautiful Balinese dress. The work seemed at once so noble and so sensual that it seemed right to look and feel my best while engaged in it. When it came time to do the final coat—a fine earthen plaster inside and out—Forrest and I worked deftly and harmoniously together, as if we were one being. To make the rough walls smooth and all the edges softly rounded gave us great pleasure. By then we knew that every stage in the creation of this structure was paralleled by accompanying internal shifts. Forrest can tell many stories of his own awakening in this process, as can each person who has come to help us. For when every inch of wall has been trod on by human feet and lovingly sculpted with their hands, think of the hopes and dreams that have been woven into the wall.

"I have been a gardener for over twenty-five years. I have only been a cob artist for a few months. But my understanding of the sacredness of working and playing with earth has deepened incredibly as the Grotto has come into being. I will never be the same, because a new strength of character has emerged in me to equal the enduring strength of the Grotto's rock and cob wall. I hope it will provide a place of sanctuary and solace for others that will endure long after we its makers are gone."

Note: Tricia Clark McDowell is a Master Gardener and writer living in Eugene, Oregon. She is the coauthor with her husband Christopher Forrest McDowell of *The Sanctuary Garden: Creating a Place of Refuge in Your Yard or Garden.*

RELATIONSHIPS AND COB LOVE STORIES

There is something about cob's primal nature, its ability to strip people (metaphorically) naked, its joyful playing in the mud-ness—whatever it is, people begin to fall in love again with themselves, each other, and with life in general. Maybe it is the high clay content of cob that works as a bonding agent in building relationships. Often people keep on cobbing romantically into the wee hours in the moonlight.

A Match Made in Mud: Misha and Elisheva Rauchwerger

It was a match made in mud—apprentice meets workshop participant. Little did they know, three years after running through the forests as forest nymphs, covered in green clay, they would be married in the cob sanctuary where they met.

Elisheva, diagnosed with chronic fatigue syndrome and multiple chemical sensitivities from years as a flight attendant, came to the cob scene looking for nontoxic and natural building alternatives that would be affordable and accessible to a novice owner-builder. What she discovered was all this and more: a material that also has therapeutic potential, as well as the ultimate medium to express her sense of aesthetics and beauty. Ecstatic, that summer Elisheva took six workshops and immersed herself in mud!

Anxious to have a cob house of his own, within months finishing his apprenticeship at

The Cob Cottage Company, Misha set out to buy property in California's Sierra foothills in order to begin building. In nearly three years of planning and building, his cob house went through numerous transformations. First, it was envisioned to be a "huge" nine-hundred-round-foot building, then, more realistically, it was scaled down to a four-hundred-round-foot cottage. During this time, Misha's relationship with Elisheva and her son, Iain, grew; and it became obvious that the tiny one-person cottage just wouldn't do. Once

EARTH, OVENS, ART

BY KIKO DENZER

A COUPLE OF YEARS AGO I spoke to a college professor about continuing my art education. He liked my stone carving, but hesitated about the sculpted earthen ovens I'd been in building in workshops and elementary school environmental programs. He was too polite to ask, "What's it got to do with art? And why ovens?" But he did say his wife would love to have one in their new outdoor kitchen/barbecue area.

His unconsidered response answered his own well-considered question, yet the question rang in my ears. What, as a sculptor, did I have to show for myself? Workshop projects, no money (but plenty of beans and potatoes from my garden), no carving (but a cob studio in progress), and various frustrations and anxieties. As always, however, friends reminded me that faith is the way, and that questions are answers:

Where does beauty come from, if not nature? Who can be an artist without knowing nature? What is knowledge except an understanding, a feel for life, gathered as experience, and expressed, directly or abstractly, sometimes in two dimensions, sometimes in three? Indeed, what is art without earth?

Earth—saturated by rain, warmed by sun, inspired by wind—is the source of beauty. And what is bread, if not the earth we eat? The seed in the soil, brought to life by rain and sun and wind, and transformed again by water, yeast, the baker's hand, and fire. What is communion, but knowledge of God's body, and the blessing of fellowship? Beauty begins by feeding the body whose forms I seek with a chisel in stone, or a pencil on paper. Indeed, what is art without bread? So as I work toward sculpture, I also build earthen ovens, grow a garden, teach, and, when people ask, I call myself a sculptor.

"But your stone carving is better," suggests my professor friend, kindly. But what is the measure of art unless it speaks to people, and they respond? An earthen oven is both practical and beautiful, and costs little or nothing. Bread you can eat; art you cannot.

Ovens may be unsophisticated and simple, but I agree with Brancusi that "sculpture must be lovely to touch, friendly to live with, not only well-made." Sculpture should be an invitation—to sit, to rest, to talk; to cook, to make feasts and festivals.

Making an earthen oven is such an invitation, and its full, round form is indeed "lovely to touch, friendly to live with." But somehow an oven is not sculpture, despite the fact that it is mass, space, volume, form, beauty. A person who might never consider creating "art" will jump at an opportunity to build an oven. What's the difference? You could say that building an oven is craft, not art, but it seems to me that separating art from craft elevates money and status over life and beauty.

Life asks us to participate—not just to watch, not just to learn, but to create; to be whole and wholly involved, rather than apart and alone. You don't have to call yourself an "artist" to engage hands, head, and heart in the genesis of new form and relationship or to celebrate and renew yourself and the world or even just to make a mud oven so you can bake your own bread.

again, the house went into revision mode with the planned addition of a "second skin" consisting of a sunspace on the south, entryway on the east, root cellar on the north, and a bedroom for Iain on the west. But . . . plans change. Halfway through the building process, they decided that their meeting place in British Columbia was to be their future home, which necessitated scaling down the design again to get the house finished in a reasonable time.

The foundation of their two-story cottage now consists of earthen-mortared, indigenous stones, dug from the property and moved by wheelbarrow into place. The red clay soil proved to make a cob so hard it is affectionately known as "Sierra Cement." Foothill pine logs, slated to be removed from the nearby land, were acquired for simply the cost of delivery. They serve as the main posts, curved ridge beam, and lintels. The windows, door, hardwood flooring, and various timbers were salvaged from homes destined for demolition. Although a sod roof was in the original plan, it was decided that a steel roof would go up faster, be less expensive, use far less wood, have less chance of leaking or attracting termites, and still be fireproof, as well as being lighter weight in case of earthquakes. Roof insulation would be sawdust mixed with lime and borax.

The house is part of a Permaculture design with graywater systems, solar electricity, passive solar–heated water, a composting toilet, a bamboo grove, a fruit orchard, a wild perennial garden, and a large pond.

A Mud Artist Soulmate

One fall Oregon day, I was gardening at the Urban Farm at the University of Oregon at a time when I was managing the demonstration garden. A charming, graceful, gnomish man, browned from the sun of Latin America, appeared magically and asked if he could build

MISHA RAUCHWERGER

Misha and Elisheva's wedding at the cob sanctuary at Hollyhock Retreat Center, Cortez, Island, BC.

a demonstration Lorena stove that he had designed in Guatemala. How could I resist? He was offering a demonstration project, he was clearly connected to Nature in a spiritual way, and I was entranced by his natural spirit. Before I knew it, we were taking off our shoes and mud dancing, sculpting clay and sand between our toes, arm in arm, laughing. Then we began sculpting, like a sand castle, a practical cookstove out of the Lorena mix of sand and clay.

This mysterious man was Ianto. I invited him to have lunch with me. He accepted, so I took him out to the garden. I handed him a wooden bowl and a pair of chopsticks and asked him to forage for salad with me. Little did I know that here was the perfect way to win the heart of this plant lover.

I realize now, more than twenty years later, that the significance of our meeting was to share and satisfy our common, basic need to commune with Nature. When we think back to what we both remember about when

Linda and Ianto take part in Laugh Therapy in the Heart house.

we met, for me it was the mud dancing and for Ianto it was our finding rare and exotic plants while picnicking along the riverside on our way to collect goat manure for the garden. Not everyone's idea of romance, but it worked for us!

We are still dancing in the mud together, now creating homes. We built ourselves a home in a garden. This is our social and global way to meet that spiritual need for communing with Nature and helping to improve the quality of life for ourselves and future generations.

Creating and sculpting stoves, ovens, and homes out of the earth and Nature's simplest materials—clay, sand, and straw—is so sensual and satisfying, creating beauty and applying practical solutions to world problems. Our first ten years together were devoted to addressing the firewood shortage and the health concerns caused by cooking on open fires in homes, demonstrating simple cooking technologies, deconsumerism, gardening, and environmental education for children. Ianto worked in North and Central America, Africa, and Asia. Now we work daily together, helping to apply natural materials to people's need for beautiful places in which to live and work.

In earth, I met my soulmates—Ianto and Nature.

APPENDIX 1

Common Errors in Cob Construction

We've noticed from our personal experience a few serious mistakes many first-time builders are likely to make. These Class One errors are often impossible to rectify and can result in the whole project being abandoned or cause broad-scale difficulties later. You'll note that nearly all of these listed have to do with organization, preparation, and planning. More serious errors are made in siting and site preparation than all the construction blunders put together. While most construction errors can be easily remedied, most of the mistakes on this list cannot.

ORGANIZATION

Borrowing Money

Loans lead to stress. They take you away to earn the money to pay them off. Resist borrowing money; build within your current means, then expand if necessary when you can afford it.

Perfectionism

Better roughly right than precisely wrong. Self-builders sometimes get trapped in wanting everything to be perfect, then take so long to move in that they're caught paying two rents, commuting from home to site, having to make more money. At an extreme, tempers are lost, humor is overwhelmed, and marriages founder.

Unrealistic Expectations

If you don't know yourself well, get opinions from old friends: Do you have the determinations and stick-with-it to follow through with a long and complex process? Can you count on help from volunteers, family, paid help? Is your budget adequate?

Distractions from the Building Site

Building takes full attention and is best done in a single sequence. Arrange your schedule so you can be on-site, working continuous work hours. Many building projects are never finished because the owner lost focus, found a new romance, or got tired of being a weekend builder.

Beginning Too Late in the Season

First-time builders need to "think tiny" and complete a small project in one building season. Cob needs time and dry weather to dry. Wet walls can be damaged by frost, and wet straw in walls may putrefy during the winter. Wet earthen floors dry extra slowly. Working in the cold and dark is no fun! Start as early in spring as you can; the work should be half-done well before Lammas (August 1).

Underestimating the Time Budget

Everything takes twice as long and costs three times as much as you first think. Plan for contingencies, how and when the building will be enclosed and therefore heatable, and who will do the work.

THE SITE

Poor Siting

Frost hollows, north- or west-facing slopes, floodplains, sites with no long views—all are impossible to rectify. And a site uphill from the nearest delivery access can slow you down.

Insufficient Solar Access

Make sure your building gets as much sun as possible in the cool months.

Lack of Research on Neighborhood Changes

Surrounding land-use can change a place catastrophically (a new trailer park next door, freeway construction, shopping mall, forest clear-cuts . . .). Also, unsupportive neighbors can derail an experimental building project, particularly one without official permits.

DESIGN

Inadequate Design Exploration

Make a model. Invest in modeling clay or use cob, sticks, and so forth. Make sure of your plans for circulation, floor levels, roof design, doors, and major openings before you begin digging, when you will be committed to a foundation plan.

Too Big

Design your first building to be tiny to make sure you can finish the first season. You can add on later. A high proportion of first-time, amateur builders start too big and give up before the building is completed.

Thin Walls too Straight

Cob needs lateral buttressing, either by building walls thick, building walls to be curved, attaching to adjoining walls, or with structural buttresses. Build thin walls with curves and straight walls thicker.

Roof Spans too Long

If you don't want to truss your roof structure, keep spans short: 10 to 12 feet is long enough, unsupported. Remember that the stress on a horizontal beam is proportional to the square of its length.

SITE PREPARATION

Poor Drainage

Realistically project yourself into the worst possible, longest, hardest rainstorm and the highest flood. Make sure that water can flow away from the building fast enough, that surface water can't reach the walls, and that splashback will not soak the base of the walls.

Floor Levels Not Worked Out Initially

Install permanent datum stakes! Lost or damaged stakes create headaches. They should be numerous and well marked for each interior floor level, at thresholds and outside, against the building. Paint them brightly, drive them so deep nobody can get them out or accidentally drive them in farther, and put them all on a plan you won't lose.

Insufficient Protection from Weather

If you build a roof first, either temporary or permanent, then you can work unhindered by rain and sun. Make sure the roof is large enough to keep the whole site and materials dry and high enough to be out of the way of workers.

Heavy Machinery Does Damage

Mechanized equipment wrecks the site, compacts soils, damages ecology. Keep such machinery under constant supervision. Better yet, keep it off the site altogether.

Inadequate Foundation

Foundations are critical and nearly impossible to replace or add later. Be sure your foundation is not too low, discontinuous, or lacking reinforcement and that the footing trench is sufficiently compacted.

Restaking the Building Outline Bigger

When novice builders see how tiny the building seems at the layout stage, they sometimes are tempted to restake it. Resist this impulse.

MATERIALS AND TOOLS

"Clay" Turns Out to Be Silt

Make sure in advance that you have enough clay in your soil, or a clay source nearby. Don't take anyone's word. Test it, make cob bricks, dry them, load-test them, and scratch for hardness. It must be sticky enough.

Running Out of Materials

Stockpile in advance (for months, even years) all the glass, lumber, and beams you'll need. Particularly during workshops, building happens fast. Have windows and door frames on hand. Know where you will go for more soil, sand, and straw should you run out. Note in advance public holidays, when suppliers will be closed.

Running Out of Water

Store enough water for several days' work in barrels or tanks in case your well runs dry or your electricity fails, as this can set your building schedule back unexpectedly.

Concrete Stucco

Use lime-sand or earthen plaster. Portland cement can destroy a cob building.

Inadequate Tools

Don't "economize" to save a couple hundred bucks by buying mediocre tools and ruin a house worth thousands of dollars. Don't allow yourself to get frustrated and damage tools. Store your tools every night, clean and dry.

Straw Bales Got Wet

Bales are really easy to get wet and almost impossible to dry out. Make sure they are dry all the way through when you buy them. Then store them off the ground (two layers of pallets may work) and under a completely waterproof roof. Blue woven tarps nearly always leak, even new. Spikey straw pokes through polyethylene, unless it's up on a steep-pitched wooden framework above the bales. If possible, borrow a barn.

APPENDIX 2

Codes and Permits

In January 1998, for the first time, an opportunity was arranged for introducing a sustainable context to the building codes. The Planning Summit for Sustainable Building Codes was a three-day gathering of about seventy building officials and inspectors, code writers, architects, natural builders, environmental activists, lobbyists, and writers, held near Los Angeles. The summit was convened around this provocative statement:

> If building codes are intended to protect the health and safety of people from the built environment, and if the codes actually jeopardize the health and safety of everyone by encouraging the destruction of the natural systems upon which we all depend for our survival, then we are obligated to re-invent the codes from this larger perspective.
>
> —David Eisenberg, from the conference papers

Speakers at the summit noted that: (1) buildings in the United States use in their construction and maintenance 10 percent of all the energy used on Earth. (2) Globally, 40 percent of all energy use is in buildings. (3) In the United States, 75 percent of lumber production goes into buildings. (4) To be sustainable, the United States should generate no more than 1.7 pounds of carbon per person per day. At present we generate 30 pounds. (5) The industrialized countries use thirty times as much resources per person as do the nonindustrialized countries.

If, through ecologically impelled codes, the United States could drop its building energy use by 20 percent, this would cut energy use/resource destruction by 2 percent worldwide, more than the total resource budget of China and India combined. But would 20 percent be easy in a society where a 6,000-square-foot concrete mansions are being nationally paraded as environmental architecture?

At The Cob Cottage Company, we answer the phone dozens of times some days. Increasingly, the question most asked is a variation of "How do you deal with the building codes?" In general, we advise people building for themselves not to approach the government at all. There are a number of ways you can legally build with cob without involving officialdom.

As of the year 2000, there were no codes for cob building in the United States, Canada, the U.K., or South Africa. Don't confuse codes with building permits, though. You may choose to buy a building or residency permit, even in the absence of codes that would make approval of a cob design easy. Be very cautious; permits and inspections could cost a lot of work and money.

Building permits are a device for taxing new construction and regulating how construction is done. With them come permits for sewer connections, electrical wiring, planning and zoning, heating stoves, and so on, each with its own fees, inspections, and regulations. Often, application for any one of these will implicate you in the others.

Codes are pre-agreed recipes for how structures may be built, to lessen the legal liability on inspectors and bureaucrats and to reduce their need to make qualitative decisions. With a code for cob building, individual inspectors would have less responsibility to use their own powers of discrimination. Codes and standards, though they may be administratively convenient, strip us all of our creativity and reduce our ability to judge the significance of our choices. They inhibit or suppress the use of common sense. By setting fixed standards, codes demand and sanction mediocrity. They create average conditions, which constrain us in responding to the uniqueness of our situation. If codes for cob construction existed, it would be easier to buy building permits but more difficult to respond to the specific needs of site, builder, and occupant. The requirement to build in standardized ways is antithetical to the human spirit and nearly always imposes greater loads on the global ecology.

Sadly, many potential builders are clearly bothered by the thought of dire consequences. They want to know "What will happen if I don't have a permit?" and in fact we're told often "But you *need* a permit" as if there were no options. It seems ironic that in this country which separated itself from England in reaction against pettifogging bureaucrats, most of us are, by subscribing to governmental interference in our domestic process, intimidated into not providing our own healthy housing. We who can afford to do so pay outrageous rents or feed the banking industry for thirty years. When we can't, we move under a bridge or sleep in abandoned cars. It might appear that codes and permits were set up to keep the homeless homeless; if they can't afford a mortgage then they should be prevented at all cost from building their own shelter.

Some effects of building codes are: (1) Standardization of buildings, components of buildings, and dimensions of buildings, making response to local conditions impossible. Local climate, skills, and traditions, as well as site ecology and the particular needs of occupants, all become secondary to creating replicable, measurable construction. (2) Minimum rather than maximum dimensions and standards, which create unnecessarily large and expensive buildings. As a result, the economic and ecological costs of new construction and lifetime maintenance, heating, and cooling are unnecessarily high. (3) A disempowered population, who come to believe they need to leave creation of their most intimate habitat to building professionals, contractors, and architects, none of whom feels able to truly put the client's best interests foremost. (4) A cowed populace, in fear of public officials, for whom they pay four times—in taxes to maintain a bureaucracy; in fees, fines, and wasted effort; in the costs of having a building that doesn't suit them; and in compromised opportunities for their descendants whose patrimony we are all squandering.

In the short term, these four effects are annoyances and may in part be overcome. However, the longer-term results must be disastrous—the effect on global ecology and our grandchildren of such prodigal use of resources, raw materials, and energy. The ecological consequences of an "average new American home" at 2,200 square feet, with 13,000 board feet of lumber, 60 cubic yards of concrete, and 60 gallons of paint are most profound. In subscribing to building codes as they exist, we are assuring for our grandchildren global ecological wastelands and economic collapse.

Personal tragedies and the irritations of bureaucracy are trivial in comparison with these broader global impacts of building codes, many of which threaten the survival of our descendants by their insistence on gross consumption of resources. It is all the more ironic that this insistence flies the banner of health, safety, and energy savings. Indeed, to a cynic it might seem that the purpose of building codes is to stimulate profits for corporations, with an economic byproduct being the giant industry of code officials, building inspectors, materials labs, architects, engineers, plans examiners, bank loans, courts, and hearings. Codes and permits perpetuate dependence on professionals who demand the kind of buildings from which they, not the resident, benefit.

YOUR PERSONAL OPTIONS

What are your options, in terms of codes and permits? How would you be able to build the kind of house you know is possible without compromising quality for quantity? It's easiest to build in an area with a reputation of having supportive building officials, but you may not have the freedom to move.

Let's examine a range of options.

No Permits Required

Every country runs its building regulations differently. In some, there is a national standard. The United States and Canada have several thousand separate municipalities, regional districts, counties, townships, cities, states, and provinces, each with its own rules. Control by centralized government tightens over time, but many rural parts of the United States and Canada still have a relaxed attitude about regulation of buildings. Some areas require no permits at all; some want you to buy a permit but don't inspect your building. Check locally.

Inconspicuous Construction

In general, local government doesn't like to be confrontational. Building inspectors don't cruise the back roads, trolling for nonpermitted buildings. Employed officials are usually very busy and wary of barging into remote situations where they could encounter opposition. Quiet construction in rural areas seldom attracts official attention, but be sure that your neighbors don't turn you in. If possible build where neighbors can't see you, where the tax assessor wouldn't normally go, and where you are not obvious from public roads. It's much easier to build an alternative, nonpermitted building if a permitted, taxed structure already exists on the lot.

Quiet Accretive Building

In rural America, a standard strategy for developing a house is to begin with an old mobile home on wheels. Some areas require no permit for a trailer, or make the requirements much simpler than for building a "house." Over time, a roof gets built over the trailer, "to keep the sun off," then a shed will appear alongside. Over several years it's hard to notice the difference, but slowly more building happens, culminating with enclosure of the entire structure, then the quiet pulling out of the original mobile home, which then gets sold to a neighbor who repeats the process.

Minimum Footage

Depending upon how large a structure you want, how painstaking the inspector is with a measuring tape, and the scale of the local minimum, building tiny can be a useful strategy. Building departments generally have a minimum size limit below which no permit is required. If you are likely to be inspected, check carefully what the rule book says. Do your local building officials specify interior or exterior dimensions? Do they say 10 feet by 12 feet or 120 square feet? In some areas the dimension is defined by the roof area, as seen from outer space. Check with someone in the office, and take careful verbatim notes of what they say, in case there's later a dispute. If you already know your building inspectors, ask how they will interpret the rules. We have heard of interesting houses built using a series of separate 100-square-foot modules, each a separate room.

Minimum footage can be as low as 100 square feet, but more often is 120 or 200 square feet. In some areas, with some types of buildings, the minimum can be up to 1,000 square feet. The International Building Code (to which most local authorities in the western United States subscribe) states plainly that a "shed or playhouse" under 120 square feet in *interior* dimensions is exempt from permits. You might need to draw attention to this provision.

Permits Not Needed for Some Building Types

In many rural areas agricultural buildings need no permit. Check what constitutes an agricultural building. In some places such a structure may not have a water line or power connection, in others it must be on an approved "farm." Some agricultural buildings may include full kitchens and heating.

Nonresidential Permits

Often outbuildings have much lower permit fees. In some places you could buy a used mobile home, pay for a cheap permit for it as your "residence," and build a cob "shed," on which there is no regulation, alongside. Live in the cob and keep your chickens in the mobile home. More often, people will build a barn (permitted) with a perfectly legal upper floor (which can be used as an apartment).

Permitted Post-and-Beam

Some building officials are comfortable permitting cob buildings so long as there is a post-and-beam structure to carry the roof; others will demand assurance from an engineer or architect that it is structurally sound. Yet building post-and-beam with cob "infill" reduces the lumber savings achieved by structural cob. Further, the posts can be a physical obstruction to constructing the cob walls, and might actually weaken them. Sometimes officials will be leery of even a cob infill. One enterprising Australian architect slid plans through with the cob infill labeled "Stabilized Mineral Aggregate." His local council were sleeping or else, being unfamiliar with a technical term, out of embarrassment didn't ask.

Owner-Builder Permits

In some administrations, a special permit exempts owner-builders from some provisions of the code. Classic examples are in the Humboldt, Nevada, and Mendocino Counties in northern California, where an enraged populace, losing patience with government interference in their lives, forced changes in regulations twenty years ago. Given that codes are supposed to be for health and safety, it makes sense to regulate building for profit but go light on people creating their own homes. Enquire into owner-builder exemptions in your own area, and hunt through the local regulations carefully; your building department may not advertise such a provision.

Experimental-Buildings Permits

You need to be patient for this strategy, because it will entail a lot of paperwork and keep you under the scrutiny of the officials. Sometimes there is a restriction that nobody can live in an experimental house. This makes nonsense of the idea of testing to see how well the structure performs, and could make your eventual occupancy a little dicey.

Full Permit

Last resort. Expensive! Irritating! Be prepared for construction delays, nonsensical regulations, interference, and many more obstacles that yield no benefits. Realize also that you will need to submit detailed plans. When they are "approved," you have bought into whatever codes apply even to the most trivial details, and any changes you make to improve the building thereafter will need additional approval, probably with an extra fee. Expect also that permit fees can be extravagant: $2,000 to $5,000 in permits and inspections is not uncommon. The inspector may also demand lab tests for compaction, shear, and modulus of elasticity, which will add to your costs.

"Alternate" Methods and Materials

We know of no laws that *prohibit* earthen construction anywhere in North America, and most building regulations do include provisions for "alternate *[sic]* methods and materials." For instance section R-108.1 of the Council of American Building Officials (CABO) One and Two Family Dwelling Code states: "The provisions of this code are not intended to limit the appropriate use of materials . . . or methods of design or construction not specifically prescribed by this code, provided the building official determines that the proposed alternate *[sic]* materials . . . or methods of design or construction are at least equivalent . . . in suitability, quality, strength, effectiveness, fire resistance, durability, dimensional stability, safety and sanitation." The interpretation of this section invests a great deal of authority in individual officials. Several inspectors have told us that they eagerly await the opportunity to approve cob buildings within their jurisdictions.

Natural Renovations

Unless your place of choice to build has a building department with a reputation for supporting sustainable building practices, you could choose to naturally renovate your existing home while you explore areas with supportive officials. Natural renovations could include adding interior sculptured cob walls for thermal mass, with alcoves, niches, and door arches. Also, consider a cob addition to an existing building. In many places minor extensions are allowed without permits. Check locally.

EARTHEN BUILDINGS AND REGULATIONS

An obstacle to official acceptance of cob is the wide variability of soils from which it is made. The great strength of Oregon cob depends on customized mixes based on a careful analysis of the available soil at a site. The mix proportions depend primarily on the amount of clay in the soil, but also on the stickiness and expansiveness of that particular clay and on the ratio of different particle sizes of silt and sand. These conditions are simple to determine in the field but are hard to quantify and therefore hard to regulate.

Gradually permits are being issued for cob, both as infill and as structural walls. The regulatory situation is changing as building officials and inspectors become familiar with the technique. Many of them are very reasonable people, though constrained by the regulations under which they must work. Individual building inspectors and officials, hamstrung by their positions, enthuse unofficially about natural building in general. The Cob Cottage Company's office has had many requests from officials who want to be better informed and educated about new techniques. Several officials have confided privately that they plan to try cob building themselves.

Your local officials may be constrained by the rules they are required to dispense, yet be personally helpful to you. Take time to get to know them, be patient and understanding, and help them to help you to get what you want. Take in photos of existing buildings, and lend them books, videos, photographs. Find civil or structural engineers and architects who will be references. Best of all, you could begin by building something small (e.g., a tiny shed, a garden wall), then invite local building oficials in to inspect it when it's finished. You could, for instance, build a beautiful and complex series of outdoor enclosures using cob courtyard walls, which, being "fences," are rarely regulated at all. Walls of this type are an easy way to provide demonstrations of cob's strength, durability, and fire resistance to uninformed officials.

The history of regulating earthen buildings is interesting and revealing. In many countries a rapid expansion of the brick industry in the 19th century coincided with laws requiring new construction to be in brick. In the western United States, the lumber industry had a part in developing codes and regulations defining official recognition of building techniques. Nowhere was there a strong enough lobby to recognize earthen construction, so either by default or collusion, earth building was squeezed out of the regulatory process. Powerful industrial lobbies can't be expected to support any system of building that fails to advance their profits.

In Britain, the first law restricting building materials was the London Building Act of 1667, created in reaction to the great London fire of 1666. Since cob was not used in London because of the unsuitability of the soils, it was not mentioned as a material having suitable fire resistance. When the first National Building Act was proposed in 1842, it was based on bylaws from London and Bristol, where cob was almost unknown. Consequently, cob was no longer legally permissible even in regions where it had been safely used for many centuries.

In 1985, after 135 years of regulatory purgatory, cob was once again permitted in Britain by a new clause in the building laws stating that "any material which can be shown by experience, such as a building in use, to be capable of performing the function for which it is intended" is satisfactory. However, lacking definitive information on the proper construction of cob, regulatory authorities are often reluctant to approve its use. For many years, most "new" cob construction in England has been either nonresidential outbuildings or officially categorized as "restoration" projects. For example, to get local Building Council approval as recently as 1994, Kevin McCabe set his cob house on the foundation of an old stone barn, with two original walls left standing, although in 2002 he completed a new, 3,000-square-foot, 3-story house with a 3-foot-thick cob walls.

In North America, citizen groups and individuals need to work for immediate relaxation of regulations pertaining to owner-builders. In enlightened jurisdictions, the barriers to using more ecologically friendly local materials are gradually lowering, but what is needed is not so much less obstruction as positive encouragement and incentives for us to provide our own houses at a reasonable ecological and financial cost.

APPENDIX 3

Cob and Water

IN THE FALL OF 1994 LINDA AND IANTO went to Australia and New Zealand, to teach the first ever Oregon Cob workshops there and to research pioneer cob homes. We were gone three months, through one of the wettest periods in Oregon history, and returned for Linda's birthday, Valentine's Day 1995. Our tiny cob cottage, built on a marsh in a rain forest, had suffered more than a yard of rain in the time we were absent, and locals said the sun had hardly shown his face. Home at last! We opened the door, went in. Sniff sniff . . . "Must have been someone staying here." "Yup, it smells lived in." Well, wrong . . . nobody had been there. The stove had never been lit. But the house had that dry coziness of a well-lived-in home. Not a suggestion of damp, even in those forgotten corners under the sink and behind the cooker, where in many houses mustiness and mold accumulate. Another victory for passive solar earthen buildings.

Experiences like ours speak volumes against a prejudice commonly held in the United States that earthen buildings, especially in rainy climates, must be dirty, dismal, and dank. It simply isn't true.

Another easily debunked myth is that any exposure to moisture will cause the immediate collapse of an earthen building. On the contrary, recent testing of English cob buildings shows a "normal" moisture content of around 4 to 6 percent by weight, probably due to water retention by natural salts in the cob. Moisture contents of up to 26 percent have been measured without obvious structural problems. Ancient structures all over the world, including 600-year-old cob cottages in Britain and parts of the Great Wall of China that are well over 2,000 years old, attest to the longevity of earthen building, even in very rainy climates. Ted Howard, in *Mud and Man,* tells of an earthen house in New South Wales, Australia, that survived complete flooding. "It must have been an exceptionally well constructed building, to judge by one report from an old local resident. She saw the water flowing through the building in the flood of 1945, coursing through the doors and windows. The walls do not seem to have been affected in the slightest, even after the passage of forty years."

Nonetheless, water under the wrong conditions should be considered a serious threat, perhaps the most significant danger of all, to cob and other unbaked earth buildings. This appendix describes the conditions under which water can be harmful and then elaborates on some common strategies for preventing water damage, some of which are recommended while some emphatically are not.

CAUSES OF WATER DAMAGE

There are three quite different ways in which water can cause structural damage to a cob wall. One is by erosion, the physical wearing away of the wall by water running down its surface. The second is spalling of the surface when driven rain is followed by a freezing night.

The third is less obvious but more serious and is insidious because it is harder to detect. When cob walls become completely saturated, they lose their strength and can collapse under their own weight. This is a sadly frequent occurrence with historical cob buildings in England and New Zealand, considering that prevention is fairly simple. The causes are usually traceable to a combination of improperly maintained roofs, inadequate foundations or drainage, and the application of non-breathing cement plasters. Flooding and leaky pipes are less often the cause, but could be equally damaging.

Erosion

One of the most frequently asked questions about cob is, "Why doesn't it wash away in the rain?" The answer in most cases is quite simple: you don't let it get very wet. In rainy climates, we tend to build large roof overhangs that protect the wall from water. Sometimes, on sites where there is a great deal of wind-driven rain, we protect the outside with weather-resistant, lime-sand plaster. This is the strategy most often used by traditional cob builders in Britain.

Surprisingly, even unplastered cob shows remarkable resistance to weathering. A study by the Devon Earth Builders Association estimates that cob walls with roofs but no plaster (a common condition for barns and other outbuildings) tend to erode at a rate of approximately an inch per century. Given that British cob walls are usually between 24 and 36 inches thick, this causes little concern. The weather resistance of unplastered cob seems to be due to two factors. First, kneading and compacting the wet clay during mixing and building causes a process called "colloidal cementation," which renders the cob relatively impermeable. Furthermore, the projecting straws and large aggregate in roughly trimmed cob slow the movement of water down the walls and encourage it to drip off rather than channeling itself into erosive rivulets. Cob appears to hold up better than adobe blocks, which have a smoother texture and vertical mortar joints that can collect and concentrate runoff.

This is obviously a very climate-sensitive issue. In dry climates such as the Middle East, the Sahel, and the American Southwest, earthen buildings with earthen roofs are common. Some roofs are flat, designed to collect scarce rainwater and channel it into cisterns. Some are domed, usually providing no protection from rain to the walls below. Normally neither domed nor flat roofs have eaves to protect the walls. Such buildings survive torrential downpours with no more than surface damage, but sometimes require a fresh coat of earthen plaster after every rainy season. Unprotected earthen buildings are only found in places where the air is dry enough that the walls don't stay soaked for long after a rainstorm.

We have experimented by leaving small cob structures unprotected from the western Oregon rains. After the first winter, we found that although very little erosion loss had occurred, the straw inside the wall had begun to rot. We inferred that this could be a significant cause of failure in cob structures in rainy climates (if not adequately protected from soaking), because with the straw gone the material loses both its tensile and shear strength, and its protection from erosion. In subsequent winters we found that erosion indeed increased, and without protection, walls could deteriorate rapidly.

Freezing

Exposed earthen walls seem to erode much faster in climates that experience rapid and extreme temperature variations. For example, adobe walls in parts of New Mexico (where winter air temperatures frequently vary by 40 degrees or more between day and night) may lose up to an inch in twenty years from their vertical surfaces, about five times the erosion rate of English cob. This is due to a phenomenon known as "spalling," where water that has penetrated the wall freezes, expands, and pushes out chips of the surface. In England, cob buildings were often left unplastered only on the north and east (leeward) sides, away from driven rain and where the daily temperature change was least. Frost damage from spalling was therefore minimal.

It is inadvisable (as well as not much fun) to build with wet cob if a deep freeze is expected before the walls can dry somewhat. We have seen fresh frozen cob inflate like puff-pastry, then later crumble. This is a constraint in northern climates where the frost-free season is short. If hard frost is expected within a few days of building, some protection would be recommended.

Saturation

The most catastrophic failures in cob buildings are usually the result of wall saturation. When a certain level of moisture is reached, the clay in the material becomes plastic and the wall loses its load-bearing capacity. The critical moisture level probably varies widely, according to soil type, sand-to-clay ratio, wall height and weight, and other factors. In extreme conditions, an earthen wall can slump, deform, or collapse. Saturation can be caused by a number of situations, including leaking roofs, rising damp, and floods.

In his paper, "Common Structural Defects and Failures in Cob Buildings and their Diagnosis and Repair" (*Out of Earth II*, 1995), Barry Honeysett writes, "The base of the cob wall, just above the stone plinth, is probably the area of the wall most at risk. As well as the most highly stressed part of the cob, it is the most vulnerable from the effects of moisture. Dampness rising from the ground can be trapped and added to by rain penetration through cracked render, building up to a level at which the cob loses its strength and can eventually fail. The mode of collapse is most likely to be shear failure, with the bottom of the sound cob kicking outward."

Leaky Roofs

An obvious way for water to get into a cob wall is through a leak in the roof. Over time, even a small leak can get the top of the wall quite wet. This can go unnoticed for years, particularly if the surrounding area is stuccoed, under an eave, or in the dark of an attic. With a larger leak or in an especially rainy winter the cob can get soaked deep down. So long as a house is inhabited, roof leaks are unlikely to affect the cob's structure, because wall plaster will show damage, and detection will lead to fixing the problem before collapse of the wall.

Uninhabited buildings are much more at risk. Unless they are passive solar buildings, they lack an internal source of heat to dry out a damaged wall. The first warning to an absentee owner can be slump and collapse of part of the wall. Thinking the situation has gone too far, the owner may give up, allowing the roof to collapse also. Within a few decades, the cob will gradually wash away.

In northern Wales where Ianto grew up, a few humble cob cottages still survive on the windswept stormy Llyn Peninsula, jutting out into the Irish Sea. They were noted and photographed by Margaret Griffith in the 1970s, by which time most were uninhabited and in poor condition. Ms. Griffith took Ianto and Linda back there in December 1997. In twenty years most of the roofs had collapsed or been raided for their slate tiles. Some buildings were nothing more than a grass-covered pile of earth, in others the walls were still substantially intact though deteriorating fast. A sad situation. Wales, noted for its ancient literature and tradition of song, never has valued its native architecture, instead allowing a patrimony of remarkable and unique buildings to be lost for the sake of a few roof tiles.

Rising Damp and Splashback

Tests on old English cob buildings indicate that the moisture content is often highest at the bottom of the wall. Causes of so-called rising damp can be an inadequate foundation or rain splash from the roof.

When earthen walls are built with inadequate foundations, moisture can travel up into the bottom of the wall through capillary action, potentially causing saturation and failure. If your foundation is made of concrete, brick, or porous stone, consider using a damp-proof course of water-repellent cement, tar, or other bituminous material between the foundation and the base of the earthen wall. More importantly, there should be good drainage beneath and/or just outside the foundation perimeter (see more about drainage in chapter 10). Even with good foundations, earthen buildings sometimes suffer from rising damp when wet soil and debris build up high enough to come in contact with the earthen wall. This happens particularly on sloped barnyard sites where animals cause material to slide downhill against the building. The problem can be avoided by grading the site so that it slopes down away from the building in all directions and by occasional maintenance over the decades and centuries.

Another cause of excessive moisture at the bottom of a wall is a roof with little overhang and no effective gutter, so that rain pouring off the roof is repeatedly blown against the lower wall. In New Zealand, we saw erosion patterns that reflected perfectly the shape of the roof above, with on one occasion about 8 inches of a 14-inch-thick wall already missing, where water running off a corrugated steel roof had splashed against the bottom of the wall. Splash back from roof runoff can reach up to about a foot and a half, so the foundation stemwall needs to be at least knee-high. Longer eaves help a lot but splash can spread sideways a long way, particularly

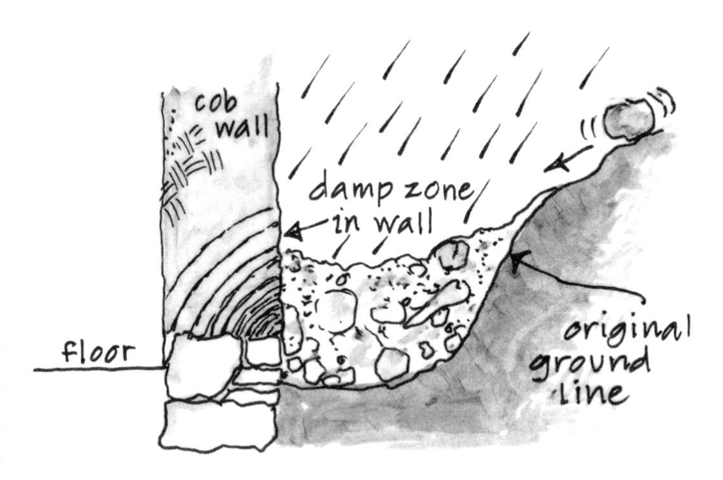

Persistent dampness in cob walls can result in loss of structural integrity and eventual collapse. Be careful.

if there is wind. Broad gutters are the most effective preventive measure, but be sure to inspect them and their downspouts seasonally for leaks and clogs.

Flooding

Despite the story from Australia cited at the beginning of this appendix, flooding to above the level of the foundation stemwall is a serious proposition. The building may survive without irreversible damage, but it also may not. Best avoid the possibility by siting your building well out of a flood plain. If there is ever a chance of standing water in the area, build a high, impermeable foundation.

Despite our protestations, one client in Buda, Texas, insisted on siting a cob cottage in a flood plain. Weeks after the building was complete, nearby Onion Creek rose to its 100-year flood level, missing the cottage by only a few vertical inches. The next monsoon season wasn't so obliging, as the river rose to its 1,000-year flood level, filling the building to a depth of five feet for about 2 days. The force of the water was sufficient to push in the door complete with frame. When the flood waters receded, part of the building was still standing, though badly damaged. It was deemed beyond repair and had to be demolished for safety reasons.

Another potential source of problems is leaky water pipes. Be extremely careful when burying pipes inside cob walls. Try to avoid buried junctions, and in serious frost areas keep water pipes just beneath the *interior* surface of the cob, well above floor height where the walls are usually coldest. Even if the pipes aren't buried in cob, an unattended burst pipe can create havoc. In the three weeks you're gone skiing, that frozen spigot can pour thousands of gallons into the kitchen, soaking the bottom of your walls. Build the foundation stemwall at least six inches above finished floor level to prevent soaking by accidental indoor flooding. Slope the floor gently towards a floor drain or doorway leading out of the building so that flood water from any source can leave the building easily.

PREVENTING WATER DAMAGE

From the discussion above, it is clear that the parts of an earthen wall most susceptible to moisture intrusion and damage are the very top and the bottom. The old English proverb states, "Give a cob house a good hat and a good pair of boots and she'll last forever." Indeed, a well-maintained roof with wide eaves and well-maintained gutters, combined with a high, impermeable foundation and good perimeter drainage, will protect most cob buildings indefinitely. Additional vulnerable points include windowsills, door reveals, and bonds between earth and other materials such as wood. Careful detailing should be used in these places to prevent water intrusion.

This level of commonsense design is extremely effective and generally sufficient. However, there are other approaches that are more complicated or questionable, namely industrial stabilizers and plasters.

Industrial Stabilizers

As David Easton points out in *The Rammed Earth House,* the simplest form of "stabilization" (defined as "the elimination of change") is to find the right proportions of clay and sand, so that the wall will not expand or contract much as it absorbs and discharges water. In the modern earth building industry, however, it has become habitual to use industrial products to make the mix more resistant to moisture. The most commonly used stabilizers in adobe and rammed earth are asphalt emulsion and Portland cement, both of which reduce absorption of water.

There has been a long and heated debate about the relative costs and benefits of industrial stabilizers. On the one hand, when properly used, they can greatly improve earth's resistance to water damage. On the other, they increase the complication, expense, toxicity, embodied energy, and environmental impact of earthen building. They can complicate building maintenance and remodeling and create a disposal problem where there was none before. Paul Graham McHenry Jr. writes in *Adobe and Rammed Earth Buildings: Design and Construction,* "Unless special conditions make waterproofing necessary, it would seem that stabilization is a costly procedure, with little benefit gained."

In our opinion, industrial stabilizers change earth into something that is not earth. They convert earthen building from a vernacular, easily accessible, and environmentally benign technology into an industrial process similar to any other. To date, we don't know of any cob builders, either in the United States or elsewhere, who have chosen to add industrial stabilizers to their mix.

Unfortunately, most building codes in the United States, where they acknowledge earthen building at all, require the addition of industrial stabilizers. Even in the heart of adobe country, where natural earthen buildings such as Taos pueblo survive in good condition after hundreds of years and more, New Mexico's code requires that asphalt emulsion or cement be added to adobe blocks, or that the walls be covered with cement stucco. When and if cob building codes are adopted in the United States, we fervently hope that the inevitable pressure from the building materials industry to mandate the addition of unnatural stabilizers can be resisted.

Stucco

Traditional earthen buildings are usually stuccoed, both for aesthetic reasons and for protection against water and abrasion. The two most common traditional kinds are made either from earth or from hydrated lime with sand or earth. Sometimes horse or cow dung is added. Both earth and lime bond well to earthen walls, and both are permeable to water vapor. This is important because it allows any moisture absorbed by the walls to escape harmlessly through evaporation. Lime and earthen plasters are said to "breathe."

Furthermore, as Nader Khalili points out in *Ceramic Houses and Earth Architecture*, clay's expansive nature makes earth plasters self-sealing; "The clay-earth absorbs water very slowly; and once it is wet, it does not let any more water pass through." But because earth plasters are made of the same basic ingredients as earthen walls (clay, straw, sand, or often dung, with various additives) they are subject to the same kinds of water damage described above, particularly erosion and freezing. One technique for reducing water damage is to make the surface extremely smooth, causing rain to run off quickly and evenly. A seemingly contradictory approach is to use lots of straw in the plaster and apply it in such a way that the straw sticks out and sheds water almost like thatch. A shaggy straw-clay stucco (without added sand) on an adobe home we know near Taos, New Mexico, withstood seven "monsoon" rainy seasons before showing significant wear, although it has no roof overhang to protect it from rain or wind. In any case, earthen stuccos will need periodic replacement.

During this century, it has become common practice to "protect" earthen buildings, both historical and newly constructed, by applying Portland cement stucco. The theory is that an impervious cement skin will prevent erosion of the wall and require less frequent maintenance than earth or lime. Unfortunately, in practice, cement stucco usually makes water problems worse.

Cement bonds poorly to earth. To prevent the thick application of up to three coats of cement stucco from falling off an earthen wall, it is standard to plaster over chicken wire fixed to the wall at frequent intervals with long nails. Cement stucco is brittle and prone to cracking, either as a result of the

Cob house collapse in Devon probably caused by cement-based stucco.

building settling or from the different rates of expansion and contraction between earth, cement, and the metal nails and mesh. These cracks allow rain to penetrate the wall. There are many other ways water can get into an earthen wall, including water vapor generated inside the house and absorbed by the wall. Unable to evaporate through the cement stucco, this moisture accumulates over time, saturating and weakening the wall, especially at the point where earth meets cement. The wet earth turns into mud and flows, leaving invisible cavities behind the plaster.

There are many dramatic examples of cement stucco causing serious damage to historical structures. The St. Francis Church in Taos, New Mexico, built in 1815 with massive adobe walls and thick buttresses, was plastered with cement stucco in 1967. In 1978, water trapped by the cement was found to have corroded the wall to a depth up to two feet. Much of the church had to be rebuilt and traditional earth plastering has resumed. Similar stories may be told about centuries-old cob buildings in the British Isles, which have suffered severe water damage after being stuccoed with cement or having nonbreathable paints or wallpaper applied to interior surfaces.

Historical buildings recently stuccoed with cement are probably at greater risk than new earthen structures with modern foundations that are plastered with cement at the time of construction. Nonetheless the nonbreathable nature of cement presents a potential hazard that can be avoided by plastering with earth or lime. Earthen plasters are sometimes stabilized by the addition of Portland cement or asphalt emulsion, with a resulting permeability somewhere between that of pure earth and pure cement.

Natural Stabilizers

There is a growing network of earth builders committed to reviving traditional plastering techniques and minimizing the use of industrial products. One of the most promising lines of experimentation involves natural additives that increase the water resistance or hardness of earthen plasters without unduly compromising "breathability." These natural stabilizers are far too many and varied to list here in full, but include natural glues such as casein; plant latexes such as prickly pear cactus juice, wheat flour paste, acacia sap, and boiled banana stem; proteins such as milk and animal blood; and plant oils including linseed, kapok, and hempseed. In Africa and elsewhere, one of the most common additives to earthen plasters is fresh manure from cows and horses, which is mixed with clay and left to "ferment" for several days. Apparently the combination of enzymatic action and microfibers produces superior plaster. In areas with freezing winters, the addition of salt to earthen plaster lowers the freezing point to prevent spalling. Much more research needs to be done in this exciting field to determine the proper usage and proportions of these additives.

Cob and Earthquakes

APPENDIX 4

All our lives we have been subject to media coverage of dramatic events, so our attention is sometimes focused on rare calamities rather than daily problems. Earthquake damage to masonry buildings is dramatic and photogenic. We have all seen newspaper photos of devastated buildings from earthquake zones, beams poking skyward out of broken walls, desolate survivors standing in rags before them.

Our fears have been hyped up, notably in the past decade or so, by vested interests who have sought to increase their sales of building materials by beefing up public buildings, and now houses too, with extra steel, concrete, and lumber, all in the name of earthquake security. The engineering profession has profited from seismic concerns, and tends to look askance at "unquantifiable" materials such as earth. But we should remember that those same freeways in San Francisco that collapsed in the 1989 Loma Prieta Earthquake had all been heavily engineered, as had many of the high-rises in Mexico City and Kobe which fell in recent quakes. Earthen buildings are not the only ones that suffer, and sometimes are successful survivors.

EARTH BUILDINGS AND EARTHQUAKES

What's the reality of earth buildings and earthquakes? Unfortunately, the answer is very complicated. Earthquakes are quirky and unpredictable, as their ground-level effect is the result of complex interacting waves, jerks, and tremors, usually originating deep, deep down. In a street of identical houses, one may collapse completely while others are left undamaged. Therefore, it's difficult to draw conclusions from limited data. Nonetheless, here are some stories that demonstrate both sides of the question.

In 1976 Guatemala suffered a series of earthquakes that left a third of the population homeless. Thirty thousand people died, many of them in collapsing adobe houses. Ianto was present in Guatemala City at the time and was conscripted by the government to study the causes of damage and to make suggestions to prevent future catastrophes. Among other effects, he noted that adobe, brick, and concrete buildings fell in almost equal proportions, and that adobe built well survived better than concrete built badly.

Another major conclusion was that adobe structures were weakened by poor mortar joints. Adobe walls are built by stacking up sun-dried mud bricks, usually placing half an inch to an inch of wet mud horizontally between each course and vertically between adjacent bricks. As the mud mortar dries, it shrinks and pulls slightly away from the adobes, leaving many microscopic cracks straight through the wall. When the earthquake hit, walls separated along these cracks, and individual adobes or whole sections of wall came loose. The most common cause of death in collapsing adobe structures was asphyxiation. Courses of adobe blocks ground against each other, powdering the mortar in between and producing clouds of fine clay dust that many people were unable to escape. Other people were crushed by falling adobe blocks or by the weight of the roof.

If weak mortar joints were indeed the major cause of earthquake damage to adobes, monolithic cob structures should perform better. In part to test that hypothesis, Ianto and Linda went to New Zealand in 1995. With a sizable stock of old cob buildings in active seismic zones, New Zealand is an excellent place to evaluate the effects of earthquakes on cob.

New Zealand was colonized by Europeans in the 19th century, largely by poor farmers from the rural parts of Ireland, Scotland, and Wales where cob had been indigenous. When they arrived at the opposite end of the earth, transport systems were minimal so building materials from off-site would have been expensive even if they had been available. In areas where timber was scarce, settlers had only the earth from their site and whatever fiber was available, initially a native grass called Tussock, and later grain straw. By 1867 there were 7,470 cob houses listed on the South Island, representing one in five of all dwellings.

In England most surviving cob buildings are well built, with substantial walls 2 to 3 feet thick, on high rock foundations. The ones that are most prominent are comfortable houses that have been continuously lived in and heated since they were built. By contrast, New Zealand's cob buildings were mostly thrown together rapidly, with walls not usually more than 16 inches thick, and some as thin as 11 inches. Foundations are minimal, in some cases completely lacking, and most of the houses are in a wet, windy climate.

For many New Zealand immigrants, a cob cottage must have been a reminder of the poverty they escaped from, so as prosperity grew their children replaced these settlers' buildings with fired-brick or lumber houses. By 1901 only 1,500 of the original cob buildings were still inhabited. The rest became animal shelters, tool sheds, and workshops or were gradually allowed to decay and collapse. Now, 150 years after they were built, many of these pioneer buildings have crumbled back into the fields. But a surprising number still exist, several hundred at least, particularly the bigger, more elaborate examples.

The most obvious causes of collapse were: cement stucco, usually dating from the mid-20th century; roof dereliction; and splash hitting the bottoms of walls. Ianto and Linda were unable to locate a single case of a cob that had fallen in an earthquake, though this is not necessarily proof of cob's resistance to earthquakes. Local people pointed out that there had been no major earthquakes since before World War II and that until recently there was little interest in cob buildings, so even people old enough to remember the big earthquakes in the 1930s would likely have had more significant worries in their lives than the fate of old, uninhabited, earthen outbuildings. Even so, earthquake-caused cracking is distinctive in its diagonal X-pattern and should be observable on surviving buildings. Curiously, it wasn't. In seventeen buildings we inspected quite carefully, we failed to find a single crack we could ascribe to an earthquake. On the other hand, there were several stories like the following.

Broadgreen is a sizable, two-story mid-Victorian mansion, built in 1855 and '56 entirely of structural cob. It is located near Nelson, at the north end of the South Island. The Nelson area has suffered two major quakes that devastated the town, one in the 1870s, the other in 1931. Many buildings fell in both quakes, including the prestigious, brick-built Boys' College, a few blocks away. As Broadgreen is now a public museum, we were able to spend most of a morning carefully inspecting it. The house is, in terms of seismic design theory, a complex example of almost everything a builder could do to ensure disaster in an earthquake, yet it stands in excellent condition.

What did the builders do that should have invited failure? They chose a site on alluvial soil. They built of silty, poor material, without much clay or coarse sand. They incorporated many loose, fist-sized stones into the mix, further weakening the continuity of the cob. The foundation was minimal, apparently of the same fist-sized rocks, in fact in places it is hard to locate. The cob was made of material from a pit directly beneath the house, which still exists as a sort of unlined basement. The walls have no taper, nor are they in any way curved or buttressed. They rise to about 25 feet at the gables, at a constant thickness of 20 inches and 16 inches. (By contrast, under the adobe code for New Mexico, which specifies a 1-to-10 width-to-height ratio, Broadgreen's walls would need to be 30 inches thick.) The facade is perforated by many openings, some of them very large, all of which surely weaken the building. The roof, instead of being ultra-lightweight, the recommendation for seismic zones, is of good Welsh slate—probably as heavy a roof as you could choose. Now 140 years old, Broadgreen stands, handsome as ever and without a single serious crack. It's interesting to speculate why, but we don't have any solid answers.

In the New Zealand cob buildings we observed, any structural cracks were almost always concentrated in the corners of the building and above windows and doors. Those cracks were not caused by earthquakes, but by settlement, leaning walls, and damp foundations. The Earth Building Association of New Zealand declares proudly: "All over New Zealand and in all seismic zones of this country, earth houses have successfully withstood earthquakes. Wellington, Nelson, and Marlborough are well known for being in our country's worst seismic zone, yet they have many examples of long-standing earth buildings."

Clearly our society's understanding of earthen buildings in earthquakes is incomplete. Research urgently needs to be done on the exact effects of major earthquakes on different types of earthen construction. We certainly know that many people have died when buildings of adobe brick have collapsed; yet conspicuous survivors of major earthquakes in California have been the missions, nearly all of which are adobe. Photos immediately after the 1906 earthquake show the Dolores Mission standing alone in a devastated San Francisco. The situation with rammed earth and cob is even less clear.

General prejudice against earth as a building material pervades reports produced by government agencies, the materials industries, engineers, and architects. We need unbiased observers with considerable experience of earthen architecture to be able to move quickly into areas recently hit by big earthquakes to do systematic analysis of what fails and why.

ANTI-SEISMIC PRECAUTIONS

In the meantime, there are a number of generally accepted precautions you can take in seismic areas. While it's impossible to guarantee any building against earthquakes, you can lessen the potential damage by careful siting, design, and construction.

In siting, avoid building on "unconsolidated sediments," meaning anything you could easily dig out with a shovel. The best sites are on solid bedrock; second best are gravels and heavy clay soils that won't liquefy when shaken. Avoid steep, unforested slopes, especially where there is evidence of past slides.

Anti-seismic design for cob buildings would include thick but tapering walls, curved in plan. There should be frequent support buttresses, especially on straight walls and unconnected ends. Intersecting walls help enormously, providing additional lateral support.

It's probably a good idea to keep the overall height down to one or two stories. However, the world's tallest earthen structures are at the extreme southern tip of Arabia, close to an intensely seismic zone. Yemen has many adobe and cob towers up to ten stories high.

They are so extravagantly in excess of anything you are likely to attempt that they indicate heights of 15 to 20 feet should be no problem at all. Probably more important than overall height is the ratio of wall width at the base to height; to be conservative, maintain a 1:10 ratio.

Keep both the size and number of openings including doors and windows to a minimum, with as much cob as possible in between. Especially avoid narrow corner columns with a door or window on either side. Over window and door openings, use stout lintels instead of arches. Let the lintels bear on at least a foot of cob on either side.

A light roof is advisable for two reasons. First, any load on top of the wall contributes to the inertial force of the earthquake. Heavy roof structures can therefore help to shake the building apart. Second, heavy falling roof members are more likely to hurt people. The best roof design for earthquakes are trusses made of bamboo, steel, or light timbers.

It may be advisable to use posts as a redundant roof support system, with ample cross-bracing between posts and roof structure to prevent racking. In the unlikely event of the cob collapsing, the posts will hold the roof up at least long enough for you to escape. Beware of burying such posts in the mass of the cob wall, or flexion by the wood could force the cob apart.

A continuous bond beam or ring beam helps to hold everything together. Pour a continuous, reinforced concrete ring beam on top of the foundation trench, and build your stemwall into it. The whole foundation should be as strong as possible, with lots of tensile reinforcement. On the top of the cob walls, use a bond beam of either concrete or bolted timbers, anchored by deadmen deep in the cob.

Compressive and shear strength are much reduced if the walls are damp, so any design feature to keep walls dry is an added precaution. See appendix 3.

Research Needed

APPENDIX 5

WHEN IANTO TELLS STRANGERS THAT he does research, they nod knowingly, not really having a clue to what he means. Perhaps a vision of hunting through giant ill-lit library stacks flitters across their minds, or they see him with lab coat and clipboard jotting down numbers, a bubbling test tube in the background. We tend to think of research as something done only by very learned people with very technical education, but in any new field, anyone with a healthy curiosity and a little time can make surprising discoveries. Natural building is so new a discipline that at The Cob Cottage Company we learn something new to *us* almost every time we go to work, and occasionally learn something that would be a new thought to most other people. To do valuable research, on a whole spread of issues, you don't need a Ph.D., a foundation grant, a microscope, or even a computer. Mostly what you need is to know the relevant questions and whether anyone has already answered them. Much of the most valuable research we have personally done has been barefoot, with our hands really dirty.

Working with cob, we have been continually impressed by how little is written, how little is known. Cob is now emerging from being an obscure historical curiosity to a potentially major building technique, but because for centuries it was a regional method known chiefly to nonliterate peasants and farmers, middle-class writers in their snobbery recorded almost nothing. We do not know of a single photograph showing traditional mixing techniques or the specifics of how European cob building was done, which is remarkable considering that traditional cob construction was contemporary with photography for almost a century.

We have a most unusual situation: a common and recent technology about which we know very little. We are forced to use archaeologists' tactics to investigate, even though thousands of practitioners of traditional British cob lived well into the 1900s.

Research is needed on a number of fronts, most of them requiring at the outset simple observations and almost no equipment. They fall into several categories.

GEOGRAPHICAL EXTENT

We still don't really know all the areas where cob buildings exist, worldwide. Where is cob still a live tradition? What can we learn from working alongside those people? In the past few years, a cob building tradition has shown up, for instance, in Denmark, the Czech Republic, Scotland, and New York. Diligent searching might uncover cob buildings in many previously unsuspected places and climates, and the world's cob buildings could be mapped, along with observations on how local conditions have affected design, materials chosen, and how the buildings have been used.

REFERENCES IN LITERATURE

Careful gleaning of major research libraries could yield very valuable information. Investigations might reveal more in languages other than English, including the languages of regions where some type of cob has been built in recent times: German, French, Czech, Danish, Slovakian, Croatian, Bosnian, Serbian, Albanian, Hungarian, Bulgarian, Ukrainian, Russian, Persian, Japanese, Chinese, Korean, and Arabic. Of special value would be photographs, or at least drawings, of people making and building with cob in traditional ways, and of cob buildings during construction.

SEISMIC TESTING AND OBSERVATION

There's a need for shake-table tests to determine how cob walls perform in earthquakes. Specifically, we should test the anti-seismic value of curved walls, different proportions of straw, arrangement of openings, and various foundation systems. Earthquake simulation yields some information but is no substitute for a rapid assessment immediately following a real earthquake. On-the-ground observers are needed to be physically present, to analyze failure and survival patterns of traditional cob versus other construction techniques immediately after major destructive earthquakes. They would need to have a succinct list of questions prepared in advance; preparing such a list would constitute important research by itself.

EARTHQUAKE-RESISTANT FOUNDATIONS

In seismically active areas, earth buildings need foundations with lots of tensile strength. Are there any good, durable alternatives to concrete and steel?

APPROPRIATE COB MIXES

How far is it safe to deviate from the ideal proportions of a cob mix? What can you do where sand is unavailable, or where long, strong straw is hard to get? What are the structural consequences of cracking due to too much clay? If clay is lacking, could you substitute lime, gypsum, or wheat paste? What is the role of silt?

MIXING TECHNIQUES

What is the efficiency of mechanical mixing using a tractor, Bobcat, or mortar mixer? When expense, environmental impact, and breakdowns are factored in, how do these methods compare with mixing by foot? Could a specialized machine be made that mixes cob more quickly and effectively than any of the options currently available?

ADMIXTURES OF OTHER MATERIALS

What happens if we add gypsum, lime, or wheat paste to the cob mix? Or to floors? With a lot of extra straw, cob becomes light clay. Is there a hybrid that is self-supporting yet lightweight, with great thermal insulation? Should cob be seen a *structural composite* rather than masonry or concrete?

POURED ADOBE

This seems to be a potential field for innovation, for instance, to create replicable units for industry or storage or to supply rectilinear or straight-walled houses. Research is needed on easy ways to prevent cracking and how to make simple, homemade forms. What would a poured adobe–light clay blend look like, centering the straw in the middle of the wall somehow, but with a solid, poured cob mix on both faces?

COB FOR DOMES AND VAULTS

There is a long tradition of earthen roofs, domes, and vaults, though mostly in adobe or unfired brick. What different and new potentials exist for using monolithic earthen construction? What techniques and what geometry should be tried? How safe are such heavy overhead structures? How are they best waterproofed?

THERMAL COMFORT OF COB

There seems to be little agreement about the insulant value of most natural materials, though nobody claims to feel cold in any of them except rock. Empirical research needs to be done on the comfort of cob in order to derive a simple explanation of the relationship between comfort and the factors of air temperature, wall and floor temperature, and airflow in buildings. The effect of contact versus radiant versus blown-air heat are also poorly understood. We don't yet have agreement on R-value for cob, though with heavy walls R-values can be misleading. R-value was allegedly invented by the fiberglass industry, to promote its products, and its utility is chiefly in measuring heat *loss* through lightweight walls. Cob retains heat in heavy walls, so R-values are an inappropriate measure. A Canadian friend observed "R-values? What's the use of R-values? It's arse-value that I care about, whether my bum's warm when I'm sitting down at home." It would be useful to have a simple table showing relationship between cob density and heat storage, rate of heat flow, and insulation value.

In very cold winter zones, cob seems to be snugger than calculations might suggest. (So much for calculations!) What does it take to keep an earthen building comfortable in, for instance, Duluth or Winnipeg? There are reportedly cob houses in Ukraine and northern China; there is a huge 19th-century cob house built as a royal retreat in the Lillehammer Valley in Norway, just south of the Arctic Circle. What precautions did the builders take, and how do people living there keep comfy?

INSULATION FOR COB

We need to develop simple ways to insulate with natural materials. In some applications the insulation should be water repellent and nonrotting, for instance, for floors and foundations. In others it should be lightweight, cheap, easy to apply, as on the outside of north-facing exterior walls or in hot places, for example, on west walls. Some forms of natural insulation would need to be fireproof, for application to built-in stoves, fireplace backs, or ovens. What about sandwiches of straw–light clay between cob skins?

ALL-NATURAL ROOFING OPTIONS

In rainy and cold climates, all-natural roofs are challenging. Traditional natural roofs (palm thatch, slate, thatch, cedar shakes) tend to leak warm air out or rainwater in, sometimes both. The seven layers of birch bark traditional in Scandinavia for sod roofs is impractical. In most parts of North America, impermeable membranes keep warm air in but are usually synthetic. What 20th- or 21st-century technologies could we employ, yet keep the materials natural?

It seems inconsistent to build an all-natural house with a synthetic membrane in the roof. Thatch is a possible remedy, yet it is expensive and a complicated craft to learn. A roof is needed that breathes yet doesn't lose large amounts of heat, and is waterproof yet made of all-natural materials. Fired terra-cotta tiles are traditional in many areas. Could they be locally produced in forest areas, using for energy the logging slash that is currently burned in giant piles?

HYBRID BUILDINGS

What are the long-term consequences of combining cob with other building systems? For example, how does water vapor travel between cob and adjacent straw bales? What are the strongest ways to attach cob walls to other wall systems?

HYPOCAUST FLOOR HEATING

Hypocausts (under-floor fire ducts) may not have been tried with earthen floors. We need to see if the mix should be adapted to high and fluctuating temperature and how deep the floor should be for comfort under different winter conditions by experimenting with different sizes of heat source and various lengths of passage. Do they need liners? Will they need cleaning, and if so, how?

ECONOMIC/ECOLOGICAL CHOICES

What are the life-cycle costs in money, time, and environmental effects of diverse designs and materials? How does the use of different mixing techniques (manual vs. mechanical), purchased ingredients, and hired labor affect the price of cob? How big are the environmental footprints of various types of earthen buildings? How do they compare with other natural building systems?

HEALTH EFFECTS OF COB OR OTHER BUILDINGS

This is a huge field, wide open for research. Common sense tells us that natural building materials should be less damaging to health, but research is needed on why—systematic explanations of the biophysics and biochemistry of healthy buildings and analysis of long-term health of people living in cob compared with other materials.

Gradually anecdotal evidence is appearing that living in earth may actually be curative. The physically and emotionally healing aspects of natural buildings need careful investigation.

PERCEPTUAL AND PSYCHOLOGICAL IMPACT

How do we experience rectilinear compared with natural geometries in homes and workplaces? What effect do these spaces have on us? Can this be measured in productivity, truancy, sickness, or absenteeism? What is the therapeutic value of natural materials and spatial designs?

Why do curved spaces feel bigger than rectilinear ones with the same measured area? How can we use this phenomenon to our advantage? Are "round feet" a reality, and if so, under what conditions is the round foot phenomenon most valuable?

COMMERCIAL OPPORTUNITIES

Many natural materials that are home-made in North America are commercially available in other countries. Examples include the New Zealand wool insulation industry and dried cob in 50 kilogram bags sold in Western Europe. In North America is there, for example, scope for small bioregional production of cob bricks, quicklime, roundwood rafter systems, or bagged clay-dung plasters?

CHEAP HOUSING

What could be the relevance of cob building to the homeless, low-income people, those deliberately deconsumerizing, and our own impoverished descendants? How about the giant refugee camps where people live for years under cardboard and plastic sheets flapping in the desert wind? Could they themselves create decent housing from the earth beneath their feet? What would it take to train a million earth building teachers? Who wants to do it?

Wildlife in the Home

APPENDIX 6

BY IANTO

IN THE VILLAGES WHERE I LIVED IN AFRICA, the boundaries between house and farm and wild places were fairly arbitrary. My neighbors' houses always seemed alive with an odd menagerie of humans, farm animals, and wildlife. The plants that grew in their compounds were often those of the surrounding forest, and sometimes these plants grew inside their houses too. It wasn't at all unusual to share a house with pigs, monkeys, songbirds, and bats. All of them trotted and fluttered and swung in and out at will, as did neighbors and family, so that it was impossible to distinguish what was intentional and what was just tolerated. In general, the householders seemed to enjoy their interspecies visitors, and after several months I realized that they directly accommodated both the domestic and the wild through the design of their buildings and yards and by daily management. I want to make a case for encouraging wildlife into your home, and I will suggest some techniques for doing that.

Why would anyone want bats in their bedroom or guinea pigs in the kitchen? Well, bats eat mosquitoes, in huge quantities; guinea pigs are living composters, snatching up vegetable scraps on the floor and making meat and fertilizer. Additionally, chickens in your house eat ticks that carry fever; snakes eat mice; and toads gobble up flies, moths, and beetles. Not to mention the cricket in the hearth, birds singing, the scent of wild herbs, or the reminders of seasonal change that deciduous plants or hibernators bring. Our own ancestors knew these truths—in medieval England and the colonial United States, the house itself was a balanced ecosystem of many species, each regulating and supporting the others.

In the overdeveloped nations we have suffered more than a century of being sanitized. The purveyors of soaps and disinfectants and the giant cleaning industry that sprang up in the 19th century conspired to persuade our grandparents to banish all signs of life from their homes.

We still carry residues of their attitude that a house is a tightly controlled territory where all visitors, human and otherwise, must be carefully selected. We feel that Nature and houses are mutually incompatible, that the perils of wildness must at all cost be kept at bay. Even domestic animals are more often prisoners than guests, and wildlife of any kind is expelled, or worse, executed without trial. Nature, increasingly, is something distant that you drive to see on the weekend, becoming significant only when we photograph it and only catching our attention if it's big.

For those of us who are trying to build with Nature, are there ways we can welcome wildlife, both plants and animals, into our daily lives? Nature needs no persuading; she is ready to occupy every available niche, immediately, no matter how humanized it seems. We don't need to "attract" wild things; merely removing barriers ensures they will move in and live with us, a daily wonder and joy in our more and more mechanical world.

The cottage I live in is an example of my own changing attitudes. Slowly coming to terms with the animals in my house, I learned to like them all for who they are, while struggling through a personal need for control. Now I open my house and heart, enjoy them for themselves, valued neighbors who were here before me, whose tribe hunted the nooks and crannies of the earth long before humans ran upright. They close the circle of life and death, remind me of my own mortality, and by comparison reinforce my humanity.

Soon it will be evening. The bats who live in my roof will remind me that it's summer by squeezing out of the eave cracks and hunting down mosquitoes. They're tame, and if I whistle in a high monotone, they come to make sure it's really me, tumbling within a foot of my face, saying bat greetings I can't hear. The open door invites them in, and most nights a furry presence flutters in, flitting around the room so fast I can hardly follow, and is gone. Mosquito patrol done for the night.

We have no fly screens. Not that there aren't flies; there are plenty, but the spiders get them. Nature is at once profligate and precisely economical, siting her flytraps with exquisite care just where they will be most effective. Daily I watch the struggles of hornets, yellow jackets, and houseflies, as spider rolls and tucks, gift-wrapping her catches for eating at leisure, later. Every night when I leave the door open, a skunk comes by. He's a wily little fellow with spots and beady eyes, and he knocks the floor as he searches methodically about the room. Tap-tap, silence; tap-tap, quiet again; tap-tap. Over the years we've learned to hide eggs or he'll eat them up, so now he comes to check on what else we have. Next comes a house mouse, good, with sounds of careful chewing, then of cleaning up the floor, crumbs, seeds, anything left around.

There's a gopher snake who comes in sometimes, sliding silently over my bare foot as I sit at the open door. He's very thorough, working his systematic way around the corners of the room, looking for an opening. Here's a knothole; in he goes, all four feet of him, inching into the wall cavity. He'll be in there for hours, checking for mouse nests, termites, mud-wasps, anything to swallow.

We've come a little further than bird feeders and potted plants, not stopping at the cute, the warm, or the furry. Now there are microponds in the garden to support frogs and garter snakes, tangles of uncut grass for the praying mantis. We've dug holes, made wet places, heaped brush piles for the birds to nest in.

But, I am asked, what about them Getting Out of Control? Look at it this way: like all biological control, the key is in understanding habitat. If you don't want cockroaches, don't feed them; if termites are eating your foundation, examine the drainage—termites dislike dry wood.

Encouraging other life-forms is more than just biological control. There are more profound reasons for associating ourselves with nonhuman life. It is no accident that the poorest families even in urban ghettoes seem to have more dogs than they can afford, that they tend window boxes and love houseplants. As cotravelers we all evolved together, we codepended for our existence on many levels, some quite unexpected. Some wild animals seem to seek out human fellowship, while in some places humans eat dogs. Once a friend watched a fox with a young bear hunting together and playing in the Oregon rain. Most significantly, other species provide lenses through which we see the world differently, constant monitors of conditions we're insensitive to, early-bird alarms like the canaries carried down coal mines or peacocks screaming at intruders. The organisms around us are continuously aware, not merely through their own senses but by being plugged in to the collective sensitivities of whole ecosystems. For us humans, this is a reservoir of information background that enriches our lives and may in fact be essential to maintaining our sanity. To be totally alive, we must be able to read our surroundings at more levels than our own individual senses can access.

Without sanctimony, I would label our need to co-associate as a spiritual need: the undefinable satisfaction we get from petting a cat, watching a spider spinning a web, growing out seeds on the kitchen windowsill, or hearing the first songbird in spring.

What might all this mean to the designer, the householder, the land-use planner, or the gardener? First, let's look at the balances of Nature, how systems self-regulate.

Self-regulating systems depend upon complexity and diversity. In other words the more different elements (complexity) in a system and the more they differ from one another (diversity), the more stability is likely. Put bluntly, Nature abhors regularity and sets about diversifying it very rapidly. By providing a wide range of habitat types we are more likely to attract and support diversity of life. The biosystems we offer, indoors or out, will be rich proportional to their variety—of materials, of scale, of texture, and of durability.

We now know three truths about the direction of Universe, three basal laws that govern everything, including ourselves as individuals and communities, and as a species. First, that the Universe constantly diversifies; second, that each thing is unique and cannot be replicated; and third, that communication develops wherever possible. Nature strives for infinite diversity, individuality, and intercommunication. Without continuous human interference, even the most barren of our creations quickly develops an ecosystem that is complex beyond our imaginings. In complete opposition to these three laws the world we have tried to create since the end of the Middle Ages has striven for regularity, uniformity, and isolation.

At the heart of the problem is a class of professionals who are among the most industrialized, affluent, and educated—architects, builders, and physical planners. I make this admission as a reformed architect, myself. We are the most brainwashed by Cartesian thinking, the most in need of reorientation. Partly because of paper design and an emphasis on the photogenic qualities of a building, we tend to see our creations as isolated, immutable products rather than evolving processes connected to everything else.

We wield tremendous power. Our decisions affect huge transfers of energy and create giant entropic reactions. We will influence whole ecosystems indefinitely into the future. But the places we build don't need to be biological deserts. We may note that in the whole of the British Isles, the greatest concentration of bird species is found in the densest concentration of architecture—within a fifteen-mile radius of Central London. Over the centuries, through a slow accretion of buildings, parks, gardens, and abandoned industry, London of all places has developed an unparalleled richness of habitat.

How should our designs actively encourage wild things? Here are a few guidelines.

- Constantly strive to understand ecology. Read a basic text, in a couple of hours you can learn a lot. Just hang out, watching and questioning.
- Don't even begin designing until you really know your site. You need to understand its existing ecology and what it is trying to become.
- Where you build, protect all life on the site, not just the big trees. Don't compact the soil, don't culvert water, leave wild places intact, and don't drain wetlands.
- Work with complexity, diversity of materials, and smallness of scale.
- In what you protect or introduce, give

preference to species native to that system, and that bioregion.

- Use natural materials; restrict artificial toxins.
- Decay is inevitable. Acknowledge mortality and change in your buildings. Choose materials that decay gracefully and feed new life.
- Choose life over death, hedges over chain link fences, grass over pavement, grazing sheep over tractor mowers, absorption swales over storm drains.
- Hard paving is always a last resort. It's irreversible, like execution.
- Relax your need for control. Leave some strategic gaps under the roof for bats; build a little sloppy.
- If you're a professional builder, educate and work with your clients, the real managers of the places you create.
- We need a constant awareness of the tilting and spinning of Earth to establish where and when we are living. Without this backdrop we can't truly know who we are. A house can be a reminder of hour and of season through its cosmic geometry and by the ever-changing life it supports.

Teaching and Learning

APPENDIX 7

The Cob Cottage Company has sown seeds to strengthen interconnectedness between human health, well-being, psychology, ecology, and architecture. The seeds have been cultivated by more than a thousand students in the United States, Britain, Mexico, Canada, New Zealand, Australia, and Denmark. We call these groundbreakers—this community of cob builders, who are also artists, architects, therapists, contractors, teachers, healthcare providers, pioneers, and entrepreneurs—the Cob Web.

The *CobWeb* is the name of our newsletter and an analogy for the weaving together of cobbers' stories, developments, discoveries, friendships, associations, and enthusiasm. Together our voices are heard. As the old English proverb says, "If you throw enough mud, some of it will stick." Or, as *we* say, "The *CobWeb* is a newsletter for people with cob stuck to their souls."

The Cob Web is a support network of people building with a sense of the magic of community. Our dreams are coming true. Natural building is having a strong, positive, life-changing effect on individuals and relationships. This large natural building family is growing fast.

The following list shows steps you can take to learn, find support, spread enthusiasm, and share skills with one another. One step leads you to the next stage of growth along your path to building your own or someone else's home.

❧ ***Read cob and natural building literature.*** This book is a good start. See our bibliography for more books, publications, and newsletters.

The *CobWeb* newsletter is a networking tool created by Michael Smith and The Cob Cottage Company in 1995, first intended for graduates of Cob Cottage workshops, a means by which they could communicate and share ideas. However, because many noncobbers wanted access to the information, we made subscriptions available to the general public. It is written by alumni and subscribers and edited and assembled and mailed by volunteers. We have a twice-annual mail-out party to which alumni come back year after year. Check the Resources for more information.

❧ ***Take a workshop*** on natural building or cob construction. Several groups now offer these all over the United States and Canada, sometimes in Europe, Australia, New Zealand, and South Africa.

❧ ***Slide shows.*** Check our Web site for a calendar of slide shows being presented by The Cob Cottage Company, our alumni and associates, or call (541) 942–2005 to request a slide show in your area.

❧ ***Cob Web sites.*** At The Cob Cottage Company Web site (www.deatech.com/cobcottage) you can find updates, our calendar of workshops, recent photos of cob buildings, and links to other Web sites.

❧ ***Open houses and cob traveling tours.*** Sponsors of Cob Cottage workshops usually are happy to show off their project. Tours can sometimes be arranged by calling in advance. But we don't encourage drop-in visits. When Ianto and Linda were on the South Island of New Zealand, they toured historical cob cottages. It's a lovely way to travel, cob hopping along the way.

❧*Work parties.* A great way to "barn raise" or cob raise your building. Keep a guest journal of visitors and helpers. Eric Hoel reported that four hundred people put their hands, feet, and hearts into his building. This process adds to the richness and character of your home. Also see Web sites for scheduled work parties. Most work parties are free.

OFFERING WORKSHOPS

BY IANTO

It was never my intention to become a teacher of mud hut building. Throughout my adult life I have taught, sporadically—architecture, placemaking, perception and awareness, systematic botany, resource conservation, design, drawing for people who can't draw. I've taught African women's self-help groups how to develop fuel-efficient cooking stoves, Peace Corps volunteers how not to be Ugly Americans. I've taught Permaculture in the desert of Brazil and Ornithology in Oregon. I've worked on islands in the Pacific, in the snowy Himalayas, in the arid mountains of Lesotho. All those years I sought a subject that is a metaphor of the broadest understanding, something of universal appeal where anyone could involve all their senses to learn at all levels simultaneously. Perhaps at last I have found it.

Teaching and learning, like cause and effect, are inextricably interdependent. We learn as we teach, we teach as we learn. We teach best what we ourselves are exploring at that very time; the best teaching is a journey of discovery between student and instructor. We all have a lifetime of experience, however young we are, whatever we have spent our time doing. In the best settings, all students are also teachers. Without dialogue, we can only learn *information*.

Try to see learning as a process that, were it visible, would look like a kind of ragged, multidimensional crochet, a fabric of understanding with loops of knowledge sticking out at the newly worked edge. Tell yourself: I can't absorb a completely new thought, it has to attach to a framework I already acknowledge. The crochet loops are the only points of attachment. As teachers we have to connect with the familiar in the other person's worldview, find a point of understanding he or she already has. Trying to throw totally new material at a person is futile.

Learning cob excites people in an almost uncanny way. Intuitively perhaps we all know we can do it and do it well, enjoying every moment. Within minutes even little children can be teaching adults, showing, correcting, explaining. Octogenarian elders who have never touched mud since childhood plunge in enthusiastically. Here is an area where the crochet of our understanding has many old loops that have been waiting for connection, in some of us all of our lives. I can't prove it, but earth building sure looks like genetic knowledge that lies dormant in all of us.

At The Cob Cottage Company we regularly evaluate how we instruct; we analyze and review each course as it ends, with all participants and then staff only. Our objective always has been to teach the most enjoyable, instructive, inspiring, smooth-running workshops people have ever been in. By 2002 we had taught more than 100 week-long cob construction workshops and several longer, so some patterns are emerging. Gradually we have been able to make some generalizations about what works best.

❧***If people want to be able to build, skillfully, without supervision, they need at least seven days, full time.*** Nine or ten is better. Shorter workshops, half a day to three days, give people some ideas, but in that length of time they can only get their feet dirty and help them decide if they want to take this seriously.

❧ ***Only accept participants who can be there full time.*** Don't let people share a course—"Can I come three days, then my brother will come the other four?" It's a sequential process, and people who are not there the entire time will exhaust you with questions already answered, irritate the rest of the group, and fragment group unity.

❧ ***Encourage whole families to involve themselves.*** Many people have forgotten the importance of working together as a family. Even tiny kids are blissfully employed when there's mud around; we've had a number of children of ages four, three, even less than two, happily involved. Sometimes spouses have never shared a building project, and it can be very healing. One man brought his 78-year-old father who taught us all how to work, rhythmically and without complaint.

❧ ***Make sure the instructor-to-student ratio is comfortable.*** We try for one instructor with up to eight people, two for sixteen students. Take on a trainee to "gopher" things you need, and hopefully free you up of logistical difficulties. Your task is to watch and listen to the trainees, to give them your entire attention. If you're instructing, don't do anything else besides instructing and building. Someone else should be making lunch, keeping the books, looking for materials, hosting visitors. Identify students early on who will benefit from doing all these things, and delegate to them. It's all an important part of learning to build.

❧ ***Don't imagine for one moment that you're only teaching how to build.*** Attendees at cob workshops come because they don't like some aspect of their current life. They are ready to learn, grow, change, on many levels. People will come to you ready to remodel their entire lives. They will want to reform their diets, stop being such consumers, simplify their operation, quit that lousy job, and improve their relationships. Look out! Anyone who is ready to consider cob building is ready for Total Change. Honoring that, knowing that these changes may be scary, make the experience as rich as you can on every level. The food should be simple but outstanding; the compost toilet should smell good and have a wonderful view. The site should be tidy and inspiring. You will need to appear patient, tolerant, generous, and relaxed, even when you're not.

❧ ***Only teach workshops to create necessary buildings that will be used and that will be demonstrations.*** "Practice building" in an empty field or industrial yard doesn't honor participants with your confidence that they can do a good job from the outset. We tell people, "Your bare hands are laying a wall that may last forty generations. Make sure that your fortieth-level grandchild coming here in 3010 will marvel at the quality of the work you do today." It's very sobering, but inspiring too, knowing that some of those thumbprints could still be there in a millennium.

❧ ***Break up the day into several 1½- to 2-hour work sessions.*** It's okay to start early and finish late, as there's a lot of mud mileage to be traveled, but stop regularly. Balance didactic talks, focused discussions, building practice, and demonstrations of technique with eating, drinking, eating, lots of tea breaks, snacktime. We learn through our stomachs: take a *long* lunch break. Save beer till day's end.

❧ ***Incorporate field trips.*** In our seven-day workshops, a midweek field trip is a memorable experience. It can be a tour of a completed cob building, a guided walk through the local ecology, a hike to a special swimming hole, a Solstice bonfire on top of the local mountain, anything to roll back and get perspective on the building site.

❧ ***When the workshop begins, start right off by demonstration and modeling.*** You don't need a little chat session to "set the scene"; that's school system, and sets people into headwork rather than bodywork mode. We try to have all the staff making cob an hour before starting time. As each person arrives, he or she gets a brief greeting handshake, then goes right into the work. "Hi, I'm Ianto. I'm making a mix of two sand and one soil, from those two piles. Take off your shoes and jump in." As more people come, we abandon the early comers, pairing them up with later arrivals. "Hi, I'm Ianto, this is Jane, she'll show you what to do." Then walk away leaving Jane spluttering " But . . . , but . . ." When we come back in ten minutes, having set up seven other pairs, Jane and Holly are in complete rhythm, Jane confidently explaining word for word "You use two parts of sand, one of this dirt right here, watch . . ."

❧ ***Be formal in giving explanations.*** When you want to explain something, call in everybody. Wait until you have everyone's attention. The first day that people are with you, stop work regularly to field questions and give instruction; ask everyone's opinion on decisions you have yet to make.

❧ ***You'll be drowned by questions if you allow it.*** Our school system trains us to remember facts for exams and emphasizes that Teacher Knows All. Your job is to encourage people to answer their own questions by observation of their own work and watching the instructors carefully, being patient that the answers will come. We tell questioners that if they still have the same question by the end of the week we will answer it then. A well-planned course should cover all a student needs to know to build and more, eventually answering everyone's questions. But we won't be answering everyone's questions in the first hour of a seven-day course. Some may hate it at first but will build their self-confidence by not relying on the teacher. Socratic teaching, sensitively done, can create quantum leaps in self-confidence. When we ask for feedback a year after a workshop, what comes back is "Empowerment!"

Somebody once commented "Don't expect a straight answer from Cob Cottage instructors, they'll tell you either 'We'll come to that later,' or 'Figure it out yourself.'" This is outright slander! In fact, we also answer questions with, "Well, what do *you* think?" or "Try it and see."

Seriously, though, when an individual asks a question privately, say, "Hold that one until the meeting, then bring it up again. I want everybody to hear the answer." Everyone will hear the answer, but more importantly hear the question, too.

People at first want definitive answers. But never say "Never." The beauty of cob is that there are almost no hard rules. As with Nature herself, the answer nearly always is that everything varies with circumstance. The trick to managing our world is in being able to accurately weigh relative significance. Slowly, students see this for themselves, by doing, observing, changing how they proceed.

❧ ***Beware of answering in numbers.*** In particular, get away from using numbers; they deprive us of understanding relative values intuitively. The best answers are often generalized templates, not specific quantifiable absolutes. "How thick should this wall be?" is a common question. I show them with my moving hands, accordion-playing style, "Oh, about this thick." "But . . ." (exasperation in the voice) "I mean how many inches?" "Like I said, about *this* thick" (waving the arms about even more generally).

Soon they stop enumerating everything, start to build by *hunch*. "But what if it falls down?" "Then you will have learned at what point it fails. When you rebuild you'll do it differently. It will be stable the second time, and you will never have to wonder."

❧***When you give instruction, don't insult people by telling them things they can easily read for themselves.*** Your value as a teacher is proportional to how well you can explain your viewpoint, how eloquently you can share your own experiences of life, how effectively you can inspire people to take charge of their own learning.

❧***On the next to last day of the workshop, hold an open house for the public and a housewarming of the cottage being built.*** During the open house the workshop participants share what they have just learned with visitors and try on their new role as teachers. During the evening we usually celebrate beside a campfire inside the cottage. Candles honor the students' sculptural details: niches, alcoves, and arches. A time capsule gets passed around the fire circle. In the container, which may be a glass jar, a special teapot, or any vessel with a lid, symbolic treasures get placed: poem, song, coin, workshop flyer, or offerings symbolic of the student's learning or experiences in the workshop. When the building finally is demolished, a thousand years from now, there will be a record of the builders, their culture and thoughts and artifacts. It's nice also to leave a conundrum for 31st-century archaeologists.

❧***Design workshops that encourage group unity, spirit, and rhythm.*** Building is bonding. The cob toss, a chain of mud-painted builders dancing and tossing cobs to the wall not only is a way to transport the materials, it's also a reminder of the joy of working together. Expect lifelong friendships, marriages, working partnerships to come out of your courses. Allow for and encourage that. Build in lots of hanging out time, especially after lunch or supper.

❧***You model how people will operate,*** whatever you do, so you better be impeccable in your behavior and honest about your inconsistencies.

WORKING WITH VOLUNTEERS

Working informally with helpers, however feeble or inexperienced they might be, is usually preferable to struggling on alone. You may be able to find people who have experience building with cob, but you may also get offers of help from enthusiastic neophytes, who will need instruction. The Cob Cottage company offers a service of introducing builders to potential volunteers and work traders.

This may be your first cob building, or you may never have taught before. Don't be timid; you still have much to share. Your lifetime of experience, in whatever ventures, is a pool of unique knowledge, and the building itself will teach everybody. Here are some ways to involve other people so that they learn as much as possible, you get the building done, and everyone has a rewarding time.

❧Begin by inviting just a few individuals to work with you. Be formal—announce a "work party," with fixed working hours and a definite starting time. Be on time yourself. Don't encourage people to come late, or you'll have to re-explain several times. Start the first day with an explanation of what the project is, how you want to do it. Maybe repeat the formal work party every Saturday, for instance, then only let graduates of that day's instruction help with construction. Casual "helpers" who show up for an hour or two can waste everyone's time, or at worst be a hazard and safety worry.

❧At first you will be co-learning, blind leading the blind. Acknowledge your real level of expertise, that you are, for instance, only three pages ahead of your helpers. That gives them permission to question, innovate, and explore.

❧Consider charging people, just a little, for their introductory session. It will formalize the fact that you're offering them the privilege to learn a lifetime skill, and discourage casual bystanders who will absorb your energy. This exchange of money raises the value of what you can offer them.

SPONSORING WORKSHOPS

In workshops, sponsors have a unique opportunity to begin their own cottage, studio, house, or retreat through hands-on instruction with a crew of twelve to twenty. In every workshop The Cob Cottage Company teaches natural building and design by creating site-specific earthen buildings, designed to bring centuries of use and pleasure. We teach all aspects of building with earth and other natural materials.

We have three main goals in presenting workshops:

- To help sponsors build themselves small, comfortable, energy-efficient, and beautiful natural structures, especially a small house or cottage, using locally available materials such as earth, straw, native stone, and wood.
- To train people, especially owner-builders and building professionals, in traditional low-impact construction techniques.
- To create public demonstrations of healthy, resource-efficient, owner-built dwellings.

What Are the Benefits to You of Sponsoring a Workshop?

When you sponsor a Cob Cottage workshop, you gain the benefit of our expertise in all aspects of building with cob. We consult with you on siting and building design. We offer you, your family, and your friends training in cob construction.

Hosting the workshop means you have a jump-start on your building, you have a large labor pool for one week (or more), and you develop lots of new ideas and a network of potential helpers in completing your project.

Finally you gain exposure to leading-edge, low-impact technology, often including specialized help from skilled and experienced participants.

Who Can Be a Sponsor?

We recommend a sponsoring team of three people or more, one of whom must have already taken a Cob Cottage workshop. There are several questions the sponsoring team should consider:

- *The physical site:* Is there good solar exposure and natural drainage? Are there local sources for sand, clay soil, straw, and running water?
- *The function of the building:* What qualities do you dream of in your building? We can work with you to develop your design well in advance of your workshop date.
- *Accommodations:* Can you accommodate up to 20 people (plus staff) on your site, with basic camping facilities, an area for meals, and a meeting space?
- *Public accessibility:* Will the building be regularly accessible to the public? Are you prepared for it to be a demonstration site as an inspiration to others?
- *Publicity:* Can you publicize widely your workshop well in advance? Could you be responsible for publicity and registrations if necessary?

We encourage sponsors to arrange for some of their workshop participants to come early for the groundbreaking experience and foundation building and stay on through the completion of the building. Call it a natural building apprenticeship and encourage an apprentice to build another cottage that he or she can live in for a year or two, or longer, as trade for finishing your house. Staying on to complete a project from start to finish is one's best calling card as an experienced cobber. This way, apprentices don't have to wait for their own land to finish and live in a cob cottage.

HELPING HANDS

The majority of students want to cob on, like some mothers who, as their baby grows, want to have another one. So it is with cobbers. When the building is complete, many are looking for the next project. The Cob Cottage Company makes available a *Helping Hands* directory of workshop graduates, apprentices, and associates, which serves to link those who want to help with those who want help, giving ongoing support to builders and new workshop graduates who want to get more experience. The *CobWeb* and its *Helping Hands* directory link dozens of building projects and hundreds of Cob Cottage workshop graduates.

Whenever you manage a work crew, you become a teacher. Many of our students and their students have become teachers. As you feel your way through cob building, you may decide that you want to teach, either informally to friends and helpers or through full-term residential workshops. Because this is unlike any other building system, teaching cob is different from other teaching, as you will quickly discover.

Glossary

Please note that many of these terms have multiple meanings. They are defined here as they are used in this manual, though many of them are in current use by builders.

adobe. A traditional earth building technique in which rectangular bricks of mud and straw are cast, dried in the sun, and stacked with mud mortar between to make walls.

adz. A cutting tool with a thin, arching blade at right angles from the handle, used for trimming and carving wood and cob.

*alis (*or *aliz).* From the Spanish "alisar," to smoothen. A traditional adobe finish made of clay, water, flour paste, and other ingredients, usually applied with a brush or sponge and burnished with a rag or sponge.

armature. An internal support framework or skeleton. An armature of nails and wire can be used to support cob sculpture.

bauge. A French term for cob.

beingin. A quality of place defined by its passive use; a space used to be in. See also *goingthru.*

berm. A long mound of earth. Berms can be used to divert or absorb surface runoff.

binder. A substance (flour paste, glue, etc.) added to a plaster or natural paint to make it more adhesive.

bioregion. A geographically distinct area recognizable by its ecology, climate, geology, etc.

Bobcat. A small tractor that has been used successfully for mixing cob. It has large wheels that can be made to spin in opposite directions.

brown coat. An intermediate plaster coat between the base or scratch coat and the finish coat.

burnish. To polish (an *alis*, for example) by rubbing with a sponge or cloth.

casein. A milk protein used in paints and glues. It can be precipitated from sour milk by heating with an acid such as vinegar.

clay. A complex chemical formed from the decomposition of the mineral feldspar. Clay expands when wet and tends to stick to itself and other materials.

cob. From the Old English, "a loaf or rounded mass." 1. (n.) A traditional form of earth building in which a moist, plastic mixture of earth, straw, water, and sometimes sand and gravel is piled on a wall while wet and worked into place. 2. (n.) A loaf of cob mix. 3. (v.) To build with cob. 4. (adj.) Made of cob.

cob toss. A technique for transporting cob loaves from the mixing area to the wall by tossing them from one person to another in a line like a bucket brigade.

cobber. 1. One who builds with cob. 2. (in Australia) a friend; buddy.

cobber's thumb. A stick or other rigid object that fits comfortably into the palm of the hand and has a blunt tip like a thumb, used to sew courses of cob together and to perforate the surface.

corbel. A projection from the face of a wall, supporting a weight on top. A special technique has been developed for corbelling with cob.

corbel cobs. Long, thin cobs with extra straw running lengthwise, used for corbelling, arching, and strapping.

course. In masonry (as for example rock, brick, or cob), a horizontal layer of material.

curtain drain. A ditch dug around the perimeter of a building and filled with drain rock, often with a drainpipe at the bottom, to divert water and keep the ground beneath the building dry. Also called a French drain.

compressive strength. Hardness; the ability to withstand pressure applied evenly, without crushing.

datum stake. A stake driven into the ground at a building site to mark the future ground or floor level.

deadman. A buried log or the like, serving as an anchor for heavy objects such as doors and roofs. Several types are used in cob; see *Gringo block*.

ecological footprint. The total area of land required to support all the activities of an individual or a group, in perpetuity.

formwork. A temporary structure used to give shape to a fluid material such as concrete or mud.

froe (or *frow*). A long, sharp blade with a short handle at right angles, used for splitting shakes.

frost heave. Upward pressure, applied by water freezing in the ground, under a building or beneath a structure.

Gaab cob. A cob building technique using large armfuls of wet cob worked into the wall by hand.

goingthru. A quality of place defined by its active use; a space that mainly permits movement through it. See also *beingin*.

grade. In construction, the level of the ground surrounding the building.

gringo block. A wooden box with no top or bottom, buried in a cob or adobe wall as an attachment point for door and window frames, counters, cabinets, etc.

gypsum. A mineral (hydrated calcium sulfate) that is processed and marketed in powder form for use as a plaster. Gypsum plasters are hard, bright, and durable, but not waterproof.

hawk. In masonry, a small board with a handle on the underside, for holding plaster or mortar.

hay. The seed heads, stems, and leaves of grasses and other plants, cut green and dried for animal fodder.

Lammas. The day halfway between summer solstice and the fall equinox, usually falling on August 1 or 2.

leather hard. Firm but not dry, as applied to cob, earth and lime plasters, etc.

lift. Height of a single application of wet cob.

lime. A building material usually made by heating limestone, coral, or seashells, and slaking it in water. This produces lime putty, which can be mixed with sand to make mortar or plaster, or with water to make limewash.

limewash (or *whitewash*). A thin, paintlike finish made of lime putty and water, sometimes with added oils or binders.

lintel. A rigid beam spanning an opening, such as a door or window, to support the wall above.

litema. (Sesotho, pronounced dee-TAY-ma). A traditional African smear plaster made of clay and animal manure.

living roof (or *sod roof*). A roofing system in which a waterproof membrane is protected by earth and/or organic materials, with plants growing on top.

loaf method. A cob building technique in which custom-sized loaves are kneaded by hand, then transported to the wall and worked into place.

machete. A very large knife, useful for trimming cob, splitting wood, chopping straw, etc.

mixing boat. A big, flat box in which plaster and mortar is mixed.

mortar. A mixture of (usually) lime, cement, or clay with water and sand, used to set and hold rocks, bricks, etc.

mortar mixer. A machine with a rotating drum with paddles inside, designed to mix lime-sand or cement-sand mortar. Sometimes suitable for cob.

mushrooming. The unintentional widening of the top of a cob wall under construction due to careless building, slow drying conditions, etc.

niche. A small alcove in a wall.

Oregon cob. A design and construction system developed by The Cob Cottage, carefully formulated for strength and combining aspects of traditional cob with other natural building techniques, passive solar design, and an emphasis on social and environmental sustainability.

outsulation. Thermal insulation wrapped around the outside of a building to protect mass walls from ambient temperatures, e.g., lime-perlite plaster or stacked plastered bales attached tightly against the outside of a cob wall.

paring iron. A traditional British tool, like a shovel, with a flat, sharp blade in line with a long handle, used to trim and shape cob walls. See *spud.*

perlite. A lightweight mineral made of puffed clay, used in concrete, etc.

perch. A measure of volume used in traditional British cob building, described as 16 feet long, 1 foot high, and as wide as the wall being built.

persuader. A heavy wooden paddle used for shaping and consolidating a cob wall during construction by standing on the wall and beating the sides of the wall with the blade.

pise. A traditional European building technique in which earth is trodden between temporary forms.

pit method. A technique for mixing cob by foot in an excavated hollow or constructed pit.

plinth. A continuous masonry foundation or stem wall.

plumb. 1. (adj.) Vertical. 2. (v.) To make vertical.

pool float. A flexible, round-ended steel trowel, ideal for earthen plasters and poured adobe floors.

Portland cement. A synthetic blend of lime, clay, and other materials, heated and packaged as a fine gray powder. When mixed with water, sand, and gravel, it forms concrete.

poured adobe. A technique for making earthen walls or floors using a runny, cob-like mixture that is poured, and for floors troweled smooth.

Pulaski. A double-headed tool with one blade like a narrow ax and the other like an adz, used for excavation and fire fighting.

pumice. A very light volcanic rock, used for insulation, landscaping, drainage, and building, sometimes mixed with clay or lime.

rammed earth. Any building system in which slightly damp soil is compacted into temporary forms such as wooden slipforms, or permanent ones such as tires or agricultural sacks.

rammed earth bags (or *superadobes).* A system of building using agricultural sacks filled with earth (or sometimes sand or gravel) that are tamped in place in overlapping courses.

rammed tires. A building system using discarded automobile or truck tires laid in interlocking courses and rammed full of earth.

ready-mix. In cob building, a soil that contains clay and coarse sand in proportions to make good building material without addition of sand or clay.

round feet. Informal measure of area in curvilinear space. A round foot feels like about two square feet.

roundwood. Unmilled parts of trees such as roots, trunks, branches, and twigs in their natural shape; often cut and peeled but not sawn or split lengthwise.

rubble trench foundation. Trench filled with drain rock for drainage and support under a plinth or stem wall.

sand. Small particles of broken rock, smaller than a pea but large enough to see individual grains.

scaffolding. A temporary platform to stand on while building.

scratch coat. One or more undercoats of plaster, used to fill in gaps and shape the wall, but left rough so that the finish coat will adhere.

shake bolt. A short log of straight-grained, rot-resistant wood from which roofing shakes can be split with a froe.

shear. A stress or strain on a building that tends to cause two parallel planes to slide relative to each other. This can be caused by winds, ground movement, etc.

shillet. Naturally occurring fractured rock, usually shale, the size of gravel (southwestern England).

shingles. Wood sawn in thin wedges, used as roofing tiles.

shouldering. Unintentional narrowing of the top of a cob wall under construction.

silt. Particles of broken rock too fine for the eye to distinguish individual grains.

soil-cement. A mixture of soil and Portland cement.

splüge. To bulge, as wet cob has the tendency to do when pressure is applied from above.

spud. Short, thick, straight steel blade on a long handle, for debarking (peeling) logs by forcefully shoving the blade along the surface of the wood. Also used to trim cob walls.

strapping. The use of specially made, long, thin, wide, strap-shaped, straw-rich cobs, especially to fix shouldering.

straw. Stems of cereal grains, harvested dry and mature with the seed removed.

straw-clay (or *light clay).* From German, "leichtlehm." A traditional European system for making medium-weight insulating infill in timber-framed walls and roofs. It is made of straw coated in a thin clay slip, and is usually rammed between forms.

swale. A ditch running on contour across a slope to slow down and absorb surface runoff.

tarp method. A technique in which cob is mixed by foot on a sheet of thin, flexible, water-resistant material such as a tarp.

tensile strength. The ability to be pulled, twisted, or bent without breaking.

thatch. A roofing system that uses the overlapping stems or leaves of plants.

tie stone. A long stone built deep across the thickness of a stone wall, sometimes appearing on both faces.

timber framing. A traditional building system using large timbers (often hewn rather than sawn) connected together with complex joinery.

torchis. A French term for rammed earth.

trestle. A braced form serving as a support for working high on a wall.

*trom (*or *trombe) wall.* A solar-heated mass wall of masonry, cob, etc., set behind a glass skin.

urbanite. Common mineral resource initially manufactured as poured concrete for sidewalks, floorslabs, etc.; later reused in building as foundations, subfloors, drainrock, etc.

vapor barrier. An impermeable membrane used to prevent the transfer of moisture through walls and roofs.

wattle and daub. A traditional building system composed of woven panels of thin, flexible sticks plastered over with sticky mud.

zabour. A name for cob construction in the Middle East.

Resources on Cob and Natural Building

EDUCATIONAL ORGANIZATIONS:

Builders Without Borders (trainings and resources; cross cultural natural building projects for the underhoused)
119 Main Street, Kingston, NH 88042;
phone: (505) 895-5400
Web site: www.builderswithoutborders.com

Cal-Earth-California Institute of Earth Art and Architecture (info and workshops on adobe, ceramic and earth bag construction)
10376 Shangri La Ave., Hesperia, CA 92345;
phone: (760) 224-0614
Web site: www.calearth.org

The Canelo Project (publications and workshops on straw bale and natural building)
HC1 Box 324, Elgin, AZ 85611;
phone: (520) 455-5548
Web site: www.caneloproject.com

*__The Cob Cottage Company__ (free phone and mail info service on cob and natural building; training and workshops; consulting)
P.O.Box 123, Cottage Grove, OR 97424;
phone: (541) 942-2005
Web site: www.deatech.com/cobcottage/

*__CobWorks__ (workshops)
RR#1, Mayne Island, BC V0N 2T0 Canada;
phone: (250) 539-5253
Web site: www.cobworks.com

CRATerre (professional trainings in earthen construction, architecture, engineering)
Maison Levrat, Rue de Lac BP 53, F-38092 Villefontaine Cedex, France;
phone: 33-474-954391
Web site: www.craterre.archi.fr

Earthwood Building School (workshops and resources on cordwood masonry, stone circles, and earth-sheltered housing)
336 Murtagh Hill Rd., West Chazy, NY, 12992; phone: (518) 493-7744
Web site: www.cordwoodmasonry.com

The Econest Building Company (design/build, workshops on light-clay, timber frame, earth floors, natural plasters)
P.O. Box 864, Tesuque, NM 87574;
phone: (505) 984-2928
Web site: www.econests.com

Emerald Earth (workshops on natural building and permaculture)
P.O. box 764, Boonville, CA 95415;
phone: (707) 895-3302
E-mail: lorax@ap.net

Fox Maple School of Traditional Building (info and workshops on timber framing, thatching, and traditional clay infill techniques)
P.O. Box 249, Corn Hill Rd., Brownfield, ME 04010; phone: (207) 935-3720
Web site: www.nxi.com/WWW/joinersquarterly

---

* indicates those specifically relevant to cob

Gourmet Adobe (Carole Crews's natural plasters and finishes, workshops)
HC 78 Box 9811, Ranchos De Taos, NM 87557; phone: (505) 758-7251

***Groundworks** (cob workshops for women)
P.O. Box 381, Murphy, OR 97523;
phone: (541) 471-3470
Web site: www.cpros.com/-sequoia/workshop.html

***The Natural Builder** (short workshops on cob and natural building)
P.O. Box 855, Montrose, CO 81402;
phone: (970) 249-8821

The North American School of Building (training in cob and natural building; apprenticeships; demonstration buildings)
P.O. Box 444, Coquille, OR 97423;
phone: (541) 396-1825

Out on Bale (un) Ltd. (straw bale info and workshops)
2509 Campbell Ave. #292, Tucson, AZ 85719; phone: (520) 622-6896
E-mail: rew@interx.com

Rammed Earth Works (on-demand workshops and consulting in rammed earth)
101 S. Coombs, Suite N., Napa, CA 94559;
phone: (707) 244-2532
E-mail: Rew@I-café.net

Solar Energy International (workshops on alternative energy and natural building)
P.O. Box 715, Carbondale, CO 81623;
phone: (970) 963-8855
Web site: www.solarenergy.org

Southwest Solaradobe School (earth building classes, especially adobe and rammed earth)
P.O. Box 153, Bosque, NM 87006;
phone: (505) 861-1255
Web site: www.adobebuilder.com

Sun Ray School of Natural Living (design/build, workshops)
1356 Janicki Rd., Sedro Wooley, WA 98284;
phone: (360) 856-5482

Women Builders (assorted natural building workshops and referrals for women)
P.O. Box4114, Tucson, AZ 85717;
phone: (520) 206-8000
E-mail: womenbuilders@theriver.com

RESOURCE CENTERS

The Black Range Lodge (books and videos on straw bale and natural building)
Star Rt. 2 Box 119, Kingston, NM 88042;
phone: (505) 895-5652
Web site: www.StrawbaleCentral.com

* **Building for Health Materials Center** (natural building supplies)
Box 113, Carbondale, CO 81623;
phone: (800) 292-4838

Center for Maximum Potential Building Systems (R&D of innovative building technologies, including recycled materials)
8604 FM 969, Austin, TX 78724;
phone: (512) 928-4786
Web site: www.cmpbs.org

Center for Resourceful Building Technology (research and demonstration of resource-efficient building materials & techniques)
P.O. Box 3866, Missoula, MT 59806;
phone: (406) 549-7678

DAWN/Out On Bale-By Mail (mail order books and resources for natural building)
6570 W. Illinois St., Tucson, AZ 85735;
phone: (520) 624-1673
Web site: www.greenbuilder.com/dawn

Development Center for Appropriate Technology (education about alternative materials, sustainability, and building codes)
P.O. Box 27513, Tucson, AZ 85726-7513;
phone: (520) 624-6628
Web site: www.dcat.net

***Devon Earth Building Association** (research and publications about cob building and restoration)
50 Blackboy Rd., Exeter EX4 6TB, UK.

***The Dirt Cheap Builder's Catalogue**
(mail order books on natural building and homesteading)
P.O. Box 375, Cutten, CA 95534;
phone: (707) 441-1632
Web site: www.dirtcheapbuilder.com

Intaba's Kitchen
Corvallis, Oregon; phone: (541) 754-6958;
Web site: www.intabas.com

Institute for Sustainable Forestry
(researches and promotes sustainable forest management)
P.O. Box 1580, Redway, CA 95560;
phone: (707) 923-4719

Northwest Eco-Building Guild
(regional info. on sustainable building and recycled materials)
217 Ninth Ave. North, Seattle, WA 98107;
phone: (206) 622-8350

***Society for the Protection of Ancient Buildings** (research and advocacy for ancient traditional buildings in Britain)
37 Spitall Square, London E1 6DY, UK

Recommended Books

NATURAL BUILDING TECHNIQUE

Abraham, Loren and Thomas Fisher. *Living Spaces*. Cologne, Germany: Konemann, 1999.

HR Bainbridge, David, Athena Swentzell Steen, and Bill Steen. *The Straw Bale House*. White River Junction, Vt.: Chelsea Green Publishing, 1994.

* Bee, Becky. *The Cob Builders Handbook*. Murphy, Oreg.: Groundworks, 1997.

Bergeron, Michel, and Paul Lacinski. *Serious Straw Bale*. White River Junction, Vt.: Chelsea Green, 2000.

Chappell, Steve. *A Timber Framer's Workshop: Joinery, Design, and Construction of Traditional Timber Frames*. Brownfield, Me.: Fox Maple Press, 1998.

Chiras, Daniel. *The Natural House*. White River Junction, Vt.: Chelsea Green, 2000.

Chiras, Daniel and Cedar Rose Guelberth. *The Natural Plaster Book*. Gabriola Island, B.C.: New Society Publishers, 2002.

Davey, Norman. *A History of Building Materials*. London: Phoenix House, 1961.

HR * Denzer, Kiko. *Build Your Own Earthen Oven*. Boldgett, Oreg.: Hand Print Press, 2001.

Edwards, Ron and L. Wei-Hao. *Mud Brick and Earth Building the Chinese Way*. Kuranda, Australia: Rams Skull Press, 1994.

Edwards, Ron. *Cob, Building in Earth*. Kuranda, Australia: Rams Skull Press, 1997.

HR * Elizabeth, Lynne, and Cassandra Adams, eds. *Alternative Construction: Contemporary Natural Building Methods*. New York: John Wiley & Sons, 2000.

Hall, Nicholas. *Thatching: A Handbook*. Essex, UK: ITDG Publishing, 1988.

Holmes, Stafford, and Michael Wingate. *Building with Lime: A Practical* Introduction. Essex, UK: ITDG Publishing, 1997.

Induni, Bruce, and Liz Induni. *Using Lime*. London: Society for the Protection of Ancient Buildings, 1990.

HR * Kennedy, Joseph F., Michael G. Smith, and Catherine Wanek, eds. *The Art of Natural Building: Design, Construction, Resources*. Philadelphia, Pa.: New Society Publishers, 2002.

HR Kern, Ken. *The Owner-Built Home*. New York: Charles Scribner's Sons, 1972.

Khalili, Nader. *Ceramic Houses and Earth Architecture*. Hesperia, Calif.: Cal-Earth Press, 1996.

Laporte, Robert. *Mooseprints: A Holistic Home Building Guide*. Natural House Building Center, 1993.

* McCann, John. *Clay and Cob Buildings*. Buckinghamshire, UK: Shire Publications.

McHenry, Paul Graham Jr. *Adobe: Build It Yourself*. Tucson: University of Arizona Press, 1985.

———. *Adobe and Rammed Earth Buildings*. Tucson: University of Arizona Press, 1989.

HR Minke, Gernot. *Lehmbau-Handbuch* (Earth-Building Handbook). Freiburg, Germany: Okobuch, 1984.

HR = Highly recommended * = Deals with cob

HR Morgan, W.E.C. et al. *The Thatcher's Craft*. London: Rural Industries Bureau, 1960.

HR Myhrman, Matts, and S.O. MacDonald. *Build It with Bales: A Step-by-Step Guide to Straw Bale Construction*, 2nd ed. Tucson, Ariz.: Out on Bale, 1997.

Reynolds, Michael. *Earthship* vols. 1–3. Taos, N. Mex.: Solar Survival Press, 1993.

Roy, Rob. *The Complete Book of Cordwood Masonry Housebuilding: The Earthwood Method*. New York: Sterling, 1992.

Schofield, Jane. *Lime in Building*. Devon, UK: Black Dog Press, 2001. (Available from Cob Cottage Co.)

HR * Smith, Michael G. *The Cobber's Companion: How to Build Your Own Earthen Home*, 3rd ed. Cottage Grove, Oreg.: Cob Cottage Company, 2001.

Steen, Athena, and Bill Steen. *Earthen Floors*. Elgin, Ariz.: Canelo Project, 1997.

Stultz, Roland, and Kiran Mukerji. *Appropriate Building Materials: A Catalog of Potential Solutions*. Essex, UK: SKAT/ITDG Publishing, 1988.

Taylor, Charmaine. *All About Lime*. Cutten, Calif.: Taylor Publishing, 2000.

Tibbets, Joseph M. *The Earthbuilders' Encyclopedia*. Bosque, N.Mex.: Southwest Solaradobe School, 1989.

Williams, Christopher. *Craftsmen of Necessity*. New York: Vintage Books, 1974.

Wojciechowska, Paulina. *Building With Earth: A Guide to Flexible-form Earth Bag Construction*. White River Junction, Vt.: Chelsea Green, 2001.

NATURAL BUILDING HISTORY, DESIGN, AND RELATED TOPICS

HR Alexander, Christopher et al. *A Pattern Language*. New York: Oxford University Press, 1977.

HR Alexander, Christopher. *The Timeless Way of Building*. New York: Oxford University Press, 1979.

HR Bourgeois, Jean-Louis, and Carollee Pelos. *Spectacular Vernacular: The Adobe Tradition*. New York: Aperture Foundation, 1989.

Brand, Stewart. *How Buildings Learn*. New York: Viking Penguin, 1994.

HR * Cob Cottage Company. *Earth Building and the Cob Revival: A Reader*. Cottage Grove, Oreg.: The Cob Cottage Company, 1996.

HR Connell, John. *Homing Instinct*. New York: Warner, 1993.

HR Day, Christopher. *Places of the Soul: Architecture and Environmental Design as a Healing Art*. London: Aquarian/Thorsons, 1993.

Dethier, Jean. *Down to Earth-Adobe Architecture: An Old Idea, a New Future*. New York: Facts on File, 1983.

* Egeland, Pamela. *Cob and Thatch*. Exeter: Devon Books, 1988.

Farrelly, David. *The Book of Bamboo*. San Francisco: Sierra Club, 1984.

* Howard, Ted. *Mud and Man: A History of Earth Buildings in Australia*, Melbourne: Earthbuild Publications, 1992. (Available from Cob Cottage Co.)

Imhoff, Dan. *Building with Vision*. Healdsburg, Calif: Watershed Media, 2001.

HR Kahn, Louis. *Shelter*. Berkeley, Calif.:Ten Speed Press, 1990.

HR Komatsu, Yoshio. *Living on Earth*. Fukuinkan-Shoten Publishers, 1999.

Lawlor, Anthony. *The Temple in the House: Finding the Sacred in Everyday Architecture*. New York: Putnam, 1994.

HR Mollison, Bill. *Permaculture: A Practical Guide for a Sustainable Future*. Washington, D.C.: Island Press, 1990.

Mollison, Bill, and Reny Mia Slay. *Introduction to Permaculture*. Tyalgum, Australia: Tagari Publications, 1991.

Pearson, David. *Earth to Spirit: In Search of Natural Architecture*. San Francisco: Chronicle Books, 1995.

———. *The New Natural House Book: Creating a Healthy, Harmonious, and Ecologically Sound Home Environment*. New York: Simon and Shuster, 1998.

HR Rudolfsky, Bernard. *Architecture Without Architects*. Albuquerque: University of New Mexico Press, 1987.

———. *The Prodigious Builders*. New York: Harcourt Brace Javanovich, 1977.

HR Scher, Les, and Carol Scher. *Finding and Buying Your Place in the Country.* Chicago: Dearborn, 2000.

HR Taylor, John S. *A Shelter Sketchbook: Timeless Building Solutions*. White River Junction, Vt.: Chelsea Green, 1997.

HR Thoreau, Henry David. *Walden*. William Rossi, ed. New York: W.W. Norton, 1992.

Wackernagel, Mathis, and William Rees. *Our Ecological Footprint*. Philadelphia, Pa.: New Society Publishers, 1996.

* Walker, B., and C. McGregor. *Earth Structures and Construction in Scotland*. Edinburgh: Historic Scotland, 1996.

Walshe, Paul, and John Miller. *French Farmhouses and Cottages*. New York: Rizzoli, 1992.

Wells, Malcolm. *Gentle Architecture*. New York: McGraw-Hill, 1981.

* Williams-Ellis, Clough and John & Elizabeth Eastwick-Field. *Building in Cob, Pisé, and Stabilized Earth*. London: Country Life Limited, 1919.

PERIODICALS

Adobe Builder, P.O. Box 153, Bosque, NM 87006. Oversized, glossy quarterly trade journal for professional adobe builders and restorationists, mostly in the Southwest.

HR **The CobWeb*. Box 123, Cottage Grove, OR 97424; phone: (541) 942-2005. Bi-annual. The only journal dedicated to cob, with technical updates, personal stories, networking, and more.

The Joiner's Quarterly. P.O. Box 249, Snowville Rd., Brownfield, ME 04010; phone: (207) 935-3720; e-mail: foxmaple@ nxi.com. Glossy, professional quarterly for timber framers, with articles ranging from highly technical descriptions of framing tools and techniques to overviews of contemporary and historical earth building methods.

HR *The Last Straw*. HC 66, Box 119, Hillsboro NM, 88042; phone: (505) 895-5400; Web site: www.strawhomes.com. Quarterly. Up-to-date technical developments, resources and networking on straw bale construction and natural building in general.

HR *The Permaculture Activist*. P.O. Box 1209, Black Mountain, NC 28711; phone: (828) 669-6336; website: www.permacultureactivist.net. Quarterly journal packed with in-depth articles on Permaculture and occasionally natural building.

SUPPLIERS OF BOOKS ON NATURAL BUILDING

Black Range, Star Rt. 2, Box 119, Kingston, NM 88042; phone: (505) 895-5652; www.StrawBaleCentral.com

Chelsea Green Publishing, P.O. Box 428, White River Junction, VT 05001; phone: (800) 639-4099; www.chelseagreen.com

Cob Cottage Company, Box 123, Cottage Grove, OR 97424; phone: (541) 942-2005.

Earthwood Building School, 366 Murtagh Hill Rd., West Chazy, NY 12992; phone: (518) 493-7744.

Taylor Publishing, Box 375, Cutten, CA; phone: (707) 441-1632, www.dirtcheapbuilder.com

Index

the politics and practice of sustainable living

CHELSEA GREEN PUBLISHING

CHELSEA GREEN has introduced a new series called "Politics of the Living," a collection of hard-hitting works by major writers exposing the global governmental and corporate assault on life.

For more than twenty years Chelsea Green has published the best books on green building, renewable energy, organic gardening and sustainable agriculture, permaculture, and eco-food. Our series of Slow Food City Guides includes *The Slow Food Guide to New York City* and *The Slow Food Guide to Chicago*, produced in partnership with Slow Food U.S.A.

For more information about Chelsea Green, or to request a free catalog, call toll-free (800) 639-4099, or write to us at P.O. Box 428, White River Junction, Vermont 05001. Visit our Web site at www.chelseagreen.com.

Guantánamo: What the World Should Know
Michael Ratner & Ellen Ray
ISBN 1-931498-64-4
$15.00

Strangely Like War: The Global Assault on Forests
Derrick Jensen & George Draffan
ISBN 1-931498-45-8
$15.00

Welcome to the Machine: Science, Surveillance, and the Culture of Control
Derrick Jensen & George Draffan
ISBN 1-931498-52-0
$18.00

High Noon for Natural Gas: The New Energy Crisis
Julian Darley
ISBN 1-931498-53-9
$18.00

the politics and practice of sustainable living

the politics and practice of sustainable living

CHELSEA GREEN PUBLISHING

CHELSEA GREEN publishes information that helps us lead pleasurable lives on a planet where human activities are in harmony and balance with Nature. Our celebration of the sustainable arts has led us to publish trend-setting books about organic gardening, solar electricity and renewable energy, innovative building techniques, regenerative forestry, local and bioregional democracy, and whole foods. The company's published works, while intensely practical, are also entertaining and inspirational, demonstrating that an ecological approach to life is consistent with producing beautiful, eloquent, and useful books, videos, and audio cassettes.

For more information about Chelsea Green, or to request a free catalog, call toll-free (800) 639-4099, or write to us at P.O. Box 428, White River Junction, Vermont 05001. Visit our Web site at www.chelseagreen.com.

the politics and practice of sustainable living

Shelter

The New Ecological Home: A Complete Guide to Green Building Options
Daniel D. Chiras
ISBN 1-931498-16-4 • $35.00

Food

The Slow Food Guide to New York City
Martins and Watson • ISBN 1-931498-27-X
The Slow Food Guide to Chicago
Gibson and Lowndes • ISBN 1-931498-61-X
$20.00 each

Planet

Limits to Growth: The 30-Year Update
Donella Meadows, Jorgen Randers, and Dennis Meadows
ISBN 1-931498-58-X • $22.50

People

This Organic Life: Confessions of a Suburban Homesteader
Joan Dye Gussow
ISBN 1-931498-24-5 • $16.95